Costin Badica, Giuseppe Mangioni, Vincenza Carchiolo,
and Dumitru Dan Burdescu (Eds.)

Intelligent Distributed Computing, Systems and Applications

Studies in Computational Intelligence, Volume 162

Editor-in-Chief

Prof. Janusz Kacprzyk
Systems Research Institute
Polish Academy of Sciences
ul. Newelska 6
01-447 Warsaw
Poland
E-mail: kacprzyk@ibspan.waw.pl

Further volumes of this series can be found on our homepage:
springer.com

Vol. 140. Nadia Magnenat-Thalmann, Lakhmi C. Jain
and N. Ichalkaranje (Eds.)
New Advances in Virtual Humans, 2008
ISBN 978-3-540-79867-5

Vol. 141. Christa Sommerer, Lakhmi C. Jain
and Laurent Mignonneau (Eds.)
The Art and Science of Interface and Interaction Design (Vol. 1),
2008
ISBN 978-3-540-79869-9

Vol. 142. George A. Tsihrintzis, Maria Virvou, Robert J. Howlett
and Lakhmi C. Jain (Eds.)
New Directions in Intelligent Interactive Multimedia, 2008
ISBN 978-3-540-68126-7

Vol. 143. Uday K. Chakraborty (Ed.)
Advances in Differential Evolution, 2008
ISBN 978-3-540-68827-3

Vol. 144. Andreas Fink and Franz Rothlauf (Eds.)
*Advances in Computational Intelligence in Transport, Logistics,
and Supply Chain Management,* 2008
ISBN 978-3-540-69024-5

Vol. 145. Mikhail Ju. Moshkov, Marcin Piliszczuk
and Beata Zielosko
Partial Covers, Reducts and Decision Rules in Rough Sets, 2008
ISBN 978-3-540-69027-6

Vol. 146. Fatos Xhafa and Ajith Abraham (Eds.)
*Metaheuristics for Scheduling in Distributed Computing
Environments,* 2008
ISBN 978-3-540-69260-7

Vol. 147. Oliver Kramer
Self-Adaptive Heuristics for Evolutionary Computation, 2008
ISBN 978-3-540-69280-5

Vol. 148. Philipp Limbourg
Dependability Modelling under Uncertainty, 2008
ISBN 978-3-540-69286-7

Vol. 149. Roger Lee (Ed.)
*Software Engineering, Artificial Intelligence, Networking and
Parallel/Distributed Computing,* 2008
ISBN 978-3-540-70559-8

Vol. 150. Roger Lee (Ed.)
*Software Engineering Research, Management and
Applications,* 2008
ISBN 978-3-540-70774-5

Vol. 151. Tomasz G. Smolinski, Mariofanna G. Milanova
and Aboul-Ella Hassanien (Eds.)
Computational Intelligence in Biomedicine and Bioinformatics,
2008
ISBN 978-3-540-70776-9

Vol. 152. Jarosław Stepaniuk
*Rough - Granular Computing in Knowledge Discovery and Data
Mining,* 2008
ISBN 978-3-540-70800-1

Vol. 153. Carlos Cotta and Jano van Hemert (Eds.)
*Recent Advances in Evolutionary Computation for
Combinatorial Optimization,* 2008
ISBN 978-3-540-70806-3

Vol. 154. Oscar Castillo, Patricia Melin, Janusz Kacprzyk and
Witold Pedrycz (Eds.)
Soft Computing for Hybrid Intelligent Systems, 2008
ISBN 978-3-540-70811-7

Vol. 155. Hamid R. Tizhoosh and M. Ventresca (Eds.)
Oppositional Concepts in Computational Intelligence, 2008
ISBN 978-3-540-70826-1

Vol. 156. Dawn E. Holmes and Lakhmi C. Jain (Eds.)
Innovations in Bayesian Networks, 2008
ISBN 978-3-540-85065-6

Vol. 157. Ying-ping Chen and Meng-Hiot Lim (Eds.)
Linkage in Evolutionary Computation, 2008
ISBN 978-3-540-85067-0

Vol. 158. Marina Gavrilova (Ed.)
*Generalized Voronoi Diagram: A Geometry-Based Approach to
Computational Intelligence,* 2008
ISBN 978-3-540-85125-7

Vol. 159. Dimitri Plemenos and Georgios Miaoulis (Eds.)
Artificial Intelligence Techniques for Computer Graphics, 2008
ISBN 978-3-540-85127-1

Vol. 160. P. Rajasekaran and Vasantha Kalyani David
Pattern Recognition using Neural and Functional Networks,
2008
ISBN 978-3-540-85129-5

Vol. 161. Francisco Babtista Pereira and Jorge Tavares (Eds.)
Bio-inspired Algorithms for the Vehicle Routing Problem, 2008
ISBN 978-3-540-85151-6

Vol. 162. Costin Badica, Giuseppe Mangioni,
Vincenza Carchiolo and Dumitru Dan Burdescu (Eds.)
Intelligent Distributed Computing, Systems and Applications,
2008
ISBN 978-3-540-85256-8

Costin Badica
Giuseppe Mangioni
Vincenza Carchiolo
Dumitru Dan Burdescu
(Eds.)

Intelligent Distributed Computing, Systems and Applications

Proceedings of the 2nd International
Symposium on Intelligent Distributed
Computing – IDC 2008, Catania, Italy, 2008

 Springer

Costin Badica
Facultatea de Automatica, Calculatoare si
Electronica
Departamentul de Inginerie Software
Universitatea din Craiova
Bvd. Decebal, Nr. 107
200440, Craiova, Dolj
Romania
E-Mail: badica_costin@software.ucv.ro

Giuseppe Mangioni
Dipartimento di Ingegneria Informatica e delle
Telecomicazioni
Universita di Catania
Viale A. Doria, 6
95125 Catania
Italy
E-Mail: Giuseppe.Mangioni@diit.unict.it

Vincenza Carchiolo
Dipartimento di Ingegneria Informatica e delle
Telecomicazioni
Universita di Catania
Viale A. Doria, 6
95125 Catania
Italy
Email: Vincenza.Carchiolo@diit.unict.it

Dumitru Dan Burdescu
Facultatea de Automatica, Calculatoare si
Electronica
Departamentul de Inginerie Software
Universitatea din Craiova
Bvd. Decebal, Nr. 107
200440, Craiova, Dolj
Romania
Email: burdescu_dumitru@software.ucv.ro

ISBN 978-3-540-85256-8   e-ISBN 978-3-540-85257-5

DOI 10.1007/978-3-540-85257-5

Studies in Computational Intelligence   ISSN 1860949X

Library of Congress Control Number: 2008932399

© 2008 Springer-Verlag Berlin Heidelberg

Typeset & Cover Design: Scientific Publishing Services Pvt. Ltd., Chennai, India.

Printed in acid-free paper

9 8 7 6 5 4 3 2 1

springer.com

Preface

Intelligent Distributed Computing – IDC Symposium Series was started as an initiative of research groups from: (i) *Systems Research Institute, Polish Academy of Sciences in Warsaw, Poland* and (ii) *Software Engineering Department of the University of Craiova, Craiova, Romania*. IDC aims at bringing together researchers and practitioners involved in all aspects of intelligent and distributed computing to allow cross-fertilization and search for synergies of ideas and to enable advancement of research in these exciting sub-fields of computer science. Intelligent Distributed Computing 2008 – IDC 2008 was the second event in this series. IDC 2008 was hosted by *Dipartimento di Ingegneria Informatica e delle Telecomunicazioni, Università di Catania, Italia* during September 18-19, 2008.

This book represents the peer-reviewed proceedings of the IDC 2008. We received 58 submissions from 24 countries. Each submission was carefully reviewed by at least 3 members of the Program Committee. Acceptance and publication were judged based on the relevance to the symposium themes, clarity of presentation, originality and accuracy of results and proposed solutions. Finally 20 regular papers and 12 short papers were selected for presentation and were included in this volume, resulting in acceptance rates of 34.48 % for regular papers and 55.17 % for regular and short papers. The book contains also 3 invited papers authored by well-known researchers in the field.

The 35 contributions in this book address many topics related to intelligent distributed computing, systems and applications, including: adaptivity and learning; agents and multi-agent systems; argumentation; auctions; case-based reasoning; collaborative systems; data structures; distributed algorithms; formal modeling and verification; genetic and immune algorithms; grid computing; information extraction, annotation and integration; network and security protocols; mobile and ubiquitous computing; ontologies and metadata; P2P computing; planning; recommender systems; rules; semantic Web; services and processes; trust and social computing; virtual organizations; wireless networks; XML technologies.

We would like to thank to Prof. Janusz Kacprzyk, editor of *Studies in Computational Intelligence* series and member of the Steering Committee for their kind support and encouragement in starting and organizing the IDC Symposium Series. We would like to thank to the Program Committee members for their work in promoting the event and refereeing submissions and also to all colleagues who submitted papers to the IDC 2008. We deeply appreciate the efforts of our invited speakers (in alphabetical order): Prof. Alberto Montresor, Prof. Marcin Paprzycki, and Prof. Franco Zambonelli and thank them for their interesting lectures. Special thanks also go to Prof. Michele Malgeri, Dr. Alessandro Longheu, and Dr. Vincenzo Nicosia from Università di Catania, Italia, and to all the members of the Dipartimento di Ingegneria Informatica e delle Telecomunicazioni, Università di Catania, Italia for their help with organizing the IDC 2008 event.

Craiova, Catania   Costin Bădică
June 2008   Giuseppe Mangioni
                                              Vincenza Carchiolo
                                           Dumitru Dan Burdescu

Organization

Organizers

Dipartimento di Ingegneria Informatica e delle Telecomunicazioni, Università di Catania, Italia
Software Engineering Department, Faculty of Automation, Computers and Electronics, University of Craiova, Romania

Conference Chairs

Giuseppe Mangioni, Università di Catania, Italia
Costin Bădică, University of Craiova, Romania

Steering Committee

Costin Bădică, University of Craiova, Romania
Janusz Kacprzyk, Polish Academy of Sciences, Poland
Michele Malgeri, Università di Catania, Italia
Marcin Paprzycki, Polish Academy of Sciences, Poland

Organizing Committee

Vincenza Carchiolo, Università di Catania, Italia
Michele Malgeri, Università di Catania, Italia
Giuseppe Mangioni, Università di Catania, Italia
Alessandro Longheu, Università di Catania, Italia
Vincenzo Nicosia, Università di Catania, Italia
Dumitru Dan Burdescu, University of Craiova, Romania
Mihai Mocanu, University of Craiova, Romania
Dan Popescu, University of Craiova, Romania

Invited Speakers

Alberto Montresor, Dipartimento di Ingegneria e Scienza dell'Informazione, University of Trento, Italy
Marcin Paprzycki, Systems Research Institute, Polish Academy of Sciences, Poland
Franco Zambonelli, Dipartimento di Scienze e Metodi dell'Ingegneria, University of Modena and Reggio Emilia, Italy

Program Committee

Razvan Andonie, Central Washigton University, USA
Galia Angelova, Bulgarian Academy of Sciences, Bulgaria
Nick Bassiliades, Aristotle University of Thessaloniki, Greece
Frances Brazier, Vrije Universiteit, Amsterdam, Netherlands
Dumitru Dan Burdescu, University of Craiova, Romania
Giacomo Cabri, Università di Modena e Reggio Emilia, Italy
David Camacho, Universidad Autonoma de Madrid, Spain
Jen-Yao Chung, IBM T.J. Watson Research Center, USA
Gabriel Ciobanu, "A.I.Cuza" University of Iaşi, Romania
Valentin Cristea, "Politehnica" University of Bucharest, Romania
Paul Davidsson, Blekinge Institute of Technology, Sweden
Beniamino Di Martino, Second University of Naples, Italy
Vadim A. Ermolayev, Zaporozhye National University, Ukraine
Adina Magda Florea, "Politehnica" University of Bucharest, Romania
Chris Fox, University of Essex, UK
Maria Ganzha, Elblag University of Humanities and Economics, Poland
Adrian Giurca, Brandenburg University of Technology at Cottbus, Germany
De-Shuang Huang, Chinese Academy of Sciences, China
Axel Hunger, University of Duisburg-Essen, Germany
Mirjana Ivanović, University of Novi Sad, Serbia
Halina Kwasnicka, Wroclaw University of Technology, Poland
Ioan Alfred Leţia, Technical University of Cluj-Napoca, Romania
Alessandro Longheu, University of Catania, Italy
Heitor Silverio Lopes, Federal University of Technology - Parana, Brazil
José Machado, University of Minho, Portugal
Yannis Manolopoulos, Aristotle University of Thessaloniki, Greece
Urszula Markowska-Kaczmar, Wroclaw University of Technology, Poland
Ronaldo Menezes, Florida Institute of Technology, USA
Mihai Mocanu, University of Craiova, Romania
Alexandros Nanopoulos, Aristotle University of Thessaloniki, Greece
Viorel Negru, Western University of Timişoara, Romania
Ngoc-Thanh Nguyen, Wroclaw University of Technology, Poland
Peter Noerr, MuseGlobal, Inc., USA
Vincenzo Nicosia, University of Catania, Italy
George A. Papadopoulos, University of Cyprus, Cyprus

Contents

Part III: Short Papers

Part I

Invited Papers

Intelligent Gossip

Alberto Montresor

University of Trento, Italy
alberto.montresor@unitn.it

Summary. The gossip paradigm made its first appearance in distributed systems in 1987, when it was applied to disseminate updates in replicated databases. Two decades later, gossip-based protocols have gone far beyond dissemination, solving a large and diverse collection of problems. We believe that the story is not over: while gossip is not the panacea for distributed systems, there are still virgin research areas where it could be profitably exploited. In this paper, we briefly discuss a gossip-based "construction set" for distributed systems and we illustrate how intelligent distributed computing could benefit by the application of its building blocks. Simple examples are provided to back up our claim.

1 Introduction to Gossip

Since the seminal paper of Demers et al. [3], the idea of epidemiological (or gossip) algorithms has gained considerable popularity within the distributed systems and algorithms communities.

In a recent workshop on the future of gossip (summarized on a special issue of Operating System Review [12]), there has been a failed attempt to precisely define the concept of gossip. The reason for this failure is twofold: either the proposed definitions were too broad (including almost any message-based protocol ever conceived), or they were too strict (ruling out many interesting gossip solutions, some of them discussed in the next section).

While a formal definition seems out of reach, it is possible to describe a prototypical gossip scheme that seems to entirely cover the intuition behind gossip. The scheme is presented in Figure 1.

In this scheme, nodes regularly exchange information in periodic, pairwise interactions. The protocol can be modeled by means of two separate threads executed at each node: an active one that takes the initiative to communicate, and a passive one accepting incoming exchange requests.

In the active thread, a node periodically (every Δ time units, the *cycle* length) selects a peer node p from the system population through function *selectPeer*(); it extracts a summary of the local state through function *prepareMessage*(); and finally, it sends this summary to p. This set of operations is repeated forever. The other thread passively waits for incoming messages, replies in case of active requests, and modifies the local state through function *update*().

C. Badica et al. (Eds.): Intel. Distributed Comput., Systems & Appl., SCI 162, pp. 3–10, 2008.
springerlink.com

1: **loop**
2:     wait(Δ)
3:     $p \leftarrow selectPeer()$
4:     $s = prepareMessage()$
5:     send $\langle$REQUEST, $s\rangle$ to p
6: **end loop**

1: **loop**
2:     receive $\langle t, s_p\rangle$ from all
3:     **if** $t =$ REQUEST **then**
4:         $s = prepareMessage()$
5:         send $\langle$REPLY, $s\rangle$ to p
6:     **end if**
7:     $update(s_p)$
8: **end loop**

(a) active thread   (b) passive thread

Fig. 1. The generic gossip scheme

The scheme is still too generic and can be used to mimic protocols that are not gossip; as an example, we can map the client-server paradigm to this scheme by simply having all nodes selecting the same peer. For this reason, this scheme must be associated with a list of "rules of thumb" to distinguish gossip from non-gossip protocols:

- peer selection must be random, or at least guarantee enough peer diversity
- only local information is available at all nodes
- communication is round-based (periodic)
- transmission and processing capacity per round is limited
- all nodes run the same protocol

These features are intentionally left fuzzy: for example "limited", "local" or "random" is not defined any further.

With this informal introduction behind, we can focus on what makes gossip protocols so "cool" these days. The main reason is robustness: node failures do not cause any major havoc to the system, and can be tolerated in large quantity; message losses often cause just a speed reduction rather than safety issues. Low-cost is another plus: load is equally distributed among all nodes, in a way such that overhead may be reduced to few bytes per second per node.

The cause of such robustness and efficiency can be traced back to the inherently probabilistic nature of gossip protocols. They represent a certain "laid-back" approach, where individual nodes do not take much responsibility for the outcome. Nodes perform a simple set of operations periodically, they are not aware of the state of the entire system, only a very small (constant) proportion of it, and act based on completely local knowledge. Yet, in a probabilistic sense, the system as a whole achieves very high levels of robustness to benign failures and a favorable (typically logarithmic) convergence time.

We claim that if adopted, the gossip approach can open intelligent distributed computing to a whole variety of autonomic and self-* behaviors, bringing robustness to existing intelligent computing techniques. This claim will be backed up in two steps: first, by showing that several important problems in distributed systems have a robust gossip solution; second, by showing how these solutions can be integrated in existing computational intelligence techniques.

2 Gossip Lego: Fundamental Bricks

Beyond the original goal of information dissemination, gossip protocols have been used to solve a diverse collection of problems. More interestingly, it appears now that most of these solutions can be profitably combined to solve more complex problems. All together, these protocols start to look like a construction set, where protocols can be combined as Lego bricks. Figure 2 lists some of these bricks; the following subsections briefly discuss each of them.

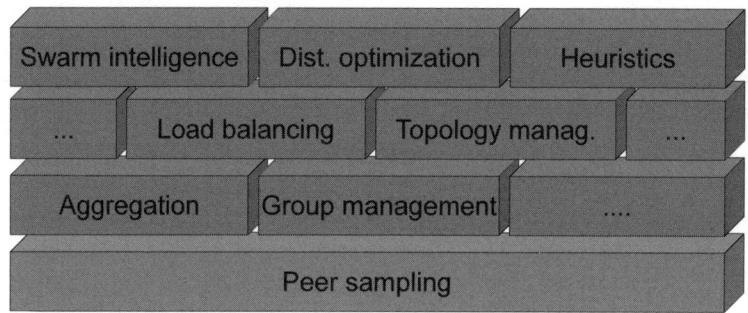

Fig. 2. Gossip Lego: Fundamental bricks

2.1 Peer Sampling

The first piece, *peer sampling*, may be seen as the green Lego baseplate where all kind of models are built. In fact, the problem it solves is at the basis of gossip: how to keep together the population of nodes that constitute the system, in such a way that it is possible implement function *selectPeer()* that selects nodes from such population.

Instead of providing each node with a global view of the system, the peer sampling service provides each node with continuously up-to-date random samples of the entire population. Higher-level gossip protocols may transparently implement *selectPeer()* by randomly choosing a per from this sample. Locally, each node only see the random node returned by *selectPeer()*; globally, the nodes and their samples define an *overlay topology*, i.e. a directed graph superimposed over the network. The graph is characterized by a *random* structure and the presence of a single strongly connected component.

An example instantiation of the peer sampling service is the NEWSCAST protocol [9], characterized by its low cost, extreme robustness and minimal assumptions. The basic idea of NEWSCAST is that each node maintains a local set of random descriptors, called the (partial) *view*. A descriptor is a pair (*node address, timestamp*). NEWSCAST is based on the same scheme as all gossip protocols. Function *selectPeer*() returns a random member of the view; function *prepareMessage*() returns the local view, plus a fresh descriptor of itself. Function *update*() keeps a fixed number of freshest descriptors (based on timestamps), selected from those locally available in the view and those contained in the received message. Nodes belonging to the network continuously inject their identifiers in the network with the current timestamp, so old identifiers are gradually removed from the system and are replaced by newer information. This feature allows the protocol to "repair" the overlay topology by forgetting information about crashed neighbors, which by definition cannot inject their identifiers.

Implementations exist in which these messages are small UDP messages containing approximately 20-30 descriptors, each composed of an IP address, a port and a timestamp. The cycle length Δ is typically long, in the range of 10s. The cost is therefore small, few tens of bytes per second, similar to that of heartbeats in many distributed architectures. The protocol provides high quality (i.e., sufficiently random) samples not only during normal operation (with relatively low churn), but also during massive churn and even after catastrophic failures (up to 70% nodes may fail), quickly removing failed nodes from the local views of correct nodes.

2.2 Decentralized Aggregation

Aggregation is a common name for a set of functions that provide a summary of some global system property. In other words, they allow local access to global information in order to simplify the task of controlling, monitoring and optimizing distributed applications. Examples of aggregation functions include network size, total free storage, maximum load, average uptime, location and intensity of hotspots, etc.

An example of gossip-based aggregation algorithm is contained in [8]. The algorithm assumes that each node maintains a local approximation of the value to be aggregated, initially equal to the value of the local property. Function *selectPeer*() exploit the underlying peer-sampling protocol to return a random node. Function *prepareMessage*() returns the current local approximate value, while function *update*() modifies the local approximate value based on some aggregation-specific and strictly local computation based on the previous values. This local pairwise interaction is designed in such a way that all approximate values in the system will quickly converge to the desired aggregate value. For example, in case of average aggregation, at the end of an exchange both nodes install the average of their current local approximate values; after each exchange, the global average will not change, while variance is reduced. It can be proved that at each cycle, the expected reduction is equal to $(2\sqrt{e})^{-1}$, independently of the size of the network.

2.3 Load Balancing

The aggregation protocol described above is *proactive*, meaning that all nodes participating in the computation are made aware of the final results. This suggests a simple improvement of a well-known load balancing protocol, as well as showing how simple protocol pieces can be combined together [7].

The load balancing scheme we want to improve works as follows: [1, 13]: given a set of tasks that must be executed by a collection of nodes, nodes periodically exchange tasks in a gossip fashion, trying to balance the load in the same fashion as our average aggregation protocol. The problem with this approach is that tasks may be costly moved from one overloaded node to another overloaded node, without really improving the situation - nodes remain overloaded.

Our idea is based on the concept that moving information about tasks is cheaper than moving tasks. For this reason, we use our aggregation service to compute the average load, and then later we put in contact - through a specialized peer sampling service - nodes that are underloaded with nodes that are overloaded. By avoiding overloaded-to-overloaded exchanges, this algorithm guarantees that an optimal number of transfers are performed.

2.4   Slicing

Once collected all nodes in the same basket through peer sampling, one may want to start to differentiate among them, creating sub-groups of nodes that are assigned to specific tasks. This functionality is provided by a *slicing service*, where the population of nodes is divided into groups (slices) which are maintained, in a decentralized way, in spite of failures.

The composition of slices may be defined based on complex conditions based on both node and slice features; example of possible slice definitions include the following:

- nodes with at least 4GB of RAM;
- not more than 10.000 machines, each of them with ADSL connection or more;
- the group composed by the 10% most performant machines;
- a group of nodes whose free disk space sums up to 1PB.

Several protocols have been devised to solve these problems [6, 4, 14]; all of them are based on special versions of peer sampling. For example, if only nodes with special characteristics (e.g., RAM greater than 4GB) are allowed to insert their node descriptor in exchanged messages, we quickly obtain a sub-population that only contains the desired nodes. By using count aggregation, you can limit the size to a specified value; by ranking values, you can select the top 10%; by using sum aggregation, you can obtain the desired disk space.

2.5   Topology Maintenance

Once you have your slice of nodes, it could be required to organize them in a complex structured topology. T-MAN is a gossip-based protocol scheme for the construction of several kinds of topologies in logarithmic time, with high accuracy [5].

Each node maintains a partial view; as in peer sampling, views are periodically updated through gossip. In a gossip step, a node contacts one of its neighbors, and the two peers exchange a subset of their partial views. Subsequently both participating nodes update their lists of neighbors by merging the the received message.

The difference form peer sampling is how to select peers for a gossip step (function *selectPeer*()), and how to select the subset of neighbors to be sent (function *prepareMessage*()).

In T-MAN, *selectPeer*() and *prepareMessage*() are biased by a *ranking function* that represents an order of preference in the partial views. The ranking function of T-MAN is a generic function and it can capture a wide range of topologies from rings to binary trees, to n-dimensional lattices. For example, in an ordered ring, the preference goes to immediate successors and predecessors over the ring itself. It is possible to demonstrate that several different topologies can be achieved in a logarithmic time.

3   Towards Intelligent Gossip

The bricks presented so far are all dedicated to simple tasks, mostly related to the management of the gossip population itself. You can keep together the entire population though peer sampling, select a group of nodes that satisfies a specific condition through slicing, build a particular topology through T-MAN, and finally monitor the resulting system through aggregation.

But gossip is not limited to this. Recent results suggest a path whereby results from the optimization community might be imported into distributed systems and architected to operate in an autonomous manner. We briefly illustrate some of these results.

3.1   Particle Swarm Optimization

PSO [11] is a nature-inspired method for finding global optima of a function f of continuous variables. Search is performed iteratively updating a small number n of random "particles", whose status information includes the current position vector x_i, the current speed vector v_i, together with the optimum point p_i and its *fitness* value $f(p_i)$, which is the "best" solution the particle has achieved so far. Another "best" value that is tracked by the particle swarm optimizer is the global best position g, i.e. the best fitness value obtained so far by any particle in the population.

After finding the two best values, every particle updates its velocity and position based on the memory of its current position, the best local positions and the best global position; the rationale is to search around positions that have proven to be good solutions, avoiding at the same time that all particles ends up in exactly the same positions.

Nothing prevents the particle swarm to be distributed among a collection of nodes [2]. At each node p, a sub-swarm of size k is maintained; slightly departing from the standard PSO terminology, we say that each swarm of a node p is associated to a *swarm optimum* g^p, selected among the particles local optima.

Clearly, different nodes may know different swarm optima; we identify the best optimum among all of them with the term *global optimum*, denoted g.

Swarms in different nodes are coordinated through gossip as follows: periodically, each node p sends the pair $\langle g^p, f(g^p) \rangle$ to a peer node q, i.e. its current swarm optimum and its evaluation. When q receives such a message, it compares the swarm optimum of p with its local optimum; if $f(g^p) < f(g^q)$, then q updates its swarm optimum with the received optimum ($g^q = g^p$); otherwise, it replies to p by sending $\langle g^q, f(g^q) \rangle$.

Simulations results [2] show that this distributed gossip algorithm is effective in balancing the particles load among nodes; furthermore, the system is characterized by extreme robustness, as the failure of nodes has the only effect of reducing the speed of the system.

3.2 Intelligent Heuristics

Many heuristic techniques can be used to approximately solve complex problems in a distributed way. For example, in [10] the problem of placing servers and other sorts of superpeers is considered. The particular goal of this paper is to situate a superpeer close to each client, and create enough superpeers to balance the load. The scheme works in a gossip way, as follows. Nodes randomly take the role of superpeers, and clients are associated to them; then, nodes dissatisfied with the service their receive, start to gossip, trying to elect superpeers that could provide better service. Dissatisfaction could be motivated by the overload of the superpeer, of by the excessive distance between the client and the server. While this heuristics scheme could be stuck in local optima, stills it reasonably improve the overall satisfaction of nodes, especially considering that nodes work in the absence of complete data.

Acknowledgments

Work supported by the project CASCADAS (IST-027807) funded by the FET Program of the European Commission.

References

1. Barak, A., Shiloh, A.: A Distributed Load Balancing Policy for a Multicomputer. Software Practice and Experience 15(9), 901–913 (1985)
2. Biazzini, M., Montresor, A., Brunato, M.: Towards a decentralized architecture for optimization. In: Proc. of the 22nd IEEE International Parallel and Distributed Processing Symposium (IPDPS 2008), Miami, FL, USA (April 2008)
3. Demers, A., et al.: Epidemic Algorithms for Replicated Database Management. In: Proc. of 6th ACM Symp. on Principles of Dist. Comp. (PODC 1987), Vancouver (August 1987)
4. Fernandez, A., Gramoli, V., Jimenez, E., Kermarrec, A.-M., Raynal, M.: Distributed slicing in dynamic systems. In: Proceedings of the 27th International Conference on Distributed Computing Systems (ICDCS 2007), p. 66. IEEE Computer Society Press, Washington (2007)

5. Jelasity, M., Babaoglu, O.: T-Man: Gossip-based overlay topology management. In: Brueckner, S.A., Di Marzo Serugendo, G., Hales, D., Zambonelli, F. (eds.) ESOA 2005. LNCS (LNAI), vol. 3910, pp. 1–15. Springer, Heidelberg (2006)
6. Jelasity, M., Kermarrec, A.-M.: Ordered slicing of very large-scale overlay networks. In: Peer-to-Peer Computing, pp. 117–124 (2006)
7. Jelasity, M., Montresor, A., Babaoglu, O.: A Modular Paradigm for Building Self-Organizing P2P Applications. In: Di Marzo Serugendo, G., Karageorgos, A., Rana, O.F., Zambonelli, F. (eds.) ESOA 2003. LNCS (LNAI), vol. 2977, pp. 265–282. Springer, Heidelberg (2004)
8. Jelasity, M., Montresor, A., Babaoglu, O.: Gossip-based aggregation in large dynamic networks. ACM Trans. Comput. Syst. 23(1), 219–252 (2005)
9. Jelasity, M., Voulgaris, S., Guerraoui, R., Kermarrec, A.-M., van Steen, M.: Gossip-based peer sampling. ACM Transactions on Computer Systems 25(3), 8 (2007)
10. Jesi, G.P., Montresor, A., Babaoglu, O.: Proximity-aware superpeer overlay topologies. IEEE Transactions on Network and Service Management 4(2), 74–83 (2007)
11. Kennedy, J., Eberhart, R.: Particle swarm optimization. In: Proceedings of IEEE International Conference on Neural Networks 1995. vol. 4 (1995)
12. Kermarrec, A.-M., van Steen, M.: Gossiping in distributed systems. Operating Systems Review 41(5), 2–7 (2007)
13. Kok, P., Loh, K., Hsu, W.J., Wentong, C., Sriskanthan, N.: How Network Topology Affects Dynamic Load Balancing. IEEE Parallel & Distributed Technology 4(3) (September 1996)
14. Montresor, A., Zandonati, R.: Absolute slicing in peer-to-peer systems. In: Proc. of the 5th International Workshop on Hot Topics in Peer-to-Peer Systems (HotP2P 2008), Miami, FL, USA (April 2008)

Infrastructure for Ontological Resource Matching in a Virtual Organization

Michał Szymczak[1], Grzegorz Frąckowiak[1], Maria Ganzha[1], Marcin Paprzycki[1], Sang Keun Rhee[2], Myon Woong Park[2], Yo-Sub Han[2], Young Tae Sohn[2], Jihye Lee[2], and Jae Kwan Kim[2]

[1] Systems Research Institute Polish Academy of Sciences, Warsaw, Poland
    marcin.paprzycki@ibspan.waw.pl
[2] Korea Institute of Science and Technology, Seoul, Korea
    greyrhee@kist.re.kr

Summary. In our earlier work we have outlined general approach to ontological matchmaking in an agent-based virtual organization. The aim of this paper is to describe in details how matchmaking is to take place within the system under construction. The Grant Announcement application is used to illustrate the proposed approach. Questions concerning efficiency of matchmaking will be addressed and in this context a distinction between asynchronous and synchronous matchmaking will be proposed.

1 Introduction

As the amount of available information increases rapidly, sometimes the efficient searching method alone is not enough to obtain necessary information in timely manner. Therefore support is needed to share the burden of searching for and filtering information. In the era of ubiquitous computing, computer systems existing everywhere should be able to proactively provide information just in time. Resource matching is essential in order to develop system searching and recommending information required for a user in a specific context. This paper describes the infrastructure and methodology of resource matching in the environment of a Research Institute where most of paperwork is carried out through an intranet. System requirements which are set by the Research Institute include facilitating user specific suggestion based on the knowledge model, actual data and geospatial information about objects. In the following sections we discuss how these factors can be included in a single suggestion request processing. In order to do so, we specify matching functionality, its possible processing modes and system building blocks which allow to realize the requirements. Resource matching utilized in forwarding notices about new research grants to appropriate users is used to illustrate the proposed approach. This work can be generalized and expanded to become a kernel of a smart information provider.

C. Badica et al. (Eds.): Intel. Distributed Comput., Systems & Appl., SCI 162, pp. 11–21, 2008.
springerlink.com   © Springer-Verlag Berlin Heidelberg 2008

2 Defining Matching

Let us start by defining matchmaking in the context of our work. By *matching* we mean establishing closeness between ontology class instances (object(s)) and (an)other selected object(s) for which certain *Matching Criteria* are met. *Calculating Relevance* is a method for finding a degree of relevance (closeness) between objects which are related through their properties. Our approach to calculating relevance was outlined in [9]. In the case considered thus far in our work, *Matching Criteria* is an ordered quadruple $\{x, q, a, g\}$, where:

- x is the selected ontology class instance (the *source object*)
- q is a SPARQL query [5] which defines a subset of objects that are considered relevant and will be matched against the source object x
- $a > 0$ is the *relevance threshold*—value above which objects will be considered relevant
- g is a sub-query processed by the GIS subsystem; this part of the system is responsible for finding cities which are located close to others (part of the *Duty Trip Support* application, see [7]). This sub-query is a triple $\{gc, gr, ga\}$, where:
  - gc is an URI of a city demarcated with the *City* class properties of the system ontology
  - gr is an operator which allows to either limit returned number of cities of interest (*AMOUNT* condition) or to limit the maximum distance between the gc and the returned cities (*RADIUS* condition)
  - ga is the parameter of the gr operator, it either limits the number of cities that can be returned or the maximum distance between the gc and the returned cities

To make the idea of the GIS sub-query clear we can consider its following two instances:

$$\{gc, gr, ga\} = \{geo{:}WarsawCity,\ RADIUS, 100\}$$
$$\{gc, gr, ga\} = \{geo{:}WarsawCity,\ AMOUNT, 50\}$$

As a result of the first query the GIS module should return all cities which are located not further than 100 km from the city of Warsaw (represented by the RDF resource *geo:WarsawCity*). On the other hand, the second request means that a maximum of 50 cities should be found (that are located closest to Warsaw).

Note that, in general, the GIS sub-query can be omitted, or replaced with a different criterion (or a group of criteria). Therefore, due to the lack of space, it will be left to be discussed in more details in subsequent publication.

2.1 Grant Announcement-Based Matching Example

In our earlier work [7] we have provided an example of a scientist employed in a Science Institute in North-east Asia.

When represented in the system, this sample employee (Prof. Chan), can have several profiles assigned. Below we depict an example of the general *Employee* profile, which consists of a *Personal Profile* and an *Experience Profile*:

```
: Employee\#1 a onto : ISTPerson ;
  onto : id "1234567890"^^xsd : string ;
  onto : hasProfile (: Employee\#1PProfile ,: Employee\#1EProfile ),
  onto : belongsToOUs (:GOU ).
: ResearchOU a onto : OrganizationUnit ;
  onto : name ''Researchers Organization Unit''^^xsd : string .
```

In this example the *Employee#1PProfile—Personal Profile*, which describes the "human resource properties" of an employee. In what follows we use basic properties (however, our system supports a complete list of needed HR-related properties): *fullname*, *gender* and *birthday*. Furthermore, the *belongsToOUs* property indicates Prof. Chan's appointment in the organization.

```
: Employee\#1PProfile a onto : ISTPersonalProfile ;
  onto : belongsTo : Employee\#1;
  person : fullname ''Yao Chan''^^xsd : string ;
  person : gender person : Male ;
  person : birthday ''1982−01−01T00 :00 :00 ''^^xsd : dateTime .
```

The second possible profile of *Employee#1* (Prof. Chan) is an *Experience Profile*. It demarcates human resource specialization in terms of fields of knowledge and project experience. Note that codes for the specification of fields of knowledge originate from the KOSEF (Korea Science and Engineering Foundation) [3].

```
: Employee\#1EProfile a onto : ISTExperienceProfile ;
  onto : belongsTo : Employee\#1;
  onto : doesResearchInFields
      scienceNamespace : Volcanology −13105,
      scienceNamespace : Paleontology −13108,
      scienceNamespace : Geochronology −13204;
  onto : knowsFields
[a onto : Knowledge ;
  onto : knowledgeObject scienceNamespace : Volcanology −13105;
  onto : knowledgeLevel "0.75"^^xsd : float] ,
[a onto : Knowledge ;
  onto : knowledgeObject scienceNamespace : Paleontology −13108;
  onto : knowledgeLevel "0.40"^^xsd : float] ,
[a onto : Knowledge ;
  onto : knowledgeObject scienceNamespace : Geochronology −13204;
  onto : knowledgeLevel "0.90"^^xsd : float] ;
  onto : managesProjects (: Project1 ).
```

According to the *Experience Profile*, Prof. Chan specializes in *Volcanology, Paleontology* and *Geochronology*. While the level of knowledge in each of these areas is expressed as a real number from the interval $(0, 1)$, respectively: 0.75, 0.4, 0.9, without loss of generality, we omit influence of this factor on the matching process. Additionally, *Employee#1* who is described with that profile manages a project *Project1*. It is a scientific project in *Volcanology*, which has its own profile:

```
: Project1  a  onto : ISTProject ;
   onto : managedBy   : Employee\#1;
   onto : period
      [a  onto : Period ;
      onto : from  "2008−06−01T00 :00:00 "^^xsd : dateTime ;
      onto : to  "2009−05−31T00 :00:00 "^^xsd : dateTime ] ;
   onto : fieldsRef  scienceNamespace : Volcanology −13105;
   onto : projectTitle  ''Very  Important  Volcanology
      Scientific  Project ''^^xsd : string .
```

The listings introduced above set a *context* within which we place a member of an organization. In order to make the matching definition clearer, let us now introduce an instance of a *SampleGrant*. Note that the sample grant could be replaced by any resource that is delivered to the organization and information about which has to be delivered to the right employees (e.g. a book, or a transport of copy paper). Profile of the proposed *SampleGrant* specifies its domain as *Geochemistry*. Obviously a resource could be demarcated using more complicated structure of covered areas and the proposed approach would work as well.

```
: SampleGrant  a  onto : ISTAnnouncement;
   onto : hasDescription
      ''Description  of  the  exemplary  grant  announcement.
      It  should  be  really  interesting .''^^xsd : string ;
   onto : refScientificFields  (<scienceNamespace :
                          Geochemistry −13200>).
```

In order to match the *SampleGrant* announcement with a human resource represented by the *ISTPerson* class instance, the following process has to be executed:

1. Construct a set of *Matching Criteria* x, q, a, g:
   a) $x =: SampleGrant$ (comparing against :*SampleGrant*)
   b) $q =$

      PREFIX onto :
      <http :// rossini . ibspan .waw. pl / Ontologies /KIST/KISTVO>
      SELECT ?person
      WHERE {
         ?person **isa** onto : ISTPerson .
         ?profile **isa** onto : ExprienceProfile .
         ?person onto : hasProfile ?profile .
         ?person onto : belongsToOU : ResearchOU}

   c) $a = \frac{1}{40}$ (sample value)
   d) $g = NULL$, since *Grant Announcement* scenario does not require any geo-spatial support
2. Execute thus generated SPARQL query. In the case of the *Grant Announcement* scenario [8, 7] this query is going to filter employee' experience profiles

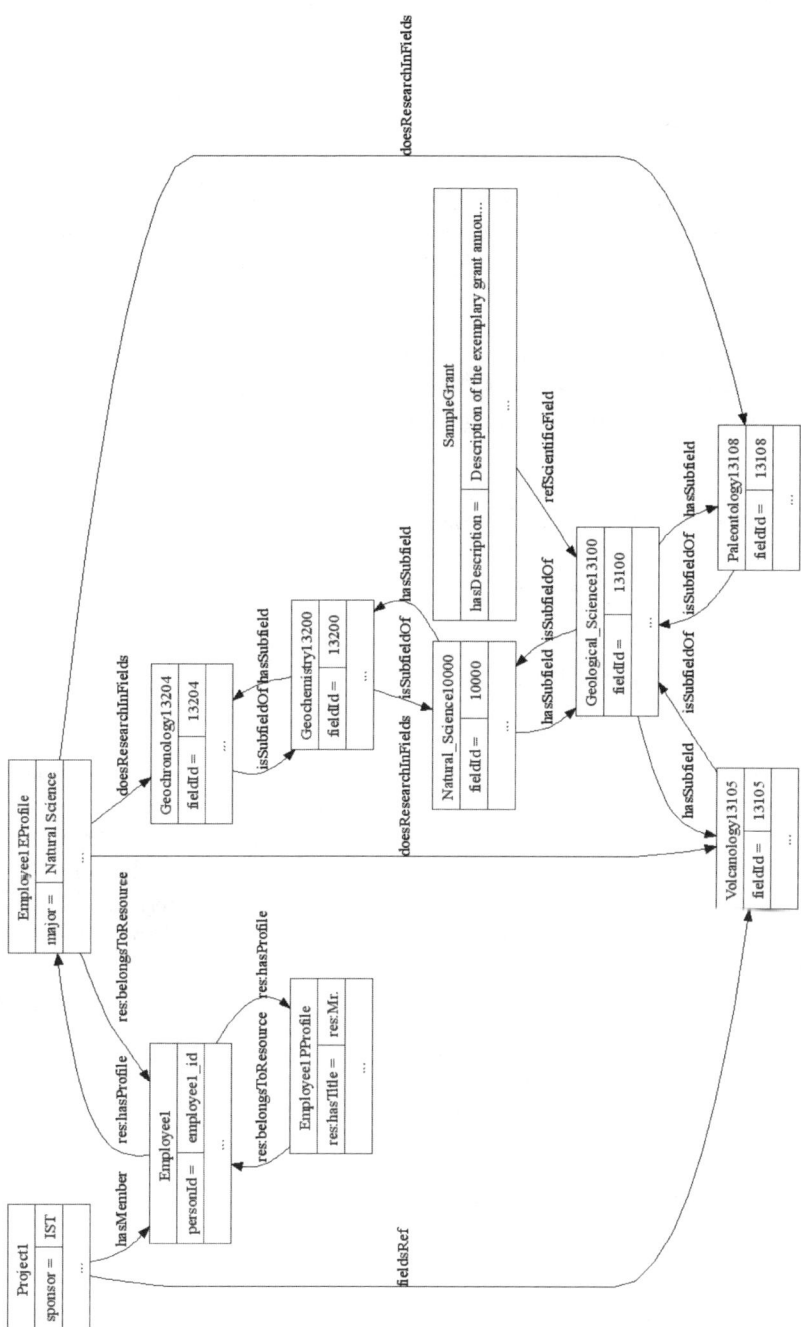

Fig. 1. Relations between sample Employee and sample Grant Announcement objects

leaving only researchers, i.e. employees that belong to the *Organization Unit* which is specific to all researchers (more on *Organization Units*, their role in the knowledge base and examples can be found in [7, 12, 9]). In our example, the result of the query is the *Employee#1EProfile*.

3. Perform *Relevance Calculations* for (in our example):
   a) source object $URI =: SampleGrant$
   b) target objects $URI's = [: Employee\#1EProfile]$

In Figure 1 we depict relations between *Employee#1* and *SampleGrant* objects. Unfortunately, due to the complexity of the example, listings including all relations between these objects would make it illegible. However, note that all objects are linked with properties which are included in the listings and refer to ontology classes and properties described in [9, 7, 12]. For instance, path between the *SampleGrant* and the *Employee1* is composed through the following intermediate nodes: *GeologicalScience13100*, *Volcanology13105* and *Employee1EProfile*. Additionally, following property weights are set in the ontology (see, [12]):

$$voPropertyWeight(doesResearchInFields) = 2$$
$$voPropertyWeight(isSubfieldOf) = 5$$
$$voPropertyWeight(hasSubfield) = 3$$
$$voPropertyWeight(refScientificField) = 2$$
$$voPropertyWeight(invRefScientificField) = 8$$
$$voPropertyWeight(hasProfile) = 1$$
$$voPropertyWeight(belongsToResource) = 1$$

These weights are given just sample values. Thus far we have not designed the mechanism for setting them up. Based on the weights and relations presented in Figure 1, closeness between the two objects can be computed. The computed relevance, according to the algorithm proposed in [10, 11], is $\frac{1}{36}$. Depending on the threshold used in the application, this degree of closeness may or ma not be considered "close enough." Since we specified the threshold value as $a = \frac{1}{40}$, this grant information will be delivered to Mr. Chan.

3   Matching Request Processing

Matching Criteria defined in the previous section require adequate computations to be performed by:

- GIS Subsystem (mostly omitted in this paper)
- SPARQL engine
- Relevance Graph matching

All these operations are by default heavily resource consuming. Therefore we distinguish two basic matching modes to be supported by the system: *synchronous* and *asynchronous*.

3.1   Synchronous Matching Request Processing

Synchronous method of matching request processing might be highly valuable for applications or even single components that require short response time. Possible usages of this method include but are not limited to (1) matching in the case of a process which has to deliver the result to a web page in a synchronous mode, or (2) requesting a single result based on the current state of objects stored in the system.

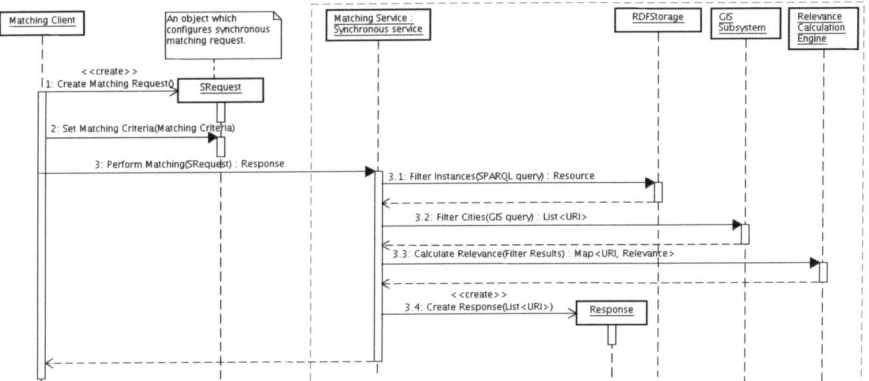

Fig. 2. Synchronous matching request processing

This approach has been represented in Figure 2 for the agent-based information processing. Specifically we can see there that:

1. A *Matching Client* (a role which can be realized by any agent capable of performing it) creates a synchronous request *SRequest* object.
2. The *Matching Client* fills *Matching Criteria* of the matching request.
3. The service which is responsible for processing synchronous matching requests (due to the fact that it extends abstract *SynchronousService* class) receives the new request and orders the *RDF Storage* (e.g. Jena) to provide results of SPARQL matching request part (*SPARQL Filtering*).
4. *GIS Subsystem* may be requested to perform *Cities Filtering* based on the GIS request.
5. Steps 3 and 4 filter objects which meet SPARQL (and GIS) *Matching Criteria*. These objects are processed by the *Relevance Calculation Engine* which is based on the *Relevance Graph* (see below).
6. Results of relevance calculation are wrapped in the *Response* object and sent back to the requesting client.

3.2   Asynchronous Matching Request Processing

Asynchronous method of request processing might be more suitable for low priority relevance calculations and for calculation which have to be repeated within

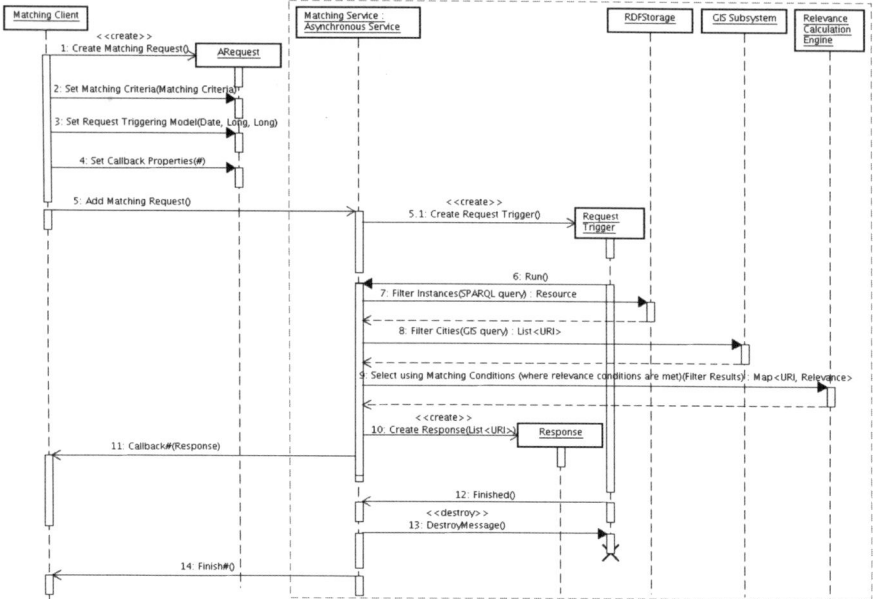

Fig. 3. Asynchronous matching request processing

a certain time frame. For instance, consider search of employees who should be informed about a *Grant Announcement* (GA). Here, we can assume in that the *GA* is valid for some predefined time (until its deadline) and during that time there may appear "new" individuals who meet the *Matching Criteria*; e.g. due to their profile update. Since the asynchronous mode supports repeating request, results which include resource with changed profiles will be returned.

Figure 3 represents the sequence diagram of the asynchronous matching process, which can be described as follows:

1. A *Matching Client* (a role similar to the *Matching Client* described in the section above) creates an asynchronous request *ARequest* object.
2. The client fills *Matching Criteria* of the matching request.
3. The client sets request triggering conditions.
4. The client sets request callback properties.
5. The service which is responsible for processing asynchronous matching requests (service which extends abstract *AsynchronousService* class) receives the new request and creates a trigger object which is set in accordance with triggering conditions specified in the *ARequest* object.
6. Each time the trigger executes its *Run* function, it starts a process which is similar to the *Synchronous Request Processing*. Function calls 6.1–6.3 correspond to the function calls 4.1–4.3 presented in Figure 2.

7. Results of relevance calculation are wrapped in the *Response* object and sent back to the client using the callback function defined in the *A Request* object (callback properties).

8. If the trigger notifies that it has finished the scheduled work, the service is informed and similar request processing finish notification is sent to the client using the same callback settings.

4   System Building Blocks

Let us now describe in some details the main building block involved in relevance calculations.

4.1   Relevance Calculation Engine

The main role of the *Relevance Calculation Engine* (*RCE*) is, given a resource, produce a list of related resources with their relevance values. This module is designed in Java, with additional libraries from the Jena API [6] for ontology model handling, and the Structure Package [2] for dealing with the graph structure. The relevance measure algorithm applied here was first introduced in [10], and its initial application was described in [11].

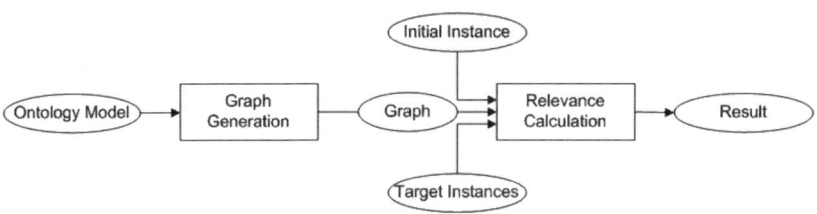

Fig. 4. Relevance calculation process

The process illustrated in Figure 4 includes objects that play key roles in calculating relevance between the *Initial Instance* and the *Target Instances*.

Relevance Graph

The core of the *RCE* is based on a graph structure that represents the underlaying Jena Ontology Model. The Model is interpreted as a directed graph $G = (V, E)$ so that:

V : set of nodes, representing all instances (or individuals)

E : set of edges, representing all object properties.

Note that reflexive relations are ignored. Upon creating edges, the value of the annotation property *voPropertyWeight* of each object property becomes the label of the edge, representing the distance between two nodes. The relevance value between two nodes is the inverse of the distance value. Graph creation function results in creating a weighted, directed graph on ontology resources and their object properties.

Relevance Calculation Interface

The *Relevance Calculation Engine* establishes closeness between a specific (*Initial*) resource and a given list of (*Target*) resources. The result is returned as an instance of the Java *Map⟨Key, Value⟩* interface, where *Key*s are the URI's of resources and *Value*s are the relevance results for these resources and the *Initial* node computed by the engine.

4.2   GIS Sub-system

In [8, 7] we have outlined utilization of the GIS module. The state of the art research shown that we can provide reliable geospatial backend for our system by using the following components: (1) GeoMaker [1] for collecting geographic coordinates of cities in the world, (2) PostgreSQL database [4] for storing that information and calculating distance between cities on demand, finally for caching the result, (3) Java for interfacing the GIS module with the rest of the system. As the GIS based calculation details were omitted in the text, we only sketch the description of the GIS. We will provide detailed reports on the architecture and efficiency of the GIS module in our subsequent publications.

5   Concluding Remarks

In the text we have outlined how the resource matching and relevance calculations will be facilitated in our agent-based virtual organization. We are currently implementing the proposed approach. Our subsequent immediate research goals are: property weights setup, stress and performance tests, matching including time frame constraints.

Acknowlegement

Work was partially sponsored by the KIST-SRI PAS "Agent Technology for Adaptive Information Provisioning" grant.

References

1. Geomaker.
   http://pcwin.com/Software_Development/GeoMaker/index.htm.
2. Java structure.
   http://www.cs.williams.edu/~bailey/JavaStructures/Software.html.

3. Korea science and engineering foundation.
   http://www.kosef.re.kr/english_new/index.html.
4. Postgis:home.
   http://postgis.refractions.net/.
5. Sparql query language for rdf. http://www.w3.org/TR/rdf-sparql-query.
6. Jena—a semantic framework for java. http://jena.sourceforge.net, 2008.
7. G. Frackowiak, M. Ganzha, M. Gawinecki, M. Paprzycki, M. Szymczak, M.-W. Park, and Y.-S. Han. *Considering Resource Management in Agent-Based Virtual Organization*. LNCS. Springer, 2008. in press.
8. G. Frackowiak, M. Ganzha, M. Gawinecki, M. Paprzycki, M. Szymczak, M.-W. Park, and Y.-S. Han. On resource profiling and matching in an agent-based virtual organization. In *Proceedings of the ICAISC'2008 conference*, LNCS. Springer, 2008.
9. M. Ganzha, M. Gawinecki, M. Szymczak, G. Frackowiak, M. Paprzycki, M.-W. Park, Y.-S. Han, and Y. Sohn. Generic framework for agent adaptability and utilization in a virtual organization—preliminary considerations. In J. Cordeiro et al., editors, *Proceedings of the 2008 WEBIST conference*, pages ISâĂŚ17âĂŚ–ISâĂŚ25. INSTICC Press, 2008. to appear.
10. S. Rhee, J. Lee, and M.-W. Park. Ontology-based semantic relevance measure. *CEUR-WS*, 294(1613-0073), 2007.
11. S. Rhee, J. Lee, and M.-W. Park. Riki: A wiki-based knowledge sharing system for collaborative research projects. In *Proceedings of the APCHI 2008 Conference*, LNCS. Springer, 2008.
12. M. Szymczak, G. Frackowiak, M. Gawinecki, M. Ganzha, M. Paprzycki, M.-W. Park, Y.-S. Han, and Y. Sohn. Adaptive information provisioning in an agent-based virtual organization—ontologies in the system. In N. Nguyen, editor, *Proceedings of the AMSTA-KES Conference*, volume 4953 of *LNAI*, pages 271–280, Heidelberg, Germany, 2008. Springer.

Architecture and Metaphors for Eternally Adaptive Service Ecosystems

Franco Zambonelli[1] and Mirko Viroli[2]

[1] DISMI – Universita' di Modena e Reggio Emilia
42100 Reggio Emilia, Italy
franco.zambonelli@unimore.it
[2] DEIS – University of Bologna
47023 Cesena (FC), Italy
mirko.viroli@unibo.it

Summary. In this paper, we first motivate the need for innovative open service frameworks that ensure capability of self-adaptability and long-lasting evolvability (i.e., eternity). On this basis, we discuss how such frameworks should get inspiration from natural ecosystems, by enabling modelling and deployment of services as autonomous individuals in an ecosystem of other services, data sources, and pervasive devices. A reference architecture is presented to clarify the concepts expressed, and then several possible approaches to realise the idea are surveyed and critically analyzed.

1 Motivations

In the near future, pervasive sensing and actuating devices will densely populate our everyday environments, will be tightly integrated with current Telecom and Internet networks, and will eventually contribute to blur the distinction between Telecom and Internet networks [5, 9].

In this context of tight convergence and integration, a single innovative open software platform will have to be provided to host and orchestrate in an integrated and self-managing way the execution of general-purpose pervasive Telecom/Web services and the organization of large masses of contextual data. Also, such an infrastructure should take into account the increasingly diverse and demanding needs of users (which will also seamlessly act as consumers and producers of data and services) [14], and must be able to flexibly tolerate evolutions over time without requiring significant re-engineering to incorporate innovations and changing needs.

Recently, a great deal of research activity has been devoted to produce solutions to match the emerging characteristics of future networks [7] and to solve problems related to, e.g., increasing dependability, reducing management efforts via self-* features, enforcing context-awareness and adaptability, tolerating evolution over time and eventually ensure that the overall service framework (if not all services within) can be highly adaptive and very long-lasting, even in the absence of explicit management actions. Unfortunately, most of the solutions so far are proposed in terms of "add-on", one-of solutions to be integrated in existing frameworks. The result of this process is often an

C. Badica et al. (Eds.): Intel. Distributed Comput., Systems & Appl., SCI 162, pp. 23–32, 2008.
springerlink.com

increased complexity of current frameworks and the emergence of contrasting trade-off between different solutions.

For instance, while the strict layering of network architectures and protocols enables services to easily survive changes at the device and communication level, it prevents information about the current execution context of services from freely flowing in the network, limiting the adaptability of services. As another example, a huge amount of research proposes application-level and self-organizing overlay networks as a mechanism to exchange information in several distributed and dynamic scenarios. However, such overlays are typically conceived to serve specific classes of application problems in specific network scenarios, and cannot tolerate adaptations with regard to changes in their usage or in the characteristics of the underlying network.

In our opinion, there is need for tackling all the above problems by reformulating their foundation, and by trying to answer the following question: *Is it possible to conceive a radically new way of modelling integrated network services and their execution environments, such that the apparently diverse issues of enabling pervasiveness, context-awareness, dependability, openness, diversity, flexible and robust evolution, can all be uniformly addressed once and for all?* In other words, in our opinion, the way towards the realisation of eternally adaptive services is to tackle the problem form the foundation, and start from a total deconstruction of current service architectures and models. We should no longer see services as localised "loci" of data and functionalities, whose activities are to be orchestrated and synchronised according to specific patterns, with the support of middleware services such as discovery services, routing services, data and context services, and where self-adaptability and evolvability are enforced via the introduction of autonomic managers [11]. Rather, we should start taking inspiration from natural systems [16, 8], where adaptability and eternal evolvability are there because of the basic "rules of the game".

No matter whether one thinks at physical systems, at chemical systems, at biological systems, as well as at general ecological systems. In all these systems, you can always recognise the following characteristics: Above a common environmental substrate (defining the basic "laws of nature" and the ground on which individuals can live), individuals of different kinds (or species) interact, compete, and combine with each other (in respect of the basic laws of nature), so as to serve their own individual needs as well as the sustainability and the evolvability of the overall system.

Although such considerations apply whether you think at a physical system, at chemical systems, or at biological system, let us try to better elaborate the idea by referring at biological evolution and at the dynamics of life on earth. Life is, to most extents (i.e., apart from planetary catastrophes) sort of eternal and eternally adaptive: mechanisms, protocols, and the basic "infrastructure" of life do not change and have never changed since its first appearance. Simply, new individuals get on appearing, finding their own way in the overall system, and possibly leading to the emergence of new reactions and new combinations of elements, in the end possibly leading to the emergence of new life forms and new ecological dynamics. The chemistry of life is eternal, the forms under which it manifests depends on the specific characteristics of the environment, and the specific contingencies occurring in the environment, and on the specific species that populate the environment. Also, very important, it is not individuals that evolve, but the ecosystem as a whole.

This is the sort of endeavour that we think one should assume towards the realisation of "Eternally adaptive service ecosystems": conceiving services and data components as individuals in an open ecosystem in which they interact accordingly to a limited set of "eco-laws" to serve their own individual purposes in respect of such laws, and where self-adaptation and eternity are inherent, endogenous properties of the ecosystem rather than peculiar characteristics of its individuals [10].

Against this background, the remainder of this paper *(i)* tries to sketch a general reference architecture for nature-inspired service ecosystems and *(ii)* surveys and shortly analyses the possible metaphors that can be adopted for such ecosystems, and that can lead to different realisation of the reference architecture.

2 A Reference Architecture for Eternally Adaptive Service Ecosystems

Independently of the specific approach adopted, a uniform reference architecture can be adopted for open service ecosystems. A pictorial representation of such an architecture is reported in Figure 1.

At the very low level, the physical ground on which the ecosystem will be deployed is laid, which is a very dense network (ideally, a pervasive continuum) of networked computing devices and information sources. The former includes all the devices that are going to increasingly pervade all our everyday environments (e.g., PDAs, smart phones, sensors, tags), all interconnected with each other. The latter includes the increasing amount of Web data sources that already (and increasingly) collect knowledge, facts and events about nearly every aspect of the world.

At the top level, service developers, producers and consumers of services and data, access the open service framework for using/consuming data or services, as well as for producing and deploying in the framework new services and new data components. At both the lower and the top levels of the architecture openness stands: on the one hand, new devices can join/leave the system at any time; on the other hand, new users can interact with the framework and can deploy new services and data items on it. Between these two levels, the components of the ecosystem reference architecture stand.

The level of "Species" is the one in which physical and virtual devices of the pervasive system, digital and network resources of any kind, persistent and temporary knowledge/data, contextual information, events and information requests, and of course software service components, are all provided with a uniform abstract view of being the"living entities" of the system, which we refer to as *ecosystem individuals*, and which populate the world. Although such individuals are expected to be modelled (and computationally rendered) in a uniform way, they will have specific characteristics very different from each other, i.e., they will be of different "species".

In a bootstrap phase, an ecosystem is expected to be filled with a set of individuals physically deployed in the environment (physical and network resources, contextual information, initialization data and services, and so on). From then on, the ecosystem eternally lives, with the population of individuals evolving in different ways: *(i)* the

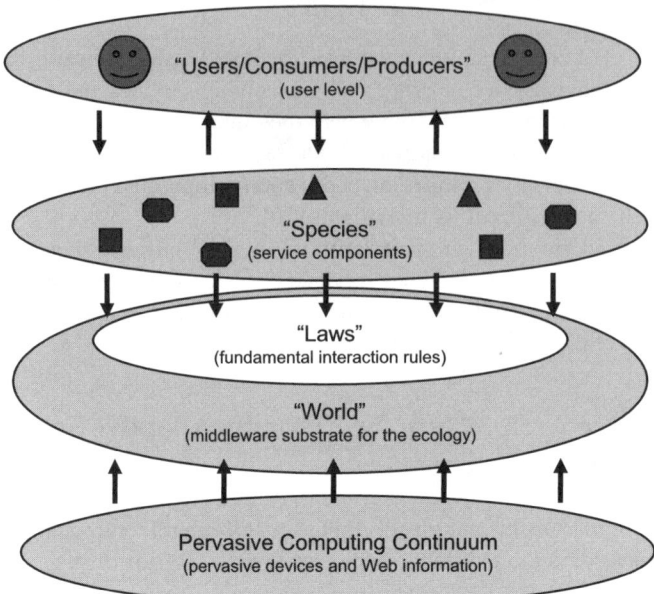

Fig. 1. Service Ecosystems Architecture

initial set of individuals is subject to changes (to tackle the physical system's mobility, faults, and evolution); *(ii)* service developers and producers inject in the system new individuals (developers insert new services and virtual devices, prosumers insert data and knowledge); and *(iii)* consumers keep observing the environment for certain individuals (inject information requests and look for certain data, knowledge, and events).

Below the level of species there is the individuals "world" level, which provides the virtual fabric that supports individuals, their activities and interactions, as well as their insertion and evolution. This overall ecological behaviour is to be enacted by a middleware substrate, a software infrastructure deployed on top of the physical deployment context (i.e., on top of the pervasive continuum), which is in charge of handling insertion and observation of individuals, their persistence and accessibility, as well as their interaction. Moreover, it should deal with the mobility and dynamism of the underlying context, properly turning it into the creation, destruction and change of individuals representing physical and network resources. More in general. the world level define the "shape" and the characteristics of the world in which individuals live.

The way in which individuals interact, compose with others, aggregate so as to form or spawn new individuals, and decay (ultimately winning or losing the natural selection process intrinsic in the ecosystem) is determined by the set of fundamental "laws" regulating the eternal service ecosystems model. Starting from the unified description of living entities—the information/service/structure they provide—and from proper matching criteria, such laws basically specify the likelihood of certain spontaneous evolutions of individuals or groups of individuals. Typical evolution patterns driven by such laws are the following: temporary data and services decay as long as they are not exploited until disappearing, and dually, they get reinforced when exploited; data, data requests, and data retrieving services might altogether match, hence spawning "data-found" events; new services can be created by aggregating existing services whose description strongly matches; an existing service can be upgraded—or corrected when faulty—by injecting a service patch that will automatically compose to it; and so on. A key consequence of the fact that all components are seamlessly seen as individuals is that ecological laws abstract away from the peculiarities of the above cases, uniformly dealing with the concepts of individuals' match-based grouping and evolution: laws of the ecology are the only part of the ecosystem that do not evolve—as happens in natural ecologies. Accordingly, services and applications built with this paradigm will never be stopped or shutdown, but simply debugged, maintained, sustained, or sacrificed by an on-the-fly evolution based on laws behaviour and on the insertion of new individuals—e.g., acting like cures, diseases, feeding resources or viruses.

3 Survey and Analysis of Possible Approaches

The key difference in the possible approaches that can be undertaken towards the realisation of eco-inspired service frameworks (as from the described reference architecture) stands in the metaphor adopted to model the ecosystem, its individuals, and its laws. In particular—without excluding the existence of other useful natural metaphors or the possibility of conceiving interesting non-natural metaphors—the main metaphors that can be adopted and have been suggested are: physical metaphors [6, 12], chemical metaphors [3], biological metaphors [2, 4, 15], together with the properly called ecological metaphors [1, 13].

A summary of the characteristics of each of these metaphors is in Figure 2. We emphasise in any case that, so far, none of these metaphors has been actually adopted to extensively study and prototype an actual, open and general-purpose service framework: either the metaphor has been applied to specific application scenarios [12, 4, 15] or its potential general adoption has been only envisioned [6, 1].

3.1 Metaphors

Let us now come to the distinguishing characteristics of each metaphor, a summary of which is in Figure 2.

Physical metaphors consider that the species of the ecosystem are sort of computational particles, living in a world of other particles and virtual computational fields, which act as the basic interaction means. In fact, all activities of particles are driven by laws that determine how particles should be influenced by the local gradients and shape of the computational field: they can change their status based on the perceived fields, and they can move or exchange data by navigating over such fields (i.e., by having particles that move following the gradient descent of a field, or by making them spread sort of data particles to be routed according to the shape of fields). The world in which such particles live and in which fields spread and diffuse can be either a simple (euclidean) metric world, or it could be a sort of relativistic world, in which shapes and distances

	Species	Laws	World
Physical	Particles (computational components) and messages (computational fields)	Navigation and activities driven by fields (gradient ascent by components)	The Universe (a network), as shaped by waves and particles.
Chemical	Atoms (semantically described) and Molecules (composed semantic descriptions)	Chemical Reactions (matching of semantic descriptions and bonding of components)	Space (localities/bags of components)
Biological	Cells (amorphous computing cells, modules of self-assembly components)	Diffusion of chemical gradients and morphogens, differentiation of behaviour and activity	Space (Abstract computational landscapes, or physical landscapes)
Ecological	Organisms (Agents) and Species (Classes) and Resources (Data)	Survive (goal-orientation), eat, produce, and reproduce	Niches (Pervasive computing environments

Fig. 2. Metaphors for Service Ecosystems

in the environment are not "inherent" but are rather shaped by fields themselves (as in gravitational space-time).

Chemical metaphors consider that the species of the ecosystem are sorts of computational atoms/molecules, with properties described by some sort of semantic descriptions which are the computational counterpart of the description of the bonding properties of physical atoms and molecules. Indeed, the laws that drive the overall behaviour of the ecosystem are sort of chemical laws, that dictates how chemical reactions and bonding between components take place (i.e., relying on some forms of pattern matching between the semantic description of components), and that can lead to both the production of aggregates (e.g., of aggregated distributed components) or of new components (e.g., of composite components). In this case, the world in which components live is typically formed by a set of localities, intended as the "solution" in which chemical reactions can occur, altough of course it is intended that components can flow/diffuse across localities to ensure globality of interactions.

Biological metaphors typically focusses on biological systems at the small scale, i.e., at the scale of individual organisms (e.g., cells and their interactions) or of colonies of simple organisms (e.g. ant colonies). The species are therefore either simple cells or very simple (unintelligent) animals, that act on the basis of very simple goal-oriented behaviours (e.g., move and eat) and that are influenced in their activities by the strength of chemical signals in their surroundings. Similarly to physical systems, in fact, components are expected (depending on their status) to be able to spread and diffuse (chemical) signals around, that can then influence the behaviour of other components. The laws of the ecosystem determines how such signals should diffuse, and how they could influence the behaviour and characteristics of components. The world in which components live is typically a virtual computational landscape that can influence the way signals diffuse and the way components can move over it.

Ecological metaphors focusses on biological systems at the level of animal species and of their interactions. The components of the ecosystem are sort of goal-oriented animals (i.e., agents) belonging to a specific species (i.e., agent classes), that are in search of "food" resources to survive and prosper (e.g., specific resources or other components). The laws of the ecosystem determine how the resulting "web of food" should be realised, that is, they determines how and in which conditions animals are allowed to search food, eat, and possibly produce and reproduce, thus influencing and ruling the overall dynamics of the ecosystem and the interaction among individuals of different species. Similarly to chemical systems, the shape of the world is typically organized around a set of localities, i.e., of ecological niches (think at a set of local pervasive computing environments), yet enabling interactions and diffusion of species across niches.

3.2   Space, Time, and Control

The analysis of which metaphor to adopt cannot abstract from the fundamental question of: what do we want to achieve with it? What features do we want our ecosystem to express?

In general terms, as already stated in the introduction, we think that the ecosystem should be able to exhibit features of self-adaptation and eternity. Such features, from a very practical viewpoint, translate in: *(i)* the capability of the ecosystem of autonomously self-organize the distributed (i.e., spatial) activities of the components, so as to autonomously adapt the overall structure and behaviour of the system to specific contingencies; *(ii)* the capability of the ecosystem of tolerating changes over time, which includes the capabilities of adaptively accommodating new species or of surviving the extinction of species, as well as the capability of accommodating very diverse and composite behaviour with the same limited set of eco-laws. In addition, since we should never forget that the service ecosystem is here to serve us, we cannot forget an additional important feature, that is, *(iii)* the need allow humans (e.g., system administrator and users) to exert control over the behaviour of the ecosystem (or of some of its parts), i.e., of directing its activities and behaviour over space and time. All of these features, of course, should be enforced without paying the price of dramatically increasing the complexity of the ecosystem (i.e., the number and complexity of eco-laws, and the structure of its components and of the world in which they live).

The analysis of the extent to which the presented metaphors are able to accommodate (and how easily and naturally) the above features, is very complex, and would require much more room than the few pages of this paper. Nevertheless, we can try at least to draw some considerations about this, as summarised in Figure 3.

Physical metaphors have been extensively studied for their spatial self-organization features, and in particular for their capability of facilitating the achievement of coherent behaviours even in large scale system (e.g., for load balancing and data distribution), and the conceptual tools available for controlling the spatial behaviour and the dynamics of such systems are well-developed. However, the physical metaphor seems to fall

	Space (self-organization)	Time (evolution and adaptation)	Control (decentralized management)
Physical	+ (global self-organizing spatial structures)	-- (no new components, always same behaviours)	++ (we know well how to build and control specific structures in physic)
Chemical	+ (mostly local self-organizing structures, sometimes global too, as in crystals)	++ (several new components can be generated under the same basic laws)	+ (reactants and catalysts can exert control over the dynamics and structure of reactions)
Biological	+ (local, morphogenesis of local shapes)	-- (limited number of new "shapes", and only local changes)	- (mechanisms of morphogenesis not fully understood)
Ecological	+ (local structures mostly, although sometimes leading to more global patterns)	++ (several new species and same laws)	-- (difficult to understand how to enforce control over ecologies of many species, at most only local centralized control)

Fig. 3. Advantages and Limitations of the Different Metaphors

short in evolution and time adaptation, in that it hardly tolerates the presence of very diverse components with very diverse behaviours (at least if we want to preserve the simplicity of the eco-laws).

Chemical metaphors, on the other hand, can effectively lead to local self-organizing structures (e.g., local composite services) and, to a more limited extent, to some sorts of global structures (e.g., networks of distributed homogeneous components, as in crystals). Real chemistry, and so chemical computational metaphors, can accommodate an incredible amount of different components and composites, yet with the same set of simple basic laws. This is an important pre-condition for facilitating evolution over time. As far as control is concerned, one can think at using sort of catalyst or reagent components to control the dynamics and the behaviour of a chemical ecosystem.

Biological metaphors appears very flexible in enabling the spatial formation of localised morphological and activity patterns, and this has been shown to have notable applications in a variety of applications to distributed systems. However, the number of patterns that can be enforced by the spread of chemical gradients and by the reactions of simple individuals seem (as it is in physical metaphors) quite limited, and this does not match with the need for time evolution and adaption. Moreover, it is quite difficult to understand how to properly control the overall behaviour of such systems (just think at the fact that, so far, the mechanisms of morphogenesis are not fully understood by scientists).

Ecological metaphors, the same as chemical ones, promises to be very suitable for local forms of spatial self-organization (think at equilibria in ecological niches), and are particularly suited for modeling and tolerating evolution over time (think at how biodiversity has increased over the course of evolution, without ever mining the health existence of life in each and every place on earth). However, unlike chemical systems, understanding how to properly control the local and global equilibria of real ecological system is a difficult task, and it would probably be very difficult also in their computational counterparts.

In summary, it is very difficult to assess once and for all which of the metaphors is the best for next generation of adaptive service ecosystems. Some exhibit suitable features for certain aspects, but fall short for others. Personally, we have a preference for using the chemical abstraction as a basis—which seems to be the most flexible one—possibly extending it with features of other metaphors: hence the correct answer is probably in some new "hybrid" metaphor, getting the best of all the above.

4 Concluding Remarks

The peculiar characteristics of emerging and future network scenarios challenge current service frameworks, calling for novel service models and associated open service frameworks capable of exhibiting properties of autonomous adaptation and long-lasting (ideally eternal) availability and effectiveness.

In this paper, we have tried to elaborate on the idea of getting inspiration from natural ecosystem, i.e., of conceiving future service frameworks as an ecology of data, services and resources. There, services are modeled and deployed as autonomous individuals in an ecosystem of other services, data sources, and pervasive devices, and their interactions takes place in the form of a natural obedience to a simple set of well-defined "laws of nature". In this way, it is possible to deliver adaptivity and eternity as inherent properties of the service framework, rather than as complicated ad-hoc solutions.

Despite the promises of the ecological approach, though, the road towards the actual deployment of usable and effective "eco-inspired" open service frameworks still requires answering to several challenging questions. What metaphor, among the many possible ones (e.g., biological, physical, chemical) should be better adopted for modeling a suitable service framework? How should we model and represent individuals, the space in which they live, and the laws of nature to which they are subject? How can such individuals and the laws of nature lead to suitable, useful, and controllable forms of spatial self-organization? How can their dynamics be controlled to ensure eternal evolvability in an open setting? What shape should be taken by an actual software infrastructure that supports the ecosystem? All of these, and many further questions we may have missed identifying, open up fascinating areas of research.

References

1. Agha, G.: Computing in pervasive cyberspace. Commun. ACM 51(1), 68–70 (2008)
2. Babaoglu, O., Canright, G., Deutsch, A., Caro, G.A.D., Ducatelle, F., Gambardella, L.M., Ganguly, N., Jelasity, M., Montemanni, R., Montresor, A., Urnes, T.: Design patterns from biology for distributed computing. ACM Trans. Auton. Adapt. Syst. 1(1), 26–66 (2006)
3. Barros, A.P., Dumas, M.: The rise of web service ecosystems. IT Professional 8(5), 31–37 (2006)
4. Beal, J., Bachrach, J.: Infrastructure for engineered emergence on sensor/actuator networks. IEEE Intelligent Systems 21(2), 10–19 (2006)
5. Castelli, G., Rosi, A., Mamei, M., Zambonelli, F.: A simple model and infrastructure for context-aware browsing of the world. In: Pervasive Computing and Communications, March 19-23, pp. 229–238 (2007)
6. Crowcroft, J.: Toward a network architecture that does everything. Commun. ACM 51(1), 74–77 (2008)
7. Dobson, S., Denazis, S., Fernández, A., Gaïti, D., Gelenbe, E., Massacci, F., Nixon, P., Saffre, F., Schmidt, N., Zambonelli, F.: A survey of autonomic communications. ACM Trans. Auton. Adapt. Syst. 1(2), 223–259 (2006)
8. Herold, S., Klus, H., Niebuhr, D., Rausch, A.: Engineering of it ecosystems: design of ultra-large-scale software-intensive systems. In: ULSSIS 2008: Proceedings of the 2nd international workshop on Ultra-large-scale software-intensive systems, pp. 49–52. ACM, New York (2008)
9. Jain, R.: Eventweb: Developing a human-centered computing system. Computer 41(2), 42–50 (2008)
10. Jazayeri, M.: Species evolve, individuals age. In: IWPSE 2005: Proceedings of the Eighth International Workshop on Principles of Software Evolution, pp. 3–12. IEEE Computer Society Press, Washington (2005)
11. Kephart, J.O., Chess, D.M.: The vision of autonomic computing. Computer 36(1), 41–50 (2003)
12. Mamei, M., Zambonelli, F.: Field-based Coordination for Pervasive Multiagent Systems. Springer, Heidelberg (2006)
13. Peysakhov, M.D., Lass, R.N., Regli, W.C.: Stability and control of agent ecosystems. In: AAMAS 2005: Proceedings of the fourth international joint conference on Autonomous agents and multiagent systems, pp. 1143–1144. ACM, New York (2005)
14. Ramakrishnan, R., Tomkins, A.: Toward a peopleweb. Computer 40(8), 63–72 (2007)
15. Shen, W.-M., Will, P., Galstyan, A., Chuong, C.-M.: Hormone-inspired self-organization and distributed control of robotic swarms. Autonomous Robots 17(1), 93–105 (2004)
16. Ulieru, M., Grobbelaar, S.: Engineering industrial ecosystems in a networked world. In: 5th IEEE International Conference on Industrial Informatics, June 23-27, pp. 1–7. IEEE, Los Alamitos (2007)

Part II

Regular Papers

An Agent Based Approach to the Selection Dilemma in CBR

Cesar Analide, António Abelha, José Machado, and José Neves

Universidade do Minho
Departamento de Informática
Braga, Portugal
{analide,abelha,jmac,jneves}@di.uminho.pt

Summary. It is our understanding that a selection algorithm in Case Based Reasoning (CBR) must not only apply the principles of evolution found in nature, to the predicament of finding an optimal solution, but to be assisted by a methodology for problem solving based on the concept of agent. On the other hand, a drawback of any evolutionary algorithm is that a solution is better only in comparison to other(s), presently known solutions; such an algorithm actually has no concept of an optimal solution, or any way to test whether a solution is optimal. In this paper it is addressed the problem of *The Selection Dilemma* in CBR, where the candidate solutions are seen as evolutionary logic programs or theories, here understood as making the core of computational entities or agents, being the test whether a solution is optimal based on a measure of the quality-of-information that stems out of them.

Terms: Algorithms, Languages, Theory.
Keywords: Case Based Reasoning, Evolutionary Computation, Extended Logic Programming, Quality-of-Information.

1 Introduction

Case Based Reasoning (CBR) [1][5] may be understood as the process of solving new problems based on the solutions of similar past ones, but also as a powerful method for computer reasoning, or a pervasive behaviour in everyday human problem solving task. On the other hand, Genetic Programming (GP) may be seen as one of the most useful, general-purpose problem solving techniques available nowadays. GP is one instance of the class of techniques called evolutionary algorithms, which are based on insights from the study of natural selection and evolution. An evolutionary algorithm solves a problem by opening or generating a large number of random problem solvers (here understood as logical programs or agents). Each problem solver is executed and rated according to a fitness metric, given beforehand. In the same way that evolution in nature results from natural selection, an evolutionary algorithm selects the best problem solvers in each generation and breeds them. GP and genetic algorithms are two different kinds of evolutionary algorithms. Genetic algorithms involve encoded strings that represent particular problem solutions. Genetic programming, when applied to *The Selection Dilemma in CBR*, the subject of this article, follows a different approach. Instead of encoding a representation of a solution, GP breeds executable computer programs.

C. Badica et al. (Eds.): Intel. Distributed Comput., Systems & Appl., SCI 162, pp. 35–44, 2008.
springerlink.com

2   The Problem

In general, given a target problem, one intends to retrieve cases from memory that are relevant to solving, where a case may consist of a problem description, its solution, and annotations about how the solution was derived. On the other hand, a genetic or evolutionary algorithm applies the principles of evolution found in nature to the problem of finding an optimal solution to a problem. In a genetic algorithm the problem is encoded in a series of bit strings that are manipulated by the algorithm; in an evolutionary algorithm the decision variables and problem functions are used directly. A drawback of any evolutionary algorithm is that a solution is better only in comparison to other(s), presently known solutions. Such an algorithm actually has no concept of an optimal solution, or any way to test whether a solution is optimal. This also means that an evolutionary algorithm never knows for certain when to stop, aside from the length of time, or the number of iterations or candidate solutions, that one may wish to explore. In this paper it is addressed the problem of using Genetic programming, when applied to The Selection Dilemma in CBR, where the candidate solutions are seen as evolutionary logic programs or theories. Indeed, the *Selection Dilemma in CBR*, when set in terms of a Genetic Programming problem, is a form of computer based search and evolution [8].

2.1   The Past

Let us consider the case where we had a series of data (e.g., pathologies) that were produced from a set of diagnosis taken over time. In this case we have the values that define the output of the function, and we can guess at some parameters which might inter-operate to produce these values - such as a measure of the fever, pain or itch of a given patient. We might like to predict how the outcome (i.e., the pathology) will fare in the future, or we may want to fill in some missing data into the series. To do so, we need to find the relationship between the parameters which will generate values as close to the observed values as possible. Therefore, our optimum will be fit to the observed values.

The GA operates upon a population of candidate solutions to the problem. These solutions can be held in the type of their parameter representation. For example, if the candidate solutions were for a function optimization problem in which the function took a fixed number of floating point parameters, then each candidate could be represented as an array of such floating point numbers.

Clearly, we want to search only the most promising search paths into the population, although we must remain aware that sometimes non-promising search paths can be the best route to the result we are looking for. In order to work out which are the most promising candidates, we evaluate each candidate solution using a user supplied evaluation function. In general, this assigns a single numeric goodness measure to each candidate, so that their relative merit is readily ascertained during the application of the genetic operators. Undoubtedly, the amount of meaning and the interpretation that can be gleaned from this single value is crucial to a successful search [9].

2.2   The Future

With respect to the computational paradigm it were considered extended logic programs with two kinds of negation, classical negation, $\neg$, and default negation, *not*. Intuitively, *not p* is true whenever there is no reason to believe p (close world assumption), whereas $\neg p$ requires a proof of the negated literal. An extended logic program (program, for short) is a finite collection of rules and integrity constraints, standing for all their ground instances, and is given in the form:

$$p \leftarrow p_1 \wedge \ldots \wedge p_n \wedge \text{ not } q_1 \wedge \ldots \wedge \text{ not } q_m; \text{ and}$$

$$?p_1 \wedge \ldots \wedge p_n \wedge \text{ not } q_1 \wedge \ldots \wedge \text{ not } q_m, (n, m \geq 0)$$

where ? is a domain atom denoting falsity, the p_i, q_j, and p are classical ground literals, i.e. either positive atoms or atoms preceded by the classical negation sign $\neg$ [7]. Every program is associated with a set of abducibles. Abducibles may be seen as hypotheses that provide possible solutions or explanations of given queries, being given here in the form of exceptions to the extensions of the predicates that make the program. These extended logic programs or theories stand for the population of candidate solutions to model the universe of discourse.

Indeed, in our approach to GP, we will not get a solution to a particular problem, but rather a logic representation (or program) of the universe of discourse to be optimized. On the other hand, logic programming enables an evolving program to predict in advance its possible future states and to make a preference. This computational paradigm is particularly advantageous since it can be used to predict a program evolution employing the methodologies for problem solving that benefit from abducibles [8], in order to make and preserve abductive hypotheses. It is on the preservation of the abductive hypotheses that our approach will be based, leading to a solution to the problem of Selection Dilemma in CBR.

Designing such a selection regime presents, still, unique challenges. Most evolutionary computation problems are well defined, and quantitative comparisons of performance among the competing individuals are straightforward. By contrast, in selecting an abstract and general logical representation or program, performance metrics are clearly more difficult to devise. Individuals (i.e., programs) must be tested on their ability to adapt to a changing environment, to make deductions and draw inferences, and to choose the most appropriate course of action from a wide range of alternatives. Above all they must learn how to do these things on their own, not by implementing specific instructions given to them by a programmer, but by continuously responding to positive and negative environmental feedback.

In order to accomplish such goal, i.e., to model the universe of discourse in a changing environment, the breeding and executable computer programs will be ordered in terms of the quality-of-information that stems out of them, when subject to a process of conceptual blending [8]. In blending, the structure or extension of two or more predicates is projected to a separate blended space, which inherits a partial structure from the inputs, and has an emergent structure of its own. Meaning is not compositional in the usual sense, and blending operates to produce understandings of composite functions or predicates, the conceptual domain, i.e., a conceptual domain has a basic structure of entities and relations at a high level of generality (e.g., the conceptual domain for journey has roles for traveller, path, origin, destination). In our work we will follow the normal view of conceptual metaphor, i.e., metaphor will carry structure from one conceptual domain (the source) to another (the target) directly.

Therefore, let i ($i \in 1, \cdots, m$) denote the predicates whose extensions make an extended logic program that model the universe of discourse, and j ($j \in 1, \cdots, n$) the attributes for those predicates. Let $x_j \in [min_j, max_j]$ be a value for attribute j. To each predicate it is also associated a scoring function $V_{ij}[min_j, max_j] \rightarrow 0 \cdots 1$, that given the score predicate i, assigns to attribute j a value in the range of its acceptable values, i.e., its domain. For the sake of simplicity, scores are kept in the interval $[0 \cdots 1]$, here given in the form:

all(attribute-exception-list, sub-expression, invariants)

This states that *sub-expression* should hold for each combination of the exceptions of the extensions of the predicates that denote the attributes in the *attribute-exception-list* and are according to the *invariants*. This is further translated by introducing three new predicates. The first predicate creates a list of all possible exception combinations (e.g., pairs, triples) as a list of sets determined by the domain size. The second predicate recurses through this list, and makes a call to the third predicate for each exception combination. The third predicate denotes sub-expression, given for each predicate, as a result, the respective score function. The Quality of the Information (QI) with respect to a generic predicate K is, therefore, given by $QI_K = 1/Card$, where $Card$ denotes the cardinality of the exception set for K, if the exception set is not disjoint. If the exception set is disjoint, the QI is given by:

$$Q_k = \frac{1}{C_1^{Card} + \cdots + C_{Card}^{Card}}$$

where C_{Card}^{Card} is a card-combination subset, with $Card$ elements.

The next element of the model to be considered, it is the relative importance that a predicate assigns to each of its attributes under observation, w_{ij}, which stands for the relevance of attribute j for predicate i (it is also assumed that the weights of all predicates are normalized [4]:

$$\forall i \sum_{j=1}^{n} w_{ij} = 1$$

It is now possible to define a predicate scoring function, i.e., for a value $x = (x_1, \cdots, n)$ in the multi dimensional space defined by the attributes domains, which is given in the form:

$$V_i(x) = \sum_{j=1}^{n} w_{ij} * V_{ij}(x_j).$$

It is now possible to measure the QI that stems from a logic program, by posting the $V_i(x)$ values into a multi-dimensional space and projecting it onto a two dimensional one. Under this procedure, it is defined a circle, as the one given in Figure 1. Here, the dashed n-parts of the circle (in this case built on the extensions of 5 (five) predicates, named as $p_1 \cdots p_5$) denote the QI that is associated with each of the predicate extensions that make the logic program P. It works out the most promising extended logic programs or theories to model the universe of discourse of the agents that make the case memory, providing the optimal solution, subject to formal proof, to the Selection Dilemma in CBR.

It is now possible to return to the case referred to above, where we had a series of data that is produced according to a set of patient attributes, being got all time along. It is therefore possible, to produce a case memory, as the one depicted below, in terms of the predicates *itch*, *fever* and *pain* [3]. The corresponding evolutionary logic programs are presented in Figures 2, 3 and 4.

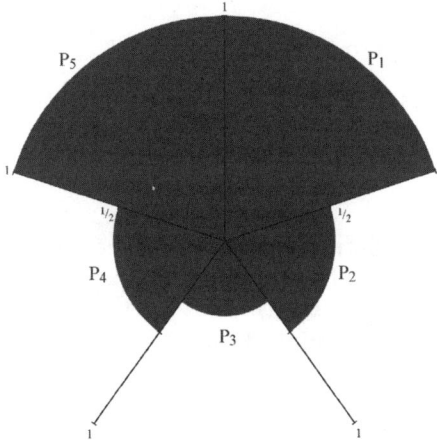

Fig. 1. A measure of the quality-of-information for logic program or theory P

The extended logic program for predicate *itch*

{

$\neg itch(X, Y) \leftarrow$ not $itch(X, Y) \wedge$ not $exception_{itch}(X, Y)$,
$exception_{itch}(X, Y) \leftarrow itch(X, itch)$,
$itch(john, itch)$,
$itch(carol, 1)$,
$exception_{itch}(kevin, 0.6)$,
$exception_{itch}(kevin, 0.8)$,
$?((exception_{itch}(X, Y) \vee exception_{itch}(X, Y)) \wedge \neg(exception_{itch}(X, Y) \wedge exception_{itch}(X, Y))$
$\}ag_{itch}$

The extended logic program for predicate *fever*

{

$\neg fever(X, Y) \leftarrow$ not $fever(X, Y) \wedge$ not $exception_{fever}(X, Y)$,

$exception_{fever}(X, Y) \leftarrow fever(X, fever)$,

$fever(kevin, fever)$,

$fever(john, 1)$,

$exception_{fever}(carol, 0.5)$,

$exception_{fever}(carol, 0.75)$,

$?((exception_{fever}(X, Y) \vee exception_{fever}(X, Y)) \wedge \neg(exception_{fever}(X, Y) \wedge exceptionfever(X, Y))$

$\}ag_{fever}$

The extended logic program for predicate *pain*

{ $\neg pain(X, Y) \leftarrow$ not $pain(X, Y) \wedge$ not $exception_{pain}(X, Y)$,

$exception_{pain}(X, Y) \leftarrow pain(X, pain)$,

$pain(carol, pain)$,

$pain(kevin, 1)$,

$exception_{pain}(john, 0.3)$,

$exception_{pain}(john, 0.45)$,

$?((exception_{pain}(X, Y) \vee exception_{pain}(X, Y)) \wedge \neg(exception_{pain}(X, Y) \wedge exception_{pain}(X, Y))$,

$\}ag_{pain}$

Now, and in order to find the relationships among the extensions of these predicates, we will evaluate the relevance of the QI, which, for patient *kevin*, will be given in the form $V_{itch}(kevin) = 0.785$; $V_{fever}(kevin) = 0$; $V_{pain}(kevin) = 1$, i.e., it is now possible to measure the QI that flows out of the logic programs referred to above (the dashed n-parts (here n is equal to 3 (three)) of the circles denote the QI for predicates *itch*, *fever* and *pain*).

It is also possible, considering what it is illustrated by Figures 5, 6 and 7, to predict not only how the outcome of the patient diagnosis will fare into the future, but also how

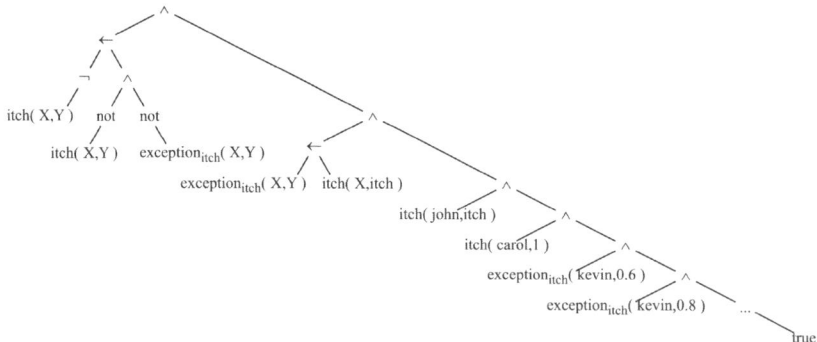

Fig. 2. The evolutionary logic program for predicate itch

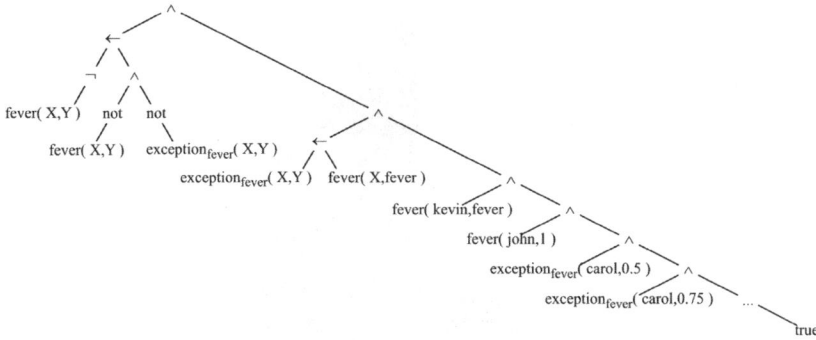

Fig. 3. The evolutionary logic program for predicate fever

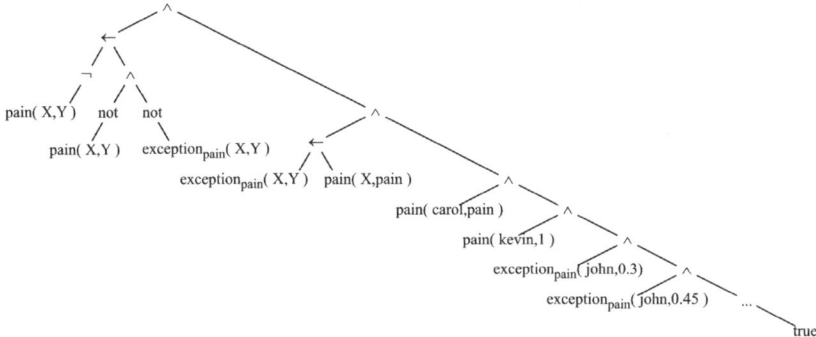

Fig. 4. The evolutionary logic program for predicate pain

to fill in some missing data into the series. To do so, we need to evolve the logic theories or logic programs, evolving the correspondent evolutionary logic programs, according to the rules of programs synthesis [10] [8]. A new predicate may be defined (the three argument predicate *pathology*), whose extension may be given in the form:

$$\{$$
$$\neg pathology(X, Y, Z) \leftarrow not\ pathology(X, Y, Z) \wedge$$
$$not\ exception_{pathology}(X, Y, Z),$$
$$pathology(john, flu, ((itch, 0), (fever, 1), (pain, 0.785))),$$
$$pathology(kevin, thrombosis, ((itch, 0.785), (fever, 0), (pain, 1))),$$
$$pathology(carol, heartattack, ((itch, 1), (fever, 0.785), (pain, 0))),$$
$$\}ag_{pathology}$$

Now, given a new case, the seriation of the pathologies is made according the percentage of overlap between the dashed areas that make the QI for the predicates in the

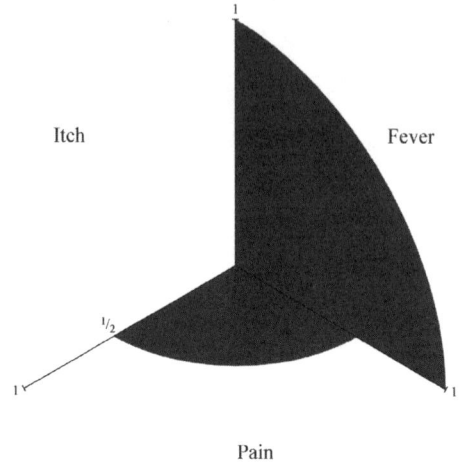

Fig. 5. A measure of the symptoms for patient John

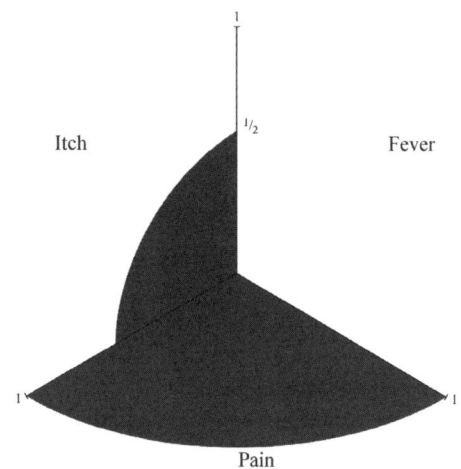

Fig. 6. A measure of the symptoms for patient Kevin

case memory, and those for the new one. For instance, if we have a case under evaluation, with the QI values depicted below:

$QI_{itch} = 0.785$
$QI_{fever} = 0$
$QI_{pain} = 0.785$

it is possible to define an order relation with respect to the pathologies referred to in the case memory, leading to:

$Thrombosis > Flu > Heartattack$

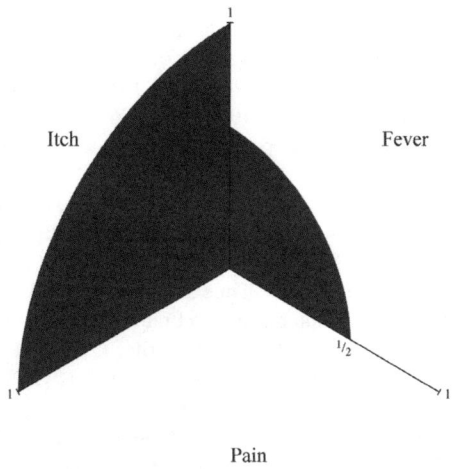

Fig. 7. A measure of the symptoms for patient Carol

3 Conclusions

This paper shows how to construct a dynamic virtual world of complex and interacting populations, entities that are built as evolutionary logic programs that compete against one another in a rigorous selection regime. It provides a solution, subject to formal proof, to the Selection Dilemma in CBR, i.e., in order to produce the optimal solution to a particular problem, one must evolve the logic program or theory that models the universe of discourse, in which its fitness is judged by one criterion alone, the owner Quality-of-Information.

Clearly, we work out:

The model, that provides a solution, subject to formal proof, to the Selection Dilemma in CBR. Indeed, a model in this context is to be understood as the composition of the extensions of the predicates that denote the objects and the relations that make the inner circle of every case in the case memory.

The parameters or attributes, that we were seeking to discover, here given in terms of the extensions of predicates of the kind just referred to above [2].

The optimal, here understood as the logic program or theory that models the universe of discourse in terms of the predicates available and maximizes their QI factors [6].

How to measure and assign values to possible solutions, which was accomplished via mechanical theorem proving and program composition [7].

References

1. Aamodt, A., Plaza, E.: Case-Based Reasoning: Foundational Issues, Methodological Variations, and System Approaches. AI Communications 7(1), 39–59 (1994)
2. Analide, C., Novais, P., Machado, J., Neves, J.: Quality of Knowledge in Virtual Entities. In: Encyclopedia of Communities of Practice in Information and Knowledge Management, pp. 436–442. Idea Group Inc. (2006)

3. Angeline, P.J.: Parse Trees. In: BŁck, T., et al. (eds.) Evolutionary Computation 1: Basic Algorithms And Operators. Institute of Physics Publishing, Bristol (2000)
4. Jennings, N.R., Faratin, P., Johnson, M.J., Norman, T.J., O'Brien, Wiegand, M.E.: Journal of Cooperative Information Systems 5(2-3), 105–130 (1996)
5. Leake, D.: Case-Based Reasoning - Experience, Lessons and Future Direction. MIT Press, Cambridge (1996)
6. Mendes, R., Kennedy, J., Neves, J.: Avoiding the Pitfalls of Local Optima: How topologies can Save the Day. In: Proceedings of the 12th Conference Intelligent Systems Application to Power Systems (ISAP 2003). IEEE Computer Society, Lemnos (2003)
7. Neves, J.: A Logic Interpreter to Handle Time and Negation in Logic Data Bases. In: Proceedings of ACM 1984 Annual Conference, San Francisco, USA, October 24-27 (1984)
8. Neves, J., Machado, J., Analide, C., Abelha, A., Brito, L.: The Halt Condition in Genetic Programming. In: Neves, J., Santos, M.F., Machado, J.M. (eds.) EPIA 2007. LNCS (LNAI), vol. 4874, pp. 160–169. Springer, Heidelberg (2007)
9. Rudolph, G.: Convergence Analysis of Canonical Genetic Algorithms. IEEE Transactions on Neural Networks, Special Issue on Evolutionary Computation 5(1), 96–101 (1994)
10. Teller, A.: Evolving programmers: The co-evolution of intelligent recombination operators. In: Kinnear, K., Angeline, P. (eds.) Advances in Genetic Programming 2, MIT, Cambridge (1996)

Modeling Interactions in Agent-Based English Auctions with Matchmaking Capabilities

Amelia Bădică and Costin Bădică

University of Craiova, Bvd.Decebal 107, Craiova, 200440, Romania
ameliabd@yahoo.com, badica_costin@software.ucv.ro

Summary. Creation of dynamic, non-trivial business relationships in agent-based trading environments requires the use of different types of middle-agents including matchmakers and arbitrators. In this note we apply the formal framework of *finite state process algebra* for modeling and analysis of complex interactions occurring in agent-based English auctions combined with matchmaking activities. In our model: i) several auctions initiated by different seller agents are carried out in parallel; ii) buyer agents have the option to register for participation only in auctions that match their goals and iii) buyers decision to what active auction to register for participation is taken dynamically.

1 Introduction

The ability of software agents to discover remote markets and to dynamically engage in commercial transactions is very important in dynamic trading environments ([8]). Connecting requester agents with provider agents in such an environment is a crucial problem (known as the *connection problem*) that requires the use of middle-agents – replacements of middlemen in a virtual environment ([7, 9]).

Typical use of middle-agents is encountered in e-commerce applications. For example, the model agent-based e-commerce system discussed in [2] uses middle-agents to connect user buyers on the purchasing side with shops on the selling side in a distributed marketplace. Each user buyer is represented by a *Client* agent and each shop is represented by a *Shop* agent. The user buyer submits an order to the system for purchasing a product via his or her *Client* agent. The *Client* agent uses a special agent called *Client Information Center – CIC* that is responsible for providing information which shop in the system sells which products. So, it can be easily noticed that *CIC* is in fact a matchmaker with respect to connecting the *Client* agent with an appropriate *Shop* agent.

Negotiations (and auctions in particular) are complex activities frequently encountered in modern e-commerce processes. They are typically characterized by tight interactions between the involved business parties ([8]). Their understanding, especially when negotiations are automatized using software agents ([10]), requires a careful analysis, usually supported by appropriate formal modeling frameworks.

For example, in the model agent-based e-commerce system presented in [2], after the process of matchmaking between the *Client* agent and the *Shop* agent, the *Client* has the possibility to create a specialized *Buyer* agent and to send him to the *Shop* site to negotiate for buying a sought-after product at an affordable price. This example clearly shows that combination of matchmaking and negotiation processes provides the *Client* agent with a flexible support for dynamically engaging in non-trivial business relationships.

In this paper we present a formal modeling of more complex agent-based e-commerce processes that integrate two types of middle-agents frequently encountered in applications: *Matchmaker* and *Arbitrator*. We have chosen an *Arbitrator* for mediating English auctions, inspired by [4] that has been extended with the ability of handling multiple auctions in parallel to allow the study of more complex models. The *Matchmaker* model has been inspired from [1]. The modeling is using the formal framework based on *finite state process algebra* – FSP, initially proposed in [3].

We start in section 2 with: (i) background on FSP formal specification language; (ii) an overview of our negotiation model; and (iii) an introduction to middle-agents focusing on *Matchmaker*. In section 3 we detail our FSP models of English auctions with matchmaking capabilities. We follow in section 4 with conclusions and proposed future work.

2 Background

2.1 Overview of FSP

FSP is an algebraic specification technique of concurrent and cooperating processes that allows a compact representation of a finite state labeled transition system (LTS hereafter), rather than describing it as a list of states and transitions.

A FSP model consists of a finite set of sequential and/or composite process definitions. Additionally, a sequential process definition consists of a sequence of one or more definitions of local processes. A process definition consists of a process name associated to a process term. FSP uses a rich set of constructs for process terms (see [11] for details). For the purpose of this paper we are using the following constructs: action prefix ($a \rightarrow P$), nondeterministic choice ($P|Q$), and process alphabet extension ($P + \{a_1, \ldots, a_n\}$) for sequential process terms and parallel composition ($P\|Q$) and relabeling ($P/\{new_1/old_1, \ldots, new_k/old_k\}$) for composite process terms.

FSP has an operational semantics given via a LTS. The mapping of a FSP term to a LTS is described in detail in [11] and it follows the intuitive meaning of FSP constructs.

The modeling that we propose here follows the general guidelines outlined in [3]. Briefly: i) agents are modeled as FSP processes and a multi-agent system is modeled as a parallel composition of processes; ii) sets $\mathcal{B}$ of buyers and $\mathcal{S}$ of sellers are assumed to be initially given and agent requests and replies are indexed with buyer and/or seller identifiers; iii) matching operation is modeled as a relation $\mathcal{M} \subseteq \mathcal{B} \times \mathcal{S}$.

2.2 Agent Negotiation Model

We understand automated negotiations as a process by which a group of software agents communicate with each other to reach a mutually acceptable agreement on some matter ([10]). In this paper we focus our attention on *auctions* – a particular form of negotiation where resource allocations and prices are determined by bids exchanged between participants according to a given set of rules ([12]).

In automated negotiations we distinguish between *protocols* (or *mechanisms*) and *strategies*. The protocol comprises public "rules of encounter" between negotiation participants by specifying the requirements that enable them to interact and negotiate. The strategy defines the private behavior of participants aiming at achieving their desired outcome ([10]).

Our negotiation model follows the generic software framework for automated negotiation proposed by [6] and it is specialized for the particular case of English auctions ([8]) following implementation details reported in [5]. This framework comprises: (1) negotiation infrastructure, (2) generic negotiation protocol and (3) taxonomy of declarative rules. The *negotiation infrastructure* defines roles of negotiation participants (eg.*Buyer* or *Seller* in an auction) and of a negotiation host. According to the *generic negotiation protocol* ([6]), participants exchange proposals (or bids) via a common space that is governed by an authoritative entity – the negotiation host. Negotiation state and intermediary information is automatically forwarded by the host to all entitled participants according to the information revealing policy of that particular negotiation ([6, 5]). *Negotiation rules* deal with the semantic constraints a particular negotiation mechanism (e.g. English auctions).

The *generic negotiation protocol* controls how messages are exchanged by the host and participants by facilitating the following negotiation activities: (1) admission to negotiation, (2) proposal (or bid) submission, (3) informing participants about the change of negotiation state, (4) agreement formation and (5) negotiation termination.

2.3 Matchmakers

A standard classification of middle-agents was introduced in the seminal work [7]. Based on assumptions about what it is initially known by the requesters, middle-agent, and providers about requester preferences and provider capabilities, authors of [7] proposed 9 types of middle-agents: *Broadcaster, Matchmaker, Front-agent, Anonymizer, Broker, Recommender, Blackboard, Introducer*, and *Arbitrator*.

A *Matchmaker* middle-agent assumes that requester preferences are initially known only to the requester, while provider capabilities are initially known to all interaction participants. This means that a provider will have to advertise its capabilities with *Matchmaker* and *Matchmaker* has responsibility to match a request with registered capabilities advertisements. However, the fact that provider capabilities are initially known also by the requester means that the result of the matching (i.e set of matching providers) is returned by *Matchmaker* to requester (so provider capabilities become thus known to the requester), and the choice of the matching provider is the responsibility of the requester. Consequently the transaction is not intermediated by *Matchmaker*, as would be the case with *Broker* or *Front-agent* ([1]).

3   FSP Model of Agent Negotiation with Matchmaking

In this section we show how the FSP model of an English auction introduced in [4] can be extended to handle multiple parallel auctions and matchmaking.

3.1 Negotiation Structure

A *negotiation structure* defines a general framework that statically constraints a given negotiation.

The *negotiation host* role orchestrates the negotiation and coordinates negotiators by employing the general negotiation protocol. We shall have a separate negotiation host for each active auction in the system. All the active negotiation hosts are managed by a *negotiation server host*.

A *negotiation participant* role describes the behavior of a negotiator that plays a certain role in the negotiation. Usually, two negotiation participant roles are defined – *buyer* and *seller*. For example, in an English auction there is a single *seller* participant and one or more *buyer* participants, while in an reverse English auction there is a single participant with role *buyer* and one or more participants with role *seller*.

A negotiation process is always initiated by a certain participant known as *negotiation initiator*. The negotiation initiator requests the initiation of a new negotiation to the negotiation server host. Usually it is required that the initiator has a given negotiation role – *negotiation initiator* role. For example, in an English auction the initiator has always role *seller*, while in a reverse English auction the initiator has always role *buyer*.

Focusing our discussion on auctions for buying and selling goods, a negotiation structure can be formally defined as follows:

Definition 1. *(Negotiation Structure) A negotiation structure is a tuple* $N = \langle ServerHost, Hosts, Seller, Buyer, Initiator \rangle$ *such that: i) ServerHost is the negotiation server host role; ii) Hosts is the set of negotiation hosts roles; this set is composed of several Host roles, each of them describing a negotiation host; iii) Seller is the seller role that defines behavior of participants selling goods in the auction; iv) Buyer is the buyer role that defines behavior of participants buying goods in the auction; iv) Initiator is the role that is allowed to initiate the auction – either buyer or seller, i.e. Initiator* $\in \{Buyer, Seller\}$.

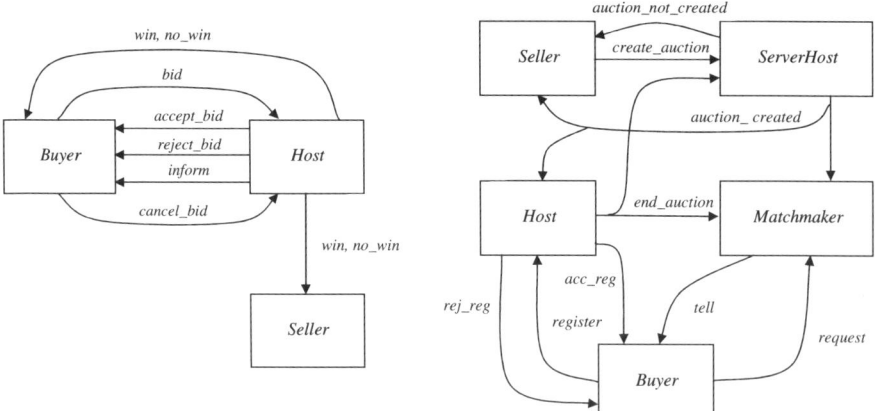

Fig. 1. Roles interaction during negotiation   **Fig. 2.** Roles interaction before and after negotiation

Interactions between the roles of a negotiation structure during and before/after negotiation are illustrated in figures 1 and 2.

Behavior of negotiation roles is described using FSP. Therefore we shall have FSP processes describing the *ServerHost*, *Host*, *Seller*, *Buyer* and *Matchmaker* roles. A participant behavior is defined by instantiating its role. Finally, the behavior of the negotiation system is defined using parallel composition of roles for each negotiation participant, including of course the negotiation server host, the negotiation hosts and the matchmaker.

3.2 Negotiation Host

A negotiation host is able to handle a single negotiation at a certain time. In other words, the negotiation host functions as a *one-at-a-time server*. In order to handle multiple negotiations concurrently, several negotiation hosts instances will be created and ran concurrently under the control of the negotiation server host.

Negotiation consists of a series of stages that, in what follows, are particularized for the case of an English auction:

i) *initiation* – the negotiation is initiated by the seller using the *init* action; note that initiation acts also as a registration of the seller agent participant; initiation is either accepted (action *accept_init*) or rejected (action *reject_init*) by the host;

ii *buyer registration* – each buyer agent must register with the negotiation using *register* action before she is allowed to submit bids; registration is granted (action *accept_registration*) or not (action *reject_registraton*) by the negotiation host;

iii) *bids submission* – each registered buyer is allowed to submit bids using *bid* action; bids are either accepted (action *accept_bid*) or not (action *reject_bid*) by the host; when a certain bid is accepted, the other registered buyer participants are notified accordingly by the host (action *inform*). additionally, a buyer may cancel submitting bids (action *cancel_bid*), in this case being deregistered from the negotiation;

iv) *agreement formation* – when the host observes a certain period of bidding inactivity, it triggers negotiation termination via action *no_bid*. This event subsequently triggers agreement formation. In this stage the host checks if an agreement can be generated. If no buyer has registered before the negotiation terminated then no agreement can be made and action *no_win* with no parameter is executed. However, if at least one buyer has successfully submitted an accepted bid then the host will decide if there is a winner (action *win*) or not (action *no_win* with parameter) depending on if the currently highest bid overbids or not the seller reservation price.

Negotiation host behavior is described as the *Host* process (see table 1). The *Host* process has a cyclic behavior and thus it runs infinitely, being able to handle an infinite sequence of negotiations, one negotiation at a time. Note that, differently from the model introduced in [4] where the decision of accepting the initiation of a new negotiation was taken by the *Host*, here this decision is taken by the *ServerHost* process (see table 2).

In a real setting, participant agents (buyers and sellers) can be created and destroyed dynamically. In our model we assume there is a given set of buyers as well as a given

Table 1. *Host* process that describes the negotiation host role for controlling a single negotiation

$$
\begin{aligned}
Host &= (init \rightarrow ServerBid(\bot, \emptyset)), \\
ServerBid(chb, Bs) &= (bid(b \in Bs) \rightarrow AnswerBid(b, chb, Bs)| \\
&\quad cancel\_bid(b' \in Bs) \rightarrow ServerBid(chb, Bs \setminus \{b'\})| \\
&\quad register(b' \notin Bs) \rightarrow AnswerReg(b', chb, Bs)| \\
&\quad no\_bid \rightarrow ServerAgreement(chb)), \\
AnswerReg(b', chb, Bs) &= (accept\_registration(b') \rightarrow ServerBid(chb, Bs \cup \{b'\})| \\
&\quad reject\_registration(b') \rightarrow ServerBid(chb, Bs)), \\
AnswerBid(b, chb, Bs) &= (accept\_bid(b) \rightarrow InformBuyers(b, Bs)| \\
&\quad reject\_bid(b) \rightarrow ServerBid(chb, Bs)), \\
InformBuyers(b, Bs) &= (inform(b_1) \rightarrow inform(b_2) \rightarrow \ldots \rightarrow \\
&\quad inform(b_k) \rightarrow ServerBid(b, Bs)), \\
ServerAgreement(\bot) &= (no\_win \rightarrow Host), \\
ServerAgreement(chb) &= (win(chb) \rightarrow Host|no\_win(chb) \rightarrow Host). \\[6pt]
Host(s \in S) &= Host/\{auction\_created(s)/init, end\_auction(s)/no\_bid, win(s)/win, \\
&\quad no\_win(s)/no\_win, register(s, b)/register(b), inform(s, b)/inform(b), \\
&\quad accept\_registration(s, b)/accept\_registration(b), \\
&\quad reject\_registration(s, b)/reject\_registration(b), \\
&\quad cancel\_bid(s, b)/cancel\_bid(b), bid(s, b)/bid(s), \\
&\quad reject\_bid(s, b)/reject\_bid(s), accept\_bid(s, b)/accept\_bid(s)\}. \\[6pt]
Hosts &= \|_{s \in S} Host(s)
\end{aligned}
$$

set of sellers that are created when the system is started. Buyers are able to dynamically register to negotiations while sellers are able to dynamically initiate negotiations.

Assuming each buyer agent has a unique name, let $\mathcal{B}$ be the set of all names of buyer agents and let S be the set of all names of seller agents that were created when the system was initiated. Let be $\bot$ a name not in $\mathcal{B}$. Definition of the *Host* process is using several indexed families of local processes:

- *ServerBid(chb, B)* such that $chb \in B \cup \{\bot\}, B \subseteq \mathcal{B}$. Here chb records the buyer associated with currently highest bid and B denotes the set of registered buyers. The condition $chb \in B \cup \{\bot\}$ means that either no buyer agent has submitted a bid in the current negotiation (when $chb = \bot$) or the buyer agent that submitted the currently highest bid must have already registered with the negotiation before the submission (i.e. $chb \in B$).
- *AnswerReg(b, chb, B)* such that $b \in \mathcal{B} \setminus B, chb \in B \cup \{\bot\}, B \subseteq \mathcal{B}$. Here b denotes the buyer that requested registration with the current negotiation, chb denotes the buyer associated with currently highest bid and B denotes the set of registered buyers. The fact that $b \in \mathcal{B} \setminus B$ means that the registration request comes from a buyer that is not yet registered with the negotiation. The fact that $chb \in B \cup \{\bot\}$ means that either the currently highest bid has not been submitted yet (i.e. $chb = \bot$) or it was submitted by a registered buyer (i.e. $chb \in B$).
- *InformBuyers(b, B)* such that $b \in B, B \subseteq \mathcal{B}$. Here b denotes the buyer that submitted an accepted bid and B denotes the set of registered buyers. The fact that $b \in B$ means that the bid that was accepted comes from a buyer that has registered with the negotiation.

Table 2. *ServerHost* process that describes the negotiation server host role

ServerHost	= *ServerHost*($\emptyset$),
ServerHost(S)	= (*end_auction*($s \in S$) → *ServerHost*($S \setminus \{s\}$)|
	create_auction($s' \notin S$) → *ServiceHost*(s', S)),
ServerHost($s' \notin Ss, Ss$)	= (*auction_not_created*(s') → *ServerHost(S)*|
	auction_created(s') → *ServerHost*($S \cup \{s'\}$))).

3.3 Negotiation Server Host

Negotiation server host manages all the active negotiations at a given time. Whenever a new auction is created, it is registered with the negotiation server host. Whenever an auction is terminated it is consequently deregistered with the negotiation server host. Note that decision of accepting or not the creation of a new auction belongs to the negotiation host server.

3.4 *Buyer* and *Seller* Roles

The *Buyer* role first queries the *Matchmaker* to find out active negotiations (action *send_request*). The *Matchmaker* responds with a set *S* of matching negotiations (action *tell*($S \subseteq S$)). The buyer agent selects a convenient matching negotiation $s \in S$ and then registers to the negotiation before starting to submit bids. If registration is granted, she can start bidding according to its private strategy – action *bid*. Here we have chosen a very simple strategy: each buyer agent submits a first bid immediately after it is granted admission to the negotiation and subsequently, whenever it gets a notification that another participant issued a bid that was accepted by the host. Additionally, each buyer participant has its own valuation of the negotiated product. If the current value that the buyer decided to bid exceeds her private valuation then the proposal submission is canceled – action *cancel_bid*, i.e. product became "too expensive". Note that after a

Table 3. *Buyer* and *Seller* processes

Buyer	= (*send_request* → *WaitReply*),
WaitReply	= (*tell*($S \subseteq S$) → **if** $S \neq \emptyset$ **then** *ContactProvider(S)* **else** *Buyer*),
ContactProvider($S \subseteq S$)	= (**while** $S \neq \emptyset$ *register*($s \in S$) → *BuyerRegister(s)*),
BuyerRegister($s \in S$)	= (*accept_registration*(s) → *BuyerBid(s)*|
	reject_registration(s) → *Buyer*),
BuyerBid($s \in S$)	= (*bid*(s) → *WaitBid(s)*|
	cancel_bid(s) → *WaitBid(s)*|
	inform(s) → *BuyerBid(s)*),
WaitBid($s \in S$)	= (*accept_bid*(s) → *Wait(s)*|
	reject_bid(s) → *BuyerBid(s)*|
	inform(s) → *BuyerBid(s)*),
Wait($s \in S$)	= (*inform*(s) → *BuyerBid(s)*|
	end(s) → *Buyer*).
Seller	= (*init* → *WaitInit*),
WaitInit	= (*accept_init* → *WaitEnd*|
	reject_init → *Seller*),
WaitEnd	= (*end* → *Seller*).

Table 4. *Matchmaker* middle-agent

$$
\begin{aligned}
Matchmaker &= Matchmaker(\emptyset), \\
Matchmaker(S \subseteq S) &= (request(b \in \mathcal{B}) \to MatchReq(b, S) \mid \\
&\quad offer(s \in S \setminus S) \to Matchmaker(S \cup \{s\}) \mid \\
&\quad withdraw(s \in S) \to Matchmaker(S \setminus \{s\})), \\
MatchReq(b \in \mathcal{B}, S \subseteq S) &= (tell(b, M(b) \cap S) \to Matchmaker(S)) + \{tell(b' \in \mathcal{B}, S' \subseteq S)\}. \\
\\
NegoMatchmaker &= Matchmaker/\{auction\_created/offer, end\_auction/withdraw\}.
\end{aligned}
$$

buyer agent submitted a bid that was accepted, she will enter a state waiting for a notification that either another successful bid was submitted or that she eventually was the last submitter of a successful bid in the current auction (i.e. a potentially winning bid, depending on if the bid value was higher than the seller reservation price) – action *end*.

The seller agent initiates the auction – action *init* and then, assuming initiation was successful, she waits for the auction to terminate – action *end*, before issuing a new initiation request.

3.5 *Matchmaker* Role

Matchmaker agent registers and deregisters active negotiations and answers *Buyer* requests for matching negotiations. The matching operation is modeled as a relation $M \subseteq \mathcal{B} \times S$. *Matchmaker* informs *Buyer* about available active negotiations (action *tell*). Note that *Buyer* is responsible to choose an appropriate matching negotiation from the available matching offers.

Note that special care should be taken in order to accurately model agents communication using FSP synchronization. *Matchmaker* model requires alphabet extension (construct $+\{tell(b' \in B, S' \subseteq S)\}$ in table 4) in order to model correctly communication between *Matchmaker* and *Buyer*.

3.6 Negotiation System

Buyer and seller agents are created by instantiating *Buyer* and respectively *Seller* roles. Note that instantiation of *Buyer* roles assumes also indexing of actions *bid*, *reject_bid*, *accept_bid*, *inform*, *cancel_bid*, *register*, *accept_registration*, *reject_registration* with buyer's name and also renaming action *end* with an indexed set of actions {*win, no_win*}. Similarly, instantiation of *Seller* role assumes renaming action *end* with a set of actions denoting various ways the auction may terminate: without a winner assuming no buyer submitted an accepted bid – *no_win*, with or without a winner assuming at least one buyer submitted an accepted bid – indexed set of actions {*win, no_win*}. Finally, instantiation of *Server* role requires no renaming, as the names of the buyer agents were supposedly known in the definition of *Server* process.

Negotiation system is defined as parallel composition of negotiation server host, negotiation hosts, seller and buyer agents processes – see table 5.

We have determined the LTS of a sample negotiation system with 2 buyers and 2 sellers using LTSA tool ([11]). The complete definition of this system is shown in the appendix. The analysis performed revealed that the system has 790 states and 2470 transitions and it is free of deadlocks.

Table 5. *System* process as parallel composition of negotiation host, buyers and seller processes

$BuyerAgent(b \in \mathcal{B})$ = $Buyer/\{bid(s,b)/bid(s), reject\_bid(s,b)/reject\_bid(s),$
$accept\_bid(s,b)/accept\_bid(s), inform(s,b)/inform(s),$
$cancel\_bid(s,b)/cancel\_bid(s), \{win(s,b), no\_win(s,b)\}/end(s),$
$register(s,b)/register(s), accept\_registration(s,b)/accept\_registration(s),$
$reject\_registration(s,b)/reject\_registration(s), tell(b,S)/tell(S)\}.$

$SellerAgent(s \in S)$ = $Seller/\{\{no\_win(s), win(s,b), no\_win(s,b)\}/end,$
$create\_auction(s)/init, auction\_created(s)/accept\_init,$
$auction\_not\_created(s)/reject\_init\}.$

$Buyers$ = $\|_{b \in \mathcal{B}} BuyerAgent(b)$

$Sellers$ = $\|_{s \in S} SellerAgent(s)$

$System$ = $(ServerHost\|Hosts\|Sellers\|Buyers\|NegoMatchmaker).$

4   Conclusions and Future Work

In this paper we applied a formal framework based on FSP process algebra for modeling a system that contains seller and buyer agents engaged in complex matchmaking and negotiation processes. We checked the resulting model with the LTSA analysis tool. As future work, we intend to: i) modeling of more complex systems containing other types of middle-agents; ii) introduction and analysis of qualitative properties of agent systems; iii) carry out of verification experiments using the proposed models.

References

1. Bădică, A., Bădică, C.: Formal Specification of Matchmakers, Front-agents, and Brokers in Agent Environments Using FSP. In: Ultes-Nitsche, U., Moldt, D., Augusto, J.C. (eds.) Proc. MSVVEIS, – 6th International Workshop on Modelling, Simulation, Verification and Validation of Enterprise Information Systems, pp. 9–18. INSTICC Press (2008)
2. Bădică, C., Ganzha, M., Paprzycki, M.: Developing a Model Agent-based E-Commerce System. In: E-Service Intelligence: Methodologies, Technologies and Applications. Studies in Computational Intelligence, vol. 37, pp. 555–578. Springer, Heidelberg (2007)
3. Bădică, A., Bădică, C., Liţoiu, L.: Middle-Agents Interactions as Finite State Processes: Overview and Example. In: Proc.16th IEEE International Workshops on Enabling Technologies: Infrastructure for Collaborative Enterprises (WETICE 2007), pp. 12–17 (2007)
4. Bădică, A., Bădică, C.: Formalizing Agent-Based English Auctions Using Finite State Process Algebra. Journal of Universal computer Science 14(7), 1118–1135 (2008)
5. Bădică, C., Ganzha, M., Paprzycki, M.: Implementing Rule-Based Automated Price Negotiation in an Agent System. Journal of Universal Computer Science 13(2), 244–266 (2007)
6. Bartolini, C., Preist, C., Jennings, N.R.: A Software Framework for Automated Negotiation. In: Choren, R., Garcia, A., Lucena, C., Romanovsky, A. (eds.) SELMAS 2004. LNCS, vol. 3390, pp. 213–235. Springer, Heidelberg (2005)
7. Decker, K., Sycara, K.P., Williamson, M.: Middle-agents for the internet. In: Proceedings of the 15th International Joint Conference on Artificial Intelligence IJCAI 1997, vol. 1, pp. 578–583. Morgan Kaufmann, San Francisco (1997)
8. Fasli, M.: Agent Technology For E-Commerce. Wiley, Chichester (2007)

9. Klusch, M., Sycara, K.P.: Brokering and matchmaking for coordination of agent societies: A survey. In: Omicini, A., Zambonelli, F., Klusch, M., Tolksdorf, R. (eds.) Coordination of Internet Agents. Models, Technologies, and Applications, pp. 197–224. Springer, Heidelberg (2001)

10. Lomuscio, A.R., Wooldridge, M., Jennings, N.R.: A classification scheme for negotiation in electronic commerce. In: Sierra, C., Dignum, F.P.M. (eds.) AgentLink 2000. LNCS (LNAI), vol. 1991, pp. 19–33. Springer, Heidelberg (2001)

11. Magee, J., Kramer, J.: Concurrency. State Models and Java Programs, 2nd edn. John Wiley & Sons, Chichester (2006)

12. McAfee, R.P., McMillan, J.: Auctions and bidding. Journal of Economic Literature 25(2), 699–738 (1987)

Output-Driven XQuery Evaluation

David Bednárek

Department of Software Engineering, Faculty of Mathematics and Physics
Charles University Prague
david.bednarek@mff.cuni.cz

Summary. When a XML document is stored in a relational or native database, its tree structure is usually dissolved into various forms of interval or Dewey indexes. Besides other advantages, these loosely-coupled structures allow parallel or distributed evaluation of XPath queries. However, when a XQuery or XSLT program produces a new XML document, its construction forms a hardly parallelizable bottleneck. In this paper, we present a method of XQuery/XSLT evaluation that directly generates Dewey-like structures representing the output of the transformation. This approach forms an output-side counterpart of Dewey-based XPath evaluation methods and makes parallel evaluation of XQuery/XSLT programs easier.

1 Introduction

Contemporary XQuery/XSLT/XPath processing and optimization techniques are usually focused on the querying side of the problem. Compared to the complexity of XPath evaluation, the generation of the output tree seems straightforward. The semantics of the XQuery/XSLT constructors suggests bottom-up in-memory construction of the tree of the output document. It works fine when the output is being concentrated into a serialized document; however, in a parallel or distributed environment, the canonical approach suffers from the following problems: First, the maintenance of in-memory tree structures in a distributed environment is difficult. Second, the bottom-up tree building approach requires that lower parts of the tree are finished before continuing to the next level; thus, this requirement may form a serial bottleneck in the distributed process.

When the output of a XQuery/XSLT program is being stored into a relational database or a native XML-database, for instance as a materialized XML-view, the document is being dissolved into a loosely-coupled structure. The elements of this structure, e.g. individual table rows in a relational database, are bound together using various forms of keys and indexes, like Dewey identifiers (see, for instance, [8]) or interval indexes ([2]). In such environment, the physical construction of the output document in the form of a tree is superfluous.

In our approach we suggest that the XQuery/XSLT processor directly generate the individual elements of the output document representation, in the form used in the target database. Instead of presenting the output document as a solid body, the elements are generated independently. Moreover, the elements may be generated at various nodes of a distributed computing environment; in the best case, at the same nodes where the elements will be stored in a distributed database.

C. Badica et al. (Eds.): Intel. Distributed Comput., Systems & Appl., SCI 162, pp. 55–64, 2008.
springerlink.com   © Springer-Verlag Berlin Heidelberg 2008

Our method is based on partial reversal of the direction of data flow in a XQuery program. In the synthetic part of the program that generates the output document, the canonic bottom-up evaluation is replaced by top-down distribution of *placeholders*. These placeholders give advice to the constructor operators on where the constructed nodes will reside in the final output document.

In the next section, we will show the principles of the reversed evaluation. In the third section, a mathematical model of the reversed evaluation will be introduced.

2 Reversed Evaluation

In the standard model of XQuery evaluation, each expression evaluates to a sequence of items; each item may be an atomic value, (a reference to) a node of an input document, or a node created by a constructor during the evaluation. In most cases, constructed nodes and trees are not navigated using XPath axes or combined using node-set operations; they are just propagated to other constructors or to the final output of the program. (Note that navigation in temporary trees was prohibited in XSLT 1.0., therefore, it is still very rare practice in XSLT 2.0 and XQuery.) In such cases, the expressions forming a XQuery program may be statically divided into two classes: Expressions returning atomic values and input document nodes and expressions returning constructed nodes.

Under the assumption that constructed trees are not navigated, only the following operators may be applied to the constructed node sequences:

- Concatenation (,) operator
- `for`-expression
- `let`-expression
- Node constructor
- Input-to-output tree conversion (implicit)
- Function call

These operators can perform only a limited degree of manipulation on the sequences. As a result, each node is just placed somewhere in the output document, without any modification to its contents. Using `where` clause, nodes may be discarded; using `let`-expression or function arguments, nodes may be copied. Thus, the total effect on a constructed node may be represented by a set of *placeholders* that represent the positions in the output document, where the node is placed (copied).

We will show that the set of placeholders may be computed in the reversed direction of the constructed node flow. The computation begins at the main expression of the program, with a single placeholder pointed at the root of the output document. The sense of the abovementioned operators is reversed – they manipulate and propagate the set of placeholders from their original output to their operands. Besides the propagation, each constructor creates corresponding nodes in the output document, at all places denoted by the placeholders that reached it. It means that these operators have side effects and the XQuery language is not evaluated as a functional language. In other words, each function returning a constructed tree sequence is replaced with a procedure with an additional parameter containing a placeholder set. Conversely, input arguments carrying constructed nodes are replaced with output arguments carrying placeholders. Similarly, a variable assigned a constructed node sequence by a let-expression is replaced by a placeholder-set collected from all usages of the variable and propagated further into the expression in the let clause.

The edge labels in the output document are generated during the evaluation of the program. In general, the labels are combined from the following partial labels:

- Fixed labels used to implement concatenation (comma) operator.
- Input node identifiers used to reflect the (implicit) document ordering of node sets produced by XPath expressions.
- Atomic values from various XML universes (xs:string, xs:integer, etc.) generated by order by clauses.

In order to protect the correct ordering and to avoid random collisions, the partial labels must not be combined by direct concatenation. Instead, *hierarchical strings* are used, whose letters are recursively composed of other hierarchical strings, labels, or atomic values. In Fig. 4, the hierarchical composition of output node labels is shown by parentheses; in our mathematical model, we will define the operator α to produce hierarchical string letters.

Output node identifiers are produced during the reversed evaluation of the XQuery program. Starting at the main expression, *(partial)* placeholders are propagated through the program; at each constructor, partial placeholders are *finalized* to produce complete output node identifiers. These identifiers, together with the properties of the constructed node, are stored into the output database.

Placeholders are generalized form of node identifiers – they allow insertion at a position marked inside an (incomplete) node identifier.

Example

As an example, we will use the Use Case TREE – Query 1 from the XQuery Test Suite [12], shown at Fig. 1. Fig. 2 shows the corresponding abstract syntax trees with arrows depicting the propagation of information. The thin dashed arrows show the propagation of input tree nodes, the thick dashed arrows carry the constructed nodes. Note that the

```
declare function local:toc($P as element()) as element()*
{
    for $X in $P/section
    return <section> {
            $X/@* , $X/title , local:toc($X)
        }    </section>
};

<toc> {
    for $S in $I/book return local:toc($S)
} </toc>
```

Fig. 1. Query 1

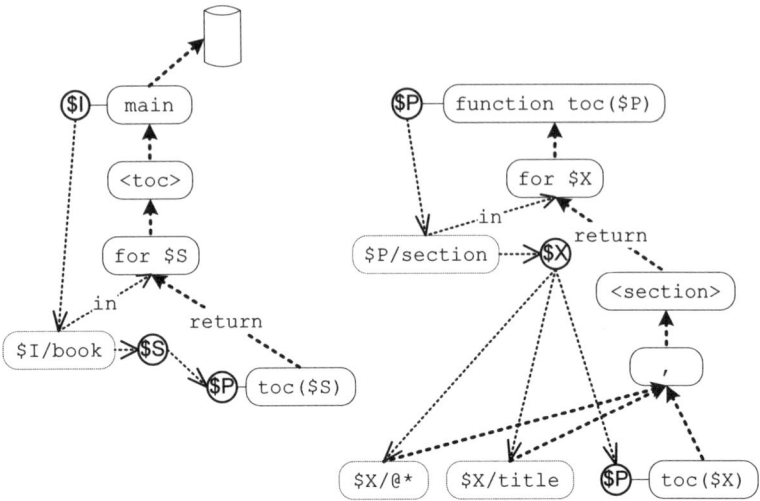

Fig. 2. Query 1 – Standard data flow

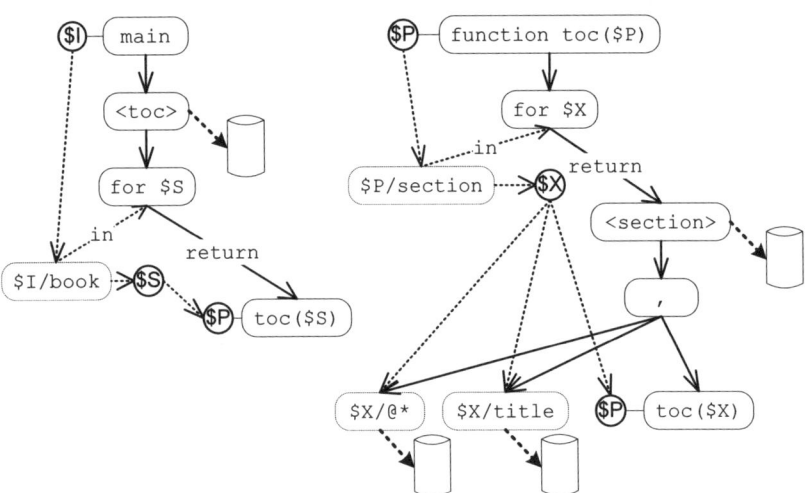

Fig. 3. Query 1 – Reversed data flow

$X/@* and $X/title expressions produce input nodes, however, their corresponding sub-trees are copied into the output document. This behavior may be regarded as an implicit conversion – by definition, it shall take place just before the <section> constructor; however, static analysis may move the conversion operation down to the XPath expressions as shown in the figure.

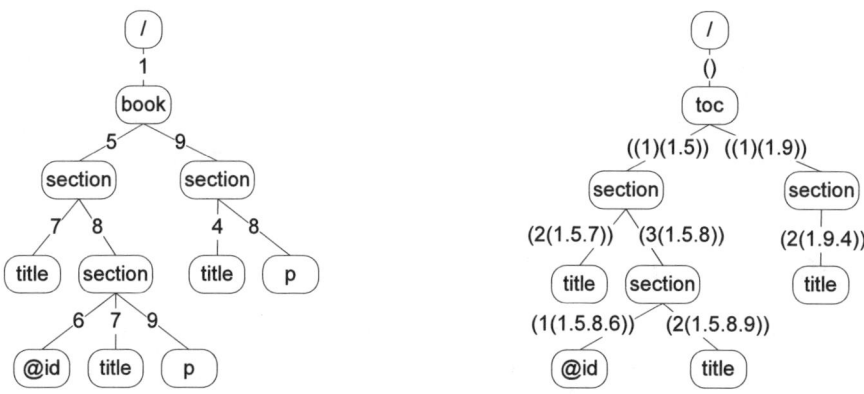

Fig. 4. Query 1 – Sample input and output documents

Fig. 4 shows an input document and the corresponding output generated by the program shown at Fig. 1. Each node is identified by the concatenation of the labels associated with the edges of the document tree; for instance, the @id node in the input document receives the identifier 1.5.8.6. The lexicographical ordering of the node identifiers corresponds to the document ordering as defined in the XML standard.

The reversed flow of placeholders is shown at Fig. 3, using solid arcs. The side effects of constructors are shown as thick dashed arrows. Implicit conversions attached to the $X/@* and $X/title expressions were transformed to side effects: Such expressions perform a Cartesian product of their new placeholder argument with the input nodes returned by their XPath expression. Each of these input nodes is (together with its subtree) copied into every place denoted by the placeholder set.

The computation starts with the *empty* placeholder (') consisting of a single hierarchical letter whose interior is empty but marked (with the apostrophe) as the insertion point. The empty placeholder reaches the <toc> constructor which finalizes the placeholder by removing the insertion point and produces a node in the form of the tuple [(),element,'toc']. The constructor operator also creates another partial placeholder ()(') to be passed down to the child expression.

A for-expression iterates through a sequence. In our example, this is a sequence of nodes serialized in the document-order; therefore, their node identifiers are used to carry the ordering information. In each iteration, the for-expression produces a new placeholder by inserting the identifier before the insertion point; in the example, the for $S expression creates the placeholder ()((1)') using the node identifier 1 from the input document. This placeholder is passed down to the function toc.

The expression for $X iterates through two input nodes having identifiers 1.5 and 1.9. Therefore, it produces these two placeholders: ()((1)(1.5)') and ()((1)(1.9)'). Subsequently, the <section> constructor is invoked twice, producing the following tuples: [()((1)(1.5)),element,'section'] and [()((1)(1.9)),element,'section'].

The concatenation (,) operator is implemented using fixed labels {1,2,3} assigned to its three operands. Thus, the sub-expression $X/title receives placeholders with the label 2 inserted, like ()((1)(1.5))(2'). The node-set computed by the XPath expression is copied to the output document. Their input node identifiers are inserted into placeholders in order to keep their document order preserved. Thus, the node identifier ()((1)(1.5))(2(1.5.7)) is generated for the first title node.

The function toc is called recursively for three times; during these calls, it receives the following placeholders: ()((1)(1.5))(3'), ()((1)(1.5))(3(1.5.8))(3') and ()((1)(1.9))(3'). However, only the first call produces any output.

3 Mathematical Model

In this section, we define a mathematical model of the reversed evaluation of a XQuery program. For brevity, we will ignore some technical details: We will not explicitly model the access to the input document(s). We will assume that XPath navigation (axes) and node-set operators are never applied to nodes constructed during the evaluation. Finally, we will not model the details of manipulation with atomic values and text nodes required by the XQuery standard, since the required concatenation may be applied additionally on the output of the transformation.

We will show the model on selected operators of the XQuery language. Since the XSLT and XQuery are related languages and the translation from XSLT to XQuery is known (see [6]), the model may be applied also to XSLT.

We assume that the XQuery was already statically analyzed to determine which expressions are used as output node sequences (such a method was already described in [1]). Those expressions will be evaluated in the reversed manner, using the placeholder sets. If an expression is used in both output and input styles (for instance, in an XPath navigation and in a constructor), the expression shall be duplicated.

Our model consists of three portions:

- *Value* model used to describe sequences of atomic values and input nodes.
- *Placeholder* model comprehending the data passed in the reversed direction.
- *Output* model used to formalize the output of the program.

We will not describe operations on the value model since almost any XPath evaluation approach may be mapped to our model. We will need the value model only at the interface between the standard and reversed evaluation regions, i.e. in the description of the semantics of for-statements and the conversion between input and output nodes.

Both value and placeholder models must include description of the context where the underlying XQuery expressions are evaluated. To define the context, we will use two stacks:

- *Call* stack storing positions in the program where functions were called on the descent to the current position. We will use ADDR to denote the set of call instructions in the program.
- *Variable* stack storing values assigned to control variables of for-statements along the descent.

All models represent sequences using a mapping from an ordered set of *sequence identifiers* to the corresponding values. Sequence identifiers, input and output node identifiers, and value stacks are represented using the same formalism – *hierarchical strings*.

Hierarchical alphabet is an (infinite) totally ordered set Σ with a function $\alpha :$ $\Sigma^* \to \Sigma$ that is a homomorphism with respect to the (lexicographical) ordering, $\alpha(u) < \alpha(w) \Leftrightarrow u < w$. Additionally, the alphabet shall contain all natural numbers, $N \subseteq \Sigma$, and all atomic value universes from the XML standard shall be mapped into the hierarchical alphabet. *Hierarchical string* is a finite word over hierarchical alphabet, a member of Σ^*. We will use λ to denote the empty string and . as concatenation operator.

Hierarchical strings may be implemented for instance by unranked ordered trees whose leaves are labeled with natural numbers or atomic values.

We will use the following operators borrowed from classical relational algebra: *Union* and *set difference*, $R \cup S$ and $R \setminus S$, *natural join*, $R \bowtie S$, *selection*, $\sigma_{P(a_1,\dots,a_n)}(R)$, based on a predicate P, *projection*, $\pi_{a_1,\dots,a_n}(R)$ (for removing attributes, we will also use the abbreviation $\pi_{\setminus a_1,\dots,a_n}(R)$ for $\pi_{A_R \setminus \{a_1,\dots,a_n\}}(R)$ where A_R is the set of attributes of R), and *rename*, $\rho_{b/a}(R)$. Additionally, we define the operator of *function application* $\Phi_{b=f(a_1,\dots,a_n)}(R)$ which adds a new attribute b to the relation R, based on the function f and the values of the attributes $a_1, \dots, a_n$.

We will use the notation $R \subseteq (a_1 : T_1, \dots, a_n : T_n)$ to declare that R is a relation with attributes $a_1, \dots, a_n$ from domains $T_1, \dots, T_n$.

A sequence of atomic values or input nodes is modeled using the following relation:

$$\mathsf{value} \subseteq (i : \mathsf{ADDR}^*, v : \Sigma^*, s : \Sigma^*, x : \Sigma^*)$$

The i and v attributes represent the call and variable stacks forming the context where the expression is evaluated. s is the sequence identifier and x is either an atomic value (mapped to a member of Σ) or an input node identifier (a member of $(\mathsf{rng}(\alpha))^*$).

A set of descriptors is expressed using the following relation:

$$\mathsf{desc} \subseteq (i : \mathsf{ADDR}^*, v : \Sigma^*, m : \Sigma^*, s : \Sigma^*)$$

The i and v attributes correspond to the call and variable stacks. The remaining hierarchical identifiers form the partial placeholder $m.\alpha(s')$ with the apostrophe marking the insertion point.

The output of the XQuery program is collected from all constructor operators throughout the program. Every contribution is modeled as the relation:

$$\mathsf{output} \subseteq (i : \Sigma^*, k : \mathsf{KIND}, n : \Sigma^*, v : \Sigma^*)$$

The i attribute is the output node identifier, k, n, and v attributes represent node kind, name and value as required by the XML document model.

The reversed evaluation of the applicable XQuery operators is described by the following equations:

- *Concatenation* – $E_0 = E_1$, E_2

$$\mathsf{desc}[E_1] = \rho_{s=u} \, \pi_{\setminus s} \, \Phi_{u=s.1} \, (\mathsf{desc}[E_0])$$
$$\mathsf{desc}[E_2] = \rho_{s=u} \, \pi_{\setminus s} \, \Phi_{u=s.2} \, (\mathsf{desc}[E_0])$$

- *Node construction* – $E_0 = $ `<a>{` E_1 `}</a>`

$$\text{output}[E_0] = \Phi_{k=\text{element}} \; \Phi_{n=a} \; \Phi_{v=\lambda} \; \pi_i \; \Phi_{i=m.\alpha(s)} \; (\text{desc}[E_0])$$
$$\text{desc}[E_1] = \Phi_{s=\lambda} \; \rho_{m/p} \; \pi_{i,v,p} \; \Phi_{p=m.\alpha(s)} \; (\text{desc}[E_0])$$

- *For expression* – $E_0 = $ `for` `$y` `in` E_1 `return` E_2

$$\text{desc}[E_2] = \pi_{\setminus j,u,w,x} \; \Phi_{v=j.\alpha(x)} \; \Phi_{s=u.\alpha(w)} \; ((\rho j/v \; \rho_{u/s} \; (\text{desc}[E_0])))$$
$$\bowtie (\rho_{w/s} \; (\text{value}[E_1])))$$

- *Order-by clause* – $E_0 = $ `for` `$y` `in` E_1 `order` `by` E_2 `return` E_3

$$\text{desc}[E_3] = \pi_{\setminus j,u,w,x} \; \Phi_{v=j.\alpha(x)} \; \Phi_{s=u.\alpha(y).\alpha(w)} \; ((\rho j/v \; \rho_{u/s} \; (\text{desc}[E_0])))$$
$$\bowtie (\rho_{w/s} \; (\text{value}[E_1])))$$
$$\bowtie (\rho j/v \; \rho_{y/x} \; \pi_{\setminus s} \; (\text{value}[E_2])))$$

4 Conclusion

We have presented a method of XQuery evaluation that reverses the standard flow of evaluation on the output, synthetic portion of the program. This approach allows to generate elements of the output document directly at their final locations in the representation of the output document, bypassing the creation and merging of temporal trees required in the standard evaluation order. The method is developed for systems where the output of the transformation is stored in a database and it may be functional if the following requirements are satisfied:

- The target database uses a kind of identifiers to link the individual elements together. (Pointer-based implementations are disqualified.)
- These identifiers must be stable and independent on the presence of sibling elements. For instance, identifier schemes based on sequential (1,2,3,...) numbering of document nodes are not suitable. Fortunately, schemes designed to support updates usually satisfy this requirement.
- The target database must tolerate temporal violation of referential integrity during the process of document generation. (For instance, children representation may be generated before their parents.) Transaction-based isolation can solve this problem; however, the cost of this approach in a distributed environment is not negligible. Higher-level management of incomplete output documents may perform better than a transsaction-based approach.
- The target database must offer an interface capable to accept individual document elements and identifiers from the XQuery/XSLT engine.

It seems that many contemporary XML storage systems, either relational (for instance [10]) or native, can support the first three requirements either immediately or with little changes. The main blocker is the fourth requirement, since the storage systems are usually built as black-boxes not capable to accept their internal identifiers from the outside. Nevertheless, advanced XML storage systems are often tightly coupled with an XPath engine on the retrieval side; therefore, the integration on the document-generation side may be realizable.

Materialized XML-views are one of the areas where such an approach may be used. Indeed, methods using identifier schemes satisfying the abovementioned requirements were already presented ([5, 4]); however, they are focused on view updates, therefore solving a different problem than our method described in this paper. In [7], related techniques were used in query reformulation above XSLT Views.

The main advantage of the presented method is the absence of in-memory structures representing the output document or its parts during the execution. This fact reduces the memory footprint of the execution. On the other hand, the amount of information generated during the execution is greater than in memory-based approaches, due to the fact that each generated node carries the Dewey identifier of the absolute location of the node. Therefore, the proposed approach may be advantageous under the following circumstances:

- Large output documents that would hardly fit into memory. Traditionally, streaming XQuery/XSLT processors are used in these cases; however, our technique allows to process programs that do not fit into streamability requirements associated with various streaming methods (see, for instance [3]).
- Shallow output documents. In this case, the absolute Dewey identifiers of the generated nodes are relatively short and the abovementioned overhead connected to their generation is lower. Studies [9] show that real XML data are rather shallow.
- A distributed environment, where the maintenance of in-memory trees is costly.
- XML messaging. Our labeling scheme may also be used as an alternative to serialized XML in tightly-coupled systems (for instance XRPC [11]) where latency is more important than bandwidth.

The exact balance of the benefits and overheads of our approach is not evaluated yet. The reversed evaluation strategy becomes a part of an experimental XML framework; experiments comparing various strategies including the reverse evaluation will be conducted in simulated environments with the abovementioned properties.

Acknowledgment

This work was supported by the program "Information society" of the Thematic program II of the National research program of the Czech Republic, No. 1ET100300419.

References

1. Bednárek, D.: Reducing Temporary Trees in XQuery. In: ADBIS 2008: Proceedings of the 12th East-European Conference on Advances id Databases and Information Systems, Pori, Finland. LNCS. Springer, Heidelberg (2008)
2. Boulos, J., Karakashian, S.: A New Design for a Native XML Storage and Indexing Manager. In: Ioannidis, Y., Scholl, M.H., Schmidt, J.W., Matthes, F., Hatzopoulos, M., Böhm, K., Kemper, A., Grust, T., Böhm, C. (eds.) EDBT 2006. LNCS, vol. 3896, pp. 755–772. Springer, Heidelberg (2006)

3. Dvořáková, J., Zavoral, F.: A Low-Memory Streaming Algorithm for XSLT Processing Implemented in Xord Framework. In: ICADIWT 2008: Proceedings of The First IEEE International Conference on the Applications of Digital Information and Web Technologies, Ostrava, Czech Republic. IEEE Computer Press, Los Alamitos (to appear, 2008)
4. El-Sayed, M., Rundensteiner, E.A., Mani, M.: Incremental Fusion of XML Fragments Through Semantic Identifiers. In: IDEAS 2005: Proceedings of the 9th International Database Engineering & Application Symposium, pp. 369–378. IEEE Computer Society, Washington (2005)
5. El-Sayed, M., Wang, L., Ding, L., Rundensteiner, E.A.: An algebraic approach for incremental maintenance of materialized XQuery views. In: WIDM 2002: Proceedings of the 4th international workshop on Web information and data management, Virginia, USA, pp. 88–91. ACM, New York (2002)
6. Fokoue, A., Rose, K., Siméon, J., Villard, L.: Compiling XSLT 2.0 into XQuery 1.0. In: WWW 2005: Proceedings of the 14th international conference on World Wide Web, China, Japan, pp. 682–691. ACM, New York (2005)
7. Groppe, S., Böttcher, S., Birkenheuer, G., Höing, A.: Reformulating XPath Queries and XSLT Queries On XSLT Views. Technical report, University of Paderborn (2006)
8. Lu, J., Ling, T.W., Chan, C.-Y., Chen, T.: From region encoding to extended Dewey: on efficient processing of XML twig pattern matching. In: VLDB 2005: Proceedings of the 31st international conference on Very large data bases, Trondheim, Norway, pp. 193–204. ACM, New York (2005)
9. Mlynkova, I., Toman, K., Pokorny, J.: Statistical Analysis of Real XML Data Collections. In: COMAD 2006: Proc. of the 13th Int. Conf. on Management of Data, New Delhi, India, pp. 20–31. Tata McGraw-Hill, New York (2006)
10. Pal, S., Cseri, I., Seeliger, O., Rys, M., Schaller, G., Yu, W., Tomic, D., Baras, A., Berg, B., Churin, D., Kogan, E.: XQuery implementation in a relational database system. In: VLDB 2005: Proceedings of the 31st international conference on Very large data bases, Trondheim, Norway, pp. 1175–1186. ACM, New York (2005)
11. Zhang, Y., Boncz, P.: XRPC: Interoperable and Efficient Distributed XQuery. In: VLDB 2007: Proceedings of the 33rd international conference on Very large data bases, Vienna, Austria, pp. 99–110. ACM, New York (2007)
12. W3C. XML Query Test Suite (November 2006)

Approximating All-to-All Broadcast in Wireless Networks

Doina Bein[1] and S.Q. Zheng[2]

[1] Department of Computer Science, Erik Jonsson School of Engineering and
Computer Science, University of Texas at Dallas, USA
`siona@utdallas.edu`
[2] Department of Computer Science, University of Texas at Dallas, USA
`sizheng@utdallas.edu`

Summary. All-to-all broadcast is a communication pattern in which every node initiates a broadcast request. In this paper we investigate the problem of building a unique cast tree, that is unoriented (unrooted) tree and has minimal total power – minimal unique cast (MUC) tree – to be used for all-to-all broadcast. We propose a polynomial-time approximation algorithm for MUC problem. The power level of a node is selected to ensure bidirectional communication with its siblings, thus broadcast and converge-cast can be performed in the tree starting at any node.

1 Introduction

Minimum-energy broadcasting has been studied extensively in the literature. Most of the work focused on finding the solution for a given source node that initiates the broadcast request (a so called *initiator* for a wireless sensor network). Convergecast is the dual of a broadcast, where the data flows back to a single node. All-to-all broadcast refers to the communication pattern in which every node is an initiator for a broadcast.

The pioneer work of Wieselthier *et al.* [14] and Stojmenovic *et al.* [12] had given a new orientation in designing broadcast trees for wireless networks. By using omnidirectional antennas at the nodes in a wireless network, a node transmission can reach multiple neighbors at the same time. For example, for node a to communicate with nodes b, c, and d, node a would need to spend at least P_{ab}, P_{ac}, and P_{ad}, to reach each individual node. But a transmission of power $P_1 = max\{P_{ab}, P_{ac}, P_{ad}\}$ will reach all the nodes a, b, and c, while a transmission of power $P_2 = max\{P_{ab}, P_{ad}\}$ will reach only nodes b and d. To send a packet at distance d, a node uses the power $P_e = d^\alpha + c_e$ where α is a constant parameter depending on the characteristics of the communication medium (typically with a value between 2 and 4), and c_e is a parameter representing an overhead due to signal processing.

For most wireless networks, there is an assumption that the transmission range of a node can be adjusted in order to minimize the overall energy used

C. Badica et al. (Eds.): Intel. Distributed Comput., Systems & Appl., SCI 162, pp. 65–74, 2008.
springerlink.com   © Springer-Verlag Berlin Heidelberg 2008

for broadcasting a message through the network. If that is the case, the node will try to minimize its transmission range, thus its power level, while in order to maintain a connected network, some nodes will have to increase their power level, to reach more nodes. Building a broadcast or multicast tree of minimum energy rooted at a given node are NP-complete problems [7, 8, 2], and several approximation algorithms to solve these problems have been proposed [14, 12, 2]. In wireless sensor networks, since the communication from the leaf nodes (sensors) back to the root node (source or initiator) is vital, the convergecast problem [3] has been analyzed from the collision detection point of view [5] and latency (total steps needed to collect the data) [6, 4].

Since the data acquired by the sensors has to be sent back to the initiator for collection and analysis, using the broadcast tree built as rooted at the initiator, the total cost of the convergecast is different from the cost of the initial broadcast, since a node will have to sent data to the parent and not to its children. For the all-to-all broadcast problem in a n-node network, considering the n individual broadcast trees, one for each possible source node, is unfeasible. A node will have to keep track of n broadcast trees. This requires a large memory space and/or large and fast processing capabilities at a node. The construction of each broadcast tree will have incur a large use of bandwidth in the preprocessing step, when the broadcast trees are computed, and also additional use of bandwidth when the network topology change (and some or all nodes have to recompute their broadcast trees). An approach that requires central coordination is to select a central node that is the closest to all the nodes that collects the data and sends it to everyone else. This can be extended to a distributed approach [9], where the network is partitioned into clusters, and selected nodes called clusterhead play the role of the central node for their clusters.

Even though the existent solutions for single (Papadimitriou and Georgiadis [11]) or source-dependent broadcast trees (Wieselthier *et al.* [14]) are not optimal, since the communication back to the initiator is not considered the energy cost is underestimated for the all-to-all broadcast or convergecast.

For the all-to-all broadcasting problem we propose a power assignment for nodes that uses a single, unoriented tree T, ensures all-to-all broadcasting in the network (two sibling nodes in T have bidirectional communication links), and is of minimal power. We propose the following heuristic (Algorithm $\mathcal{MUCT}$): For each node i in the network, determine the minimal-power broadcast tree rooted at i using some approximation algorithm for broadcast trees (see [14, 7, 2] for proposed approximation algorithms), and select the tree that has the minimum value for $power^T$.

In Section 2 we present related work. In Section 3 we define the MUC problem. and we propose an approximation algorithm for it (Section 4) In Section 5 we show the relationship between the broadcast, convergecast, and MUC problem, and we give upper and lower bounds for the total power required for MUC problem. We conclude in Section 6.

2 Related Work

Wieselthier *et al.* ([14]) had introduced the notion of wireless multicast advantage and proposed the BIP (Broadcast Incremental Power) algorithm. Stojmenovic *et al.* ([12]) proposed the concept of *internal nodes* as an alternative of clusterheads, to reduce the communication overhead of broadcasting. Li and Nikolaidis ([7]) proved that the minimum-energy broadcast problem, i.e. building a broadcast tree rooted at some given node, is NP-hard, and proposed another approximation algorithm called Iterative Maximum-Branch Minimization (IMBM). Independently, Cagali *et al.* ([2]) and Liang ([8]) showed that the problem is NP-complete. In the previous solutions to the broadcast tree problem, the total power used by some broadcast tree rooted at the specified node was to be minimized, among all the broadcast tree rooted at that node that span the network. Hence, for an n-node wireless network, each node would have had to keep track of n rooted broadcast trees, including the one rooted at itself. Papadimitrious and Georgiadis ([11]) have proposed a single broadcast tree for the entire network, on which the broadcasting initiated by any source node will take place in predetermined manner. The algorithm works for any type of weighted, general networks. They have established that the minimum spanning tree (MST) that in solvable in polynomial time sequentially and sub-polynomial time in distributed manner, is within Δ times the optimal power, where Δ is the maximum node degree in the network.

Convergecast problem was investigated for wireless sensor networks. Huang and Zhang [5] have proposed a coordinated convergecast protocol to solve the collision problem. Gandham *et al.* [4] and Kesselman and Kowalski [6] focused on minimizing the total time to complete a convergecast. They proposed a distributed convergecast scheduling algorithm for the TDMA problem, respectively a randomized distributed algorithm for the same problem.

All-to-all broadcast consumes much energy so there are very few results. The effect on energy efficiency is studied by Lindscy and Raghavendra in [9, 10]. Bauer *et al.* [1] propose a data strucure called *legend* that gathers and shares its contents with visited nodes; several traversal methods are explored.

3 Models

A fixed n-node wireless network is a pair (V, w) where V is the set of set of nodes, $|V| = n$, and w is a non-negative function, defined over $V \times V$, measuring the distance between the nodes. If the nodes have assigned certain power levels then we can model the network as a weighted digraph $G = (V, E, w)$ where E is the set of arcs, representing unidirectional communication links (the power level of some node decides to which nodes is connected). We note that if the power level of some node changes, then the set E changes. Given any two nodes i and j in G, we define the function *cost* to be $cost_{ij} = w_{ij}^{\alpha}$ if $(i, j) \in E$ or ∞ otherwise.

Since the transmission range of each node dictates whether there is an arc to some other node, and nodes do not have necessarily the same transmission

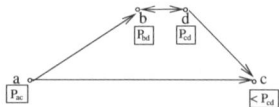

Fig. 1. Asymmetric communication between nodes

range, between two nodes u and v there can be an arc $(u, v) \in E$, or $(v, u) \in E$, or both, or neither. For example, in Figure 1, node a has enough power level to cover the distance to node c, node b to cover the distance to node d, node d to cover the distance to node c, and node c does not have enough power to reach any node (the value written in boxes under the node IDs). Assuming Euclidean distances between nodes (function w), since $w_{bd} < w_{cd}$ then there is a bidirectional communication between nodes b and d. Since $w_{bd} < w_{ab}$, then there is a unidirectional communication from a to b. Node c is isolated, since there is no node situated at a distance less than w_{cd}.

The underlying graph of G is a simple graph; by selecting $n-1$ edges to connect the nodes in G we obtained an *unoriented* (or unrooted) tree T, $T = (V, E_T)$. By choosing a node in the network as a source (root), an unoriented tree can become oriented by selecting the parent of each node as the neighboring node that has the shortest distance to the root. The total number of possible unoriented trees that can be built in a n-node network is no more than $(n + 1)^{n-1}$, which is the Cayley's tree enumeration.

Since the tree is unoriented, there is no notion of parent or children of a node. Instead, we denote by the *siblings* the set of nodes a node is connected to in the tree (neighboring nodes). For unoriented tree T, for any node i, let S_i be the set of siblings of i in T: $\forall j \in S_i, (i, j) \in E_T$. We propose to measure the energy at a node based on the distance to the farthest sibling. Given a tree $T = (V, E_T)$ rooted at some node r that spans the underlying graph of $G = (V, E)$, we propose a new measure for the power used by a node i, $power_i^T$, to be

$$power_i^T = \max_{j \in S_i} cost_{ij} \tag{1}$$

We denote by $power^T$ the total power of all the nodes in the tree T:

$$power^T = \sum_{i \in V} power_i^T \tag{2}$$

Different power assignments at a node generates different network topologies. Assuming Euclidean distance among the nodes in Figure 2, let $|xy|$ be the distance between nodes x and y, for $x, y \in \{a, b, c, d\}$. Let $|cd| < |ac| < |ab| < |bc| < |bd|$, $|ac| < |ab| < |ad|$, and $|cd| < |ad|$.

In Figure 2(a), if node c has the power P_{ca} (enough to reach nodes a and d but not node b), then the only possible spanning tree for the entire network is T_1, the one rooted at node b and has the set of arcs $\{(b, a), (a, c), (c, d)\}$. The total power used for broadcasting with node b as the source node is $P_{ba} + P_{ac} + P_{cd}$.

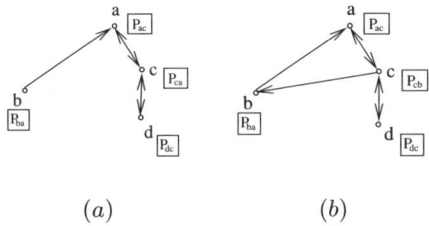

Fig. 2. Selecting the transmission power

If the nodes set their power levels as indicated by Equation (1), then $power^{T_1} = P_{ab} + P_{ba} + P_{ca} + P_{dc}$.

In Figure 2(b), if node c has the power P_{cb} (enough to reach all the nodes a, b, and d), then another tree T_2, rooted at c and with the arcs $\{(c, a), (c, b), (c, d)\}$, can be considered. The total power used for broadcasting with node c as the source node is P_{cb}. If the nodes set their power levels as indicated by Equation (1), then $power^{T_2} = P_{ac} + P_{bc} + P_{cb} + P_{dc}$.

One can observe that T_2 has the total broadcast power smaller than T_1 since $|bc| < |ba| + |ac|$ (triangle inequality), $|ba| + |ac| < |ba| + |ac| + |cd|$, and the power at a node is defined as in Equation(1). At the same time, $power^{T_1} < power^{T_2}$ since we assume $|ba| < |bc|$. Thus selecting the tree with the minimum broadcast power is necessarily a good heuristic for selecting the tree with the minimum all-to-all power.

During a single source broadcast a node may receive the same message more than once. Thus the broadcast does not induce an oriented tree, but a connected graph that it will contain an oriented tree. Thus, by eliminating the extra receipts, we can consider the induced graph as a tree. We formulate the minimal unique cast graph (MUC) problem as follows:

MUC. Given a wireless network (V, w), assign power levels to the nodes such that the corresponding unidirectional links (arcs) formed between nodes induce a strongly connected graph that can be used for all-to-all broadcast and the sum of all nodes' power is minimum.

The corresponding decision problem can be formulated as follows:

Instance: A wireless network V and a positive integer k.
Question: Does there exists a power assignment for the nodes in V such that the corresponding unidirectional links (arcs) formed between node induce a tree T that can be used for all-to-all broadcast and the sum of all nodes' power is less or equal k?

4   Approximation Algorithm for All-to-All Broadcast

Consider the following algorithm $\mathcal{MUCT}$ for a given wireless network (V, w). For each node r in V, one can construct a single-source broadcast tree T^r rooted at

r, using some approximation algorithm (see [14, 12, 13, 2] for existent approximation algorithms). For each tree T^r, consider at the nodes the power levels given by Equation (1) and the total power given by Equation (2).

We apply the following heuristic: select the tree T_o that has the minimum value for the total power $power^{T_o} = \min_{\forall r \in V} power^{T^r}$.

Algorithm 4.1. *Algorithm* $\mathcal{MUCT}$ *(Minimum Unique Cast Tree)*

Read the input: (V, w)
Initialization: Let $power_o = \infty$ and $OPT = \emptyset$.
Main Procedure:
    Forall $r \in V$ do
        Build a broadcast tree rooted at r, BT, using an approx. alg. $\mathcal{A}$
        Let T to be the unoriented tree obtained by ignore the orientation in BT.
        Define the power level of some node i as in Equation (1).
        Compute $power^T$ as in Equation (2).
        If $power^T < power_o$ then store T in OPT and $power^T$ in $power_o$.
    Endfor

The power levels at a node in the tree T_o are such that between any two siblings there is no bidirectional communication. Thus the tree can be used for all-to-all broadcast, and Algorithm $\mathcal{MUCT}$ is an approximation for the MUC problem.

5   Proofs

Let T be an oriented tree spanning the underlying graph of G. We show that by choosing the power of some node i be the maximum cost for reaching the siblings (Equation (1)), then between any two siblings in T there is a bidirectional communication link (Lemma 1). Then we show that during a broadcast from any node in the tree the total power spent is less than the value of $power^T$ given in Equation (2) (Lemma 2), and during a convergecast back to any node in the tree the total power spent is less than the value of $power^T$ (Lemma 3). This concludes a lower bound for $power^T$ (Theorem 1).

We show that the total power spent during a convergecast is greater than half of $power^T$ (Lemma 4) and during a broadcast followed by a convergecast, the total power spent is at least $power^T$ (Lemma 5). This concludes an upper bound for $power^T$ (Theorem 2).

Lemma 1. *If for any node i in the tree the power level at i is the one defined as in Equation (1) then there is a communication path between any two nodes.*

Proof. Recall that the power level of a node decides whether there is a unidirectional, bidirectional, or no communication between the node and some other node. We show that by selecting $n - 1$ edges to connect the nodes in G, edges that form an oriented tree T, and by selecting the power level of some node to be the value given in Equation (1), between any two neighboring nodes in T there will be a bidirectional communication.

Let i be some node in T and j be some sibling of i in T: $j \in S_i$. From Equation (1) it follows that $power_i^T \geq cost_{ij}$ thus there is an unidirectional communication link from i to j. Similarly, for node j $power_j^T \geq cost_{ji}$, therefore there is an unidirectional communication link from j to i. Thus between any two sibling nodes in T there is a bidirectional communication. Henceforth, there is a communication path between any two nodes in the tree.

Since between any two neighboring nodes there is a bidirectional communication link, we can consider T as a graph instead of a digraph.

For some node r, consider the orientation of the tree T with respect to r as the root node. Ancestor/descendant relationship between nodes induces a partial order. We denote by $i <_r j$ (or simply $i < j$ if r is understood) that node i is an ancestor of node j.

We denote by r-broadcasting the broadcast initiated by node r.

Lemma 2. *If any node i in the tree T has the power level as defined in Equation (1), then the total power used for a broadcast initiated by some node r along the tree T is less than the value of $power^T$ given by Equation (2).*

Proof. Some node i in T spends during the r-broadcast an amount of power, $p_i^{r,T}$, equal to

$$p_i^{r,T} = \max_{\substack{\forall j \in S_i \\ r \leq i < j}} cost_{ij} \tag{3}$$

The total power spent for r-broadcast is $p^{r,T}$

$$p^{r,T} = \sum_{i \in V} p_i^{r,T} \tag{4}$$

It can be easily observed that $p_i^{r,T} \leq power_i^T$, for any nodes i and r. Moreover, if i is a leaf node of the oriented tree T rooted at r, then $p_i^{r,T} = 0$, thus $p_i^{r,T} < power_i^T$, for any node r. It follows then that the total power spent for r-broadcast, $p^{r,T}$, given by Equation (4), is strictly smaller than the value of $power^T$, given by Equation (2).

We denote by r-convergecast the collection at node i of all the packets send by the children of i and the sending of an unique message with the aggregated data at a range that covers at least the parent of i.

Lemma 3. *If any node i in the tree T has the power level as defined in Equation (1), then the total power used for a convergecast towards some node r along the tree T is less than the value of $power^T$ given by Equation (2).*

Proof. Some node i other than r spends during the r-convergecast an amount of power, $pp_i^{r,T}$, equal to

$$pp_i^{r,T} = cost_{ik} \tag{5}$$

where $k \in S_i, r \leq k < i$, and $pp_r^{r,T} = 0$.

The total power spent for r-convergecast is $pp^{r,T}$

$$pp^{r,T} = \sum_{i \in V} pp_i^{r,T} = \sum_{(i,j) \in T, i<j} cost_{ij} \tag{6}$$

It can be easily observed that $pp_i^{r,T} \leq power_i^T$, for any nodes i and r. Moreover, if i is a leaf node in the oriented tree T rooted at r then $pp_i^{r,T} = power_i^T$. Note also that $pp_r^{r,T} = 0 < power_r^T$.

It follows then that the total power spent for r-convergecast, $pp^{r,T}$, given by Equation (6), is strictly smaller than the value of $power^T$, given by Equation (2).

We can the conclude:

Theorem 1. *For any node r in the network G and any tree T that spans G, if the power level at r is the one defined as in Equation (1) then*

$$power^T \geq max(p^{r,T}, pp^{r,T})$$

Lemma 4. *The total power used for convergecast towards some node r in the tree given by Equation (6) is greater or equal to half of $power^T$ given by Equation (2), for any node r.*

Proof. Note that

$$2pp^{r,T} = 2\sum_{(i,j) \in T} cost_{ij} = \sum_{i \in V}\sum_{j \in S(i)} cost_{ij}$$

Since for any set A, the sum of elements in A is greater or equal to the maximum element in A, it follows that for any node i,

$$\sum_{j \in S(i)} cost_{ij} \geq \max_{j \in S(i)} cost_{ij}$$

In this inequality, the right expression is $power_i^T$. Thus $2pp^{r,T} \geq power^T$.

Lemma 5. *If for any node i in the tree T, the power level at i is the one defined as in Equation (1), then the total power used for a broadcast initiated by some node r followed by a convergecast towards r along the tree T is greater or equal to the value of $power^T$ given by Equation (2).*

Proof. The power spent by a node i during r-broadcast and r-convergecast is $p_i^{r,T} + pp_i^{r,T}$.

The total power spent by all nodes during during r-broadcast and r-convergecast is

$$\sum_{i \in V}(p_i^{r,T} + pp_i^{r,T}) = p^{r,T} + pp^{r,T}$$

Since for any node i, the set $S(i)$ contains the parent of i towards r (let's called it node k) and the children of i: $S(i) = \{k\} \cup \{j \in S(i)|r \leq i < j\}$. From this and Equation (1) it follows that

$$power_i^T = \max_{\forall j \in S_i} = max(cost_{ik}, \max_{\substack{\forall j \in S_i \\ r \leq i < j}} cost_{ij})$$

Since for any two non-negative values a and b, $max(a,b) \leq a + b$, it follows that

$$max(cost_{ik}, \max_{\substack{\forall j \in S_i \\ r \leq i < j}} cost_{ij}) \leq cost_{ik} + \max_{\substack{\forall j \in S_i \\ r \leq i < j}} cost_{ij}$$

In this inequality, the left expression is $power_i^T$ (Equation (2)), and the right expression is the sum of $p_i^{r,T}$ (Equation (3)) and $pp_i^{r,T}$ (Equation (5)). Thus for any node i, $power_i^T \leq p_i^{r,T} + pp_i^{r,T}$. It follows then directly that $power^T \leq p^{r,T} + pp^{r,T}$.

We can the conclude:

Theorem 2. *For any node r in the network G and any tree T that spans G, if the power level at r is the one defined as in Equation (1) then*

$$power^T \leq pp^{r,T} + min(pp^{r,T}, p^{r,T})$$

It follows from Theorem 1 and 2 that

$$max(p^{r,T}, pp^{r,T}) \leq power^T \leq p^{r,T} + pp^{r,T}$$

6   Conclusion

We propose an approximation algorithm to build a unique cast tree to be used for all-to-all broadcast, that is unoriented and has minimal total power. Lower and upper bounds for the sum of all nodes' power are given also. Such a construction is a sufficient approximation for MUC problem, but a necessary ones. Finding a tree that has a minimal total power, that can be used for all-to-all broadcast, not necessarily using a broadcast tree as a preprocessing step, is an interesting open problem.

References

1. Bauer, N., Colagrosso, M., Camp, T.: Efficient implementations of all-to-all broadcasting in mobile ad hoc networks. Pervasive and Mobile Computing, 311–342 (2005)
2. Cagali, M., Hubaux, J.P., Enz, C.: Minimum-energy broadcast in all-wireless networks: Np-completeness and distribution issues. In: Proceedings of MOBICOM 2002, pp. 172–182 (2002)
3. Chlamtac, I., Kutten, S.: Tree-based broadcasting in multihop radio networks. IEEE Transactions on Computers 36(10), 1209–1223 (1987)
4. Gandham, S., Zhang, Y., Huang, Q.: Distributed minimal time convergecast scheduling in wireless sensor networks. In: Proccedings of the 26th IEEE International Conference on Distributed Computing Systems (ICDCS 2006), p. 50 (2006)
5. Huang, Q., Zhang, Y.: Radial coordination for convergecast in wireless sensor networks. In: Proceedings of the IEEE 1st workshop on Embedded Networked Sensors (EmNeTS-I) (2004)
6. Kesselman, A., Kowalski, D.R.: Fast distributed algorithm for convergecast in ad hoc geometric radio networks. Journal of Parallel and Distributed Computing, 578–585 (2006)
7. Li, F., Nikolaidis, I.: On minimum-energy broadcasting in all-wireless networks. In: Proceedings of LCN 2001, pp. 193–202. IEEE Computer Society Press, Los Alamitos (2001)
8. Liang, W.: Constructing minimum-energy broadcast trees in wireless ad hoc networks. In: Proceedings of MOBICOM 2002, pp. 112–122 (2002)
9. Lindsey, S., Raghavendra, C.: Energy efficient broadcasting for situation awareness in ad hoc networks. In: Proceedings of the International Conference on Parallel Processing (ICPP 2001) (2001)
10. Lindsey, S., Raghavendra, C.: Energy efficiency all-to-all broadcasting for situation awareness in ad hoc networks. Journal of Parallel and Distributed Computing, 15–21 (2003)
11. Papadimitriou, I., Georgiadis, L.: Minimum-energy broadcasting in multi-hop wireless networks using a single broadcast tree. Mobile Networks and Applications 11, 361–375 (2006)
12. Stojmenovic, I., Seddigh, M., Zunic, J.: Internal nodes based broadcasting in wireless networks. In: Proceedings of the 34th Hawaii International Conference on System Sciences (HICSS 2001) (2001)
13. Wieselthier, J.E., Nguyen, G.D., Ephremides, A.: Algorithms for energy-efficient multicasting in static ad hoc wireless networks. Mobile Networks and Applications 6 (2000)
14. Wieselthier, J.E., Nguyen, G.D., Ephremides, A.: On the construction of energy-efficient broadcast and multicast trees in wireless networks. In: Proceedings of INFOCOM 2000. IEEE Computer Press, Los Alamitos (2000)

Trusting Evaluation by Social Reputation

Vincenza Carchiolo, Alessandro Longheu, Michele Malgeri,
and Giuseppe Mangioni

Dipartimento di Ingegneria Informatica e delle Telecomunicazioni
Facoltà di Ingegneria - Università degli Studi di Catania - Italy

Summary. The increasing use of Internet for human real world activities such as e-commerce, exchange of information, advertising and several other service make the question of trust a critical issue. Today, everyone pushes information inside the net so is not easy to base trust on some, centralized authorities. many people is investigating how trust can be obtained - in some cases inspiring their investigation on social behavior - starting from some judges one may have on some others. This paper analyzes this matter, modelling trust relationship by a oriented graph and discussing some metrics useful to calculate *reputation* of a node based on others trust him/her.

1 Introduction

The stressing and increasing use of Internet for everyday life activities is harvesting more and more human real world interactions, pushing them into the virtual world of e-commerce, file sharing, on-line communities, blogs, wikies, and a plethora of other services. This transposition of social networks from a real to a virtual environment [21] makes the question of trust a critical issue, also due to the fact that in the past only few sources (public institutions, universities, corporate websites etc.) pushed their information in to the web, now anyone can spread any information across the web; actually, an authoritative source should not be considered trusted per se, however trusting has become increasingly relevant as the number of independent information sources grows.

Trust was initially defined and deeply analyzed within sociology and pysichology context. In [5], trust is given a central role: *"Trustworthiness [...] is both the constitutive virtue of, and the key causal precondition for the existence of any society"*; this relevance tends to increase within virtual environments, indeed, collecting concrete real-world evidences of virtual artifacts is often a hard question. Think for instance at whitewashing phenomenon, in real world change own's face it's not so easy. This gap leads to rely also on others mechanism and trust is a feasible solution since it aims at reproducing human behavior within social networks, where trust is achieved as a combination of personal experience (whenever available) with a person to be assigned a trustworthiness and the shared opinion (*reputation*) that person is given. In this sense, trust is used as a tool for complexity reduction [12], i.e. it provides the internal security before taking an action despite uncertainty or incomplete information (limited concrete evidences).

C. Badica et al. (Eds.): Intel. Distributed Comput., Systems & Appl., SCI 162, pp. 75–84, 2008.
springerlink.com   © Springer-Verlag Berlin Heidelberg 2008

The notion of *reputation* was mainly introduced to address the free-riding question, i.e. the malicious behavior of peer who consume resources offering limited or no participation to the network they exploit. Reputation-based systems for trust evaluation is a common approach [10, 4], generally opposed to *policy-based* systems, where the hard evidence of owned credentials is used to grant trust[1]; this is though less feasible within social networks.

Several definitions for trust can be considered, moving from social science [2][16] to more specific, computer science context [1][9]. However, trust here is not intended a synonym for authentication. The choice we make of approaching the question of trust by miming social networks behaviors not only reflects the fact that trust within computer science context still remains a social phenomenon, but it also allows to exploit some useful properties of social networks, specifically the well known *small-world* effect[20] , according to which it is often possible to find a short chain of friends of friends to anyone on the globe.

Reputation expresses the common opinion about a person and is usually combined with personal opinion in order to make a choice about that person (work together, exchange information, . . .) since it is beyond individual resources to catch all aspects of a person when making a trust decision [16]. The approach of integrating weighted contribution of both personal opinion and reputation is adopted in other works, e.g. [14][18].

Our goal is to exploit reputation for trusting evaluation; we analyze two metrics to determine the reputation of a peer node, performing some experiments on Advogato [11] data set, that is an online community dedicated to free software development that provide with a trust among people belonging to the community. In section 2 we outline the scenario in which our proposal is defined; in section 3 we introduce metrics and simulations, and in section 4 related works are considered, finally in section 5 we present some conclusions.

2   Scenario

The existing trust network is modeled as a directed labeled graph, i.e. a graph whose edges are ordered pairs ov vertices; a graph, $\mathcal{G} = (\mathcal{N}, \mathcal{E})$ where $\mathcal{N}$ is a set of nodes (representing persons in a social network) and $\mathcal{E}$ (u, v) is a set of oriented edges between nodes, i.e. $\mathcal{E} = \{(u, v) | u, v \in \mathcal{N} \wedge u \neq v\}$; a link from A to X is labelled with trustworthiness $t_X^A \in [-1, 1]$ a value that quantifies how much A trusts X; -1 means A does not trust X at all, 1 indicates that A trusts X completely, and 0 models the *indifference*, either due to a lack of information or to opposite judgements about X; note trustworthiness is not commutative, thus edge orientation is needed.

This generalized trust network can be exploited by any other application, for instance there may be a P2P network for file sharing that use trust to distinguish reliable from unrealiable peers. The trusting network itself should be intended as an overlay network placed on top of an underlying *message transportation* network; we make this hypotesis so message routing and node reachability, both used to exchange information on trustworthiness, are completely independent from trust evaluation, thus we can always suppose any node is reachable.

The next question is "How t_X^A is determined?". Whenever a person (node) A wishes to assign a trustworthiness to another person X, he combines his personal judgement with reputation, being the former assigned from his personal experience with X and the latter derived from trustworthiness X has already assigned in the network by other nodes. The role of personal experience is out of the scope of this paper; other works address this issue, e.g. [6][8].

In order to achieve a reputation value about X from the network, since we want to reproduce social behaviour, the question is *"who should A ask to get a reputation about X?"*. The most intuitive choice is to ask to all nodes that already assigned X a trustworthiness. We name such nodes as X's *judges* ($\mathcal{J}^X$); note that X could also be not aware of their existence, being not commutative the trust relationship. If A could get all X's judges opinion (trustworthiness about X), he could someway evaluate X reputation, for instance simply mediating among all values; this solution has been proposed in [22]. The main drawback with this approach is that in general X's judges may not have a trust relationship with requesting node A, thus A has no way of evaluating the *goodness* of their opinions, being then strongly subject to empowering. A refinement for this approach could be searching among X's judges who are also known (and someway trusted) by A; in this way, A is able to weight those judges' opinions, e.g. through trustworthiness he gave to them. It could also happen that A does not directly knows such judges, but some path traversing friendship relations from A exists, hence the refinement could be better defined as the search of paths from A to each of X's acquaintaces, so that A can better evaluate the opinions about X to build his reputation.

3   Trust Evaluation: Algorithms and Results

As introduced in previous section, we evaluate the reputation about a target node X exploiting information from his judges, hence we introduce a first *basic metric*, i.e. where no paths leading to the source node A are considered. To assess the correctness of reputation evaluation using this metric, we choose a generic source node A that already assigned a trustworthiness to a target X, we remove this edge from the network and try to evaluate reputation and how much it differs from trustworthiness A gave to X (A's personal judgement is neglected). This approach is known as the *leave-one-out*, a classical machine learning technique also exploited in other works (e.g. [15][17]). The formula we adopt at this stage is a simple average of X's judges trustworthiness; note that this value actually does not depend from requesting (source) node A.

To assess the effectiveness of our metrics, the Advogato data set [11] has been considered. Its properties are illustrated in table 1.

In particular, the network presents 2685 nodes with *indegree* = 0 for which any approach of evaluating a reputation clearly fails. Note however that such nodes are infrequent within social networks, since they would represent individuals who

Table 1. Advogato data set characteristics

Property	#
Nodes	7321
Edges	51660
Nodes with outdegree = 0	3253
Nodes with indegree = 0	2685
Nodes with degree = 0	1992
Average outdegree	12.70
Average indegree	11.14
Max indegree/outdegree	767

(possibly) judge people being not judged by anyone, think for example to a person who always remains inside his house, and knows about other people through the TV and Internet, but he only see movies, read blogs or websites, without directly contacting anyone, thus nobody can actually knows about his existence and cannot assign him a trustworthiness; this is clearly a remote possibility, hence such nodes are considered not significant in the remainder of our discussion. Similarly, 3253 nodes exist in the data set with *outdegree* = 0, but this also is a remote case in social network, where individuals generally interact each other thus assigning them trustworthiness; an outdegree=0 indeed would model a node that trust no one, but as soon as he lives in society, he knows people, and it is in human nature to assign a trustworthiness to a person as soon as we interact with him, thus we will neither consider such nodes in the following.

Applying the basic metrics to the Advogato data set, figure 1 illustrates the difference between the evaluated reputation and the existing trustworthiness (on x axis), obtained by varying node X over all the network (on y axis we report the number of nodes with a given percentage error). We note that most nodes are located below the 0.20 of difference, showing that the evaluated reputation can be considered an acceptable approximation for trustworthiness for the Advogato

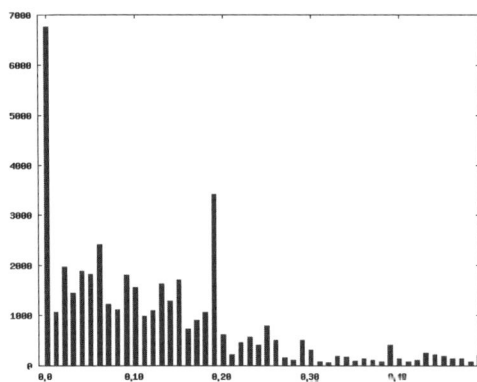

Fig. 1. # of nodes (y axis) vs. associated (reputation - trust basic metrics)

dataset the overall trust does not change. However it still remains open the possibility of cheating by judges due to the limited knowledge requesting node A has about them.

3.1 Source Centric Metric

In this section we introduce a metric aiming at reducing the problem of cheating combining trust values obtained by judges of target node X with trustworthiness source node A assigned to judges themselves.

Given a node $N \in \mathcal{J}^X$ we define $\mathcal{P}^A_{Ni}$ as the $i - th$ path joining A with N. Since several path joining a couple of node could exists we call $\mathcal{P}^A_N = \bigcup_i \mathcal{P}^A_{Ni}$ the set they belong to.

The formula used to calculate reputation about X from A's point of view - called t^A_X - depends on judges trust on X and paths from A to them, that is we use the average of such opinions weighted through paths. First of all, we define the weight of the $i - th$ path $(w_i(A, N))$ connecting A and N, where $N \in \mathcal{J}^X$;

$$w_i(A, N) = \prod_{(u,v) \in \mathcal{P}^A_{Ni}} t^u_v \tag{1}$$

where (u, v) is the oriented arc joining nodes u and v

Then, we join all $w_i(A, N)$ into a single value taking into account all paths connecting source node to one specific judge

$$w(A, N) = \frac{\sum_i w_i(A, N)}{|\{w_i(A, N)\}|} \tag{2}$$

Al last the reputation of target node, from A's point of view, is as follows:

$$t^A_X = \frac{\prod_{N \in \mathcal{J}^X} w(A, N) \cdot t^N_X}{|\mathcal{J}^X|} \tag{3}$$

The equation 2 allows us to reduce the effect of cheating because all judges are weighted with respect to source nodes. Fig 2 reports some figures obtained in the same way of those reported in figure 1 but using the source centric metric; it can be seen that results are still acceptable.

Actually, we are investigating on several approaches in order to take into account friendship chains in $w_i(A, N)$ estimation, where $N \in \mathcal{J}^X$. These approaches aim at addressing some aspects, in particular it is possible to choose between a *step-by-step* and a *cumulative* evaluation, where in the former we consider two adjacent nodes at a time, and the latter operates on the entire set of nodes belonging to the friendship chains. For instance, formula 2 is cumulative, since all trustworthiness values are multiplied in order to assess the trust node A gave to X's judge N. The step-by-step can be used to weight differently each

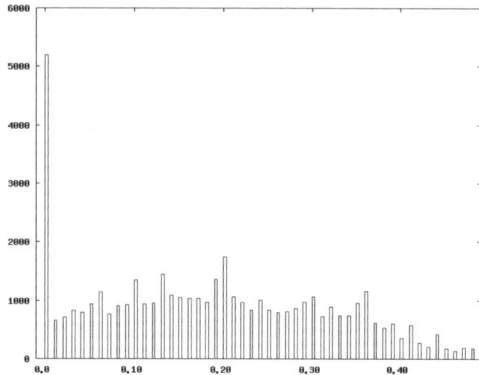

Fig. 2. # of nodes (y axis) VS associated (reputation - path centric metrics)

hop along friendship chain by focusing on each adjacent nodes pair, whereas the cumulative approach tends to consider the path as a whole.

Moreover continuous values for trustworthiness have been considered so far, however a discrete set of values (as actually the Advogato data set use) could be also adopted to better model real world trust, where often just a finite set of values (e.g. high, medium and low trust) is considered enough.

Another factor to consider is the path length, indeed in real trust networks we consider more significant an opinion if the path consists of a small set of nodes. To justify this, consider a path from A to X, nodes close to X are considered reliable due to their closeness to X (i.e. knowledge about X tends to increase) and those close to A are also reliable since A trust his friends and also friends of his friends and so on, decreasing trustworthiness as distance from A increases. From these consideration, it comes that nodes intermediate along the path would be considered the least reliable, being far away both from A and X; if the path is short enough, the presence of such nodes can be reduced.

Last but not least, the choice of the trustworthiness aggregation operator is a critical factor since it determines how much consideration has to be given to X judges. In formula 2 we used the product among all trustworthiness along the path, but other choices are possible, for instance the maximum between each pair of adjacent nodes trustworthiness allows to preserve the trust; alternatively, the minimum models a conservative approach, where we trust as least as possible.

3.2 Searching for Judges

To evaluate the reputation in an existing trust network we need to search all judges of the node to be trusted (X). In previous sections we neglect this issue, thus in in the following we address it. We point out that only local approaches are considered here, thus the existing trusting network could be not completely known a priori; this situation occurs when the network is very large and/or nodes join and leave frequently, as in social and P2P networks, hence we choose the more realistic hypothesis of a local view rather than effective but (possibly) Ir-realistic scenario of getting a global view (e.g. as in Eigentrust algorithm [19]).

Adopting a local view implies that we only suppose X is known by A (the node that is going to assign a trustworthiness to X), but X's judges have to be discovered someway since trust relationship is not commutative, hence the knowledge of X implies the knowledge of its outgoing links but this does not guarantee that those nodes are actually X's judges (a node could trust X without X knowing its existence). The simple criteria we adopt is to start from X, consider all nodes he trusts in (outgoing links), and check whether they also trust X or whether nodes they trust in also trust X; this approach is reasonable in social and P2P networks, where interaction between individuals (or their alter-ego, i.e. peers) generally results into mutual trust relationships.

In order to assess the effectiveness of judges search algorithm, we considered the same data set used to assess proposed metrics ([11]); excluding nodes with *outdegree* = 0 or *indegree* = 0 since they would model infrequent cases within social networks, as stated previously. The experiment consists of varying the role of target node X over the whole data set, applying the algorithm and comparing the average discovered indegree with the real indegree. Table 2 shows that over a total of 7320 scanned nodes, 4381 was excluded since 3523 was with outdegree=0 (the algorithm fails) and 1128 was with indegree=0 (nobody has trusted X hence reputation is unavailable). For remaining nodes it has been revealed that the average number of judges found was 2.404 over 7.057 losing 4.654 nodes, hence the experiment shows that this approach, though simple and efficient (no flooding actually occurs on the network), apparently does not seem to be very effective in finding judges.

Table 2. Judges search simulation setup and results

Description	# of nodes	avg. indegree
Total scanned nodes	7320	7.057
Nodes with outdegree=0	3253	1.275
Nodes without indegree=0	1128	1.421
Found judges per successfully request	2.404	7.057
Missed judges per successfully request	4.654	

However, a deeper analysis reveals a different result; in particular, figure 3 shows cumulative missing with respect the indegree and outdegree, i.e. x-axis is in(out) degree and y-axis is total number of missing judges of a simulation session where a randomly chosen node asks for reputation about all other nodes. We note that as indegree increases, i.e. the number of judges is higher, the number of missing judges rapidly decreases, showing that our approach tends to fail just for poorly connected nodes (an infrequent case in social networks).

The same figure also compare missed judges with outdegree, showing again that the higher is outdegree, the less is the number of missed nodes; this means that our approach of starting the search of judges from X's outgoing links, even if

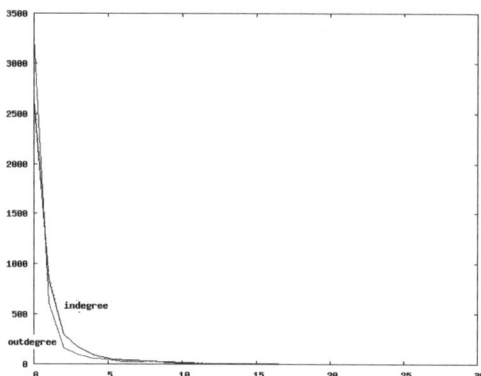

Fig. 3. # of missed judges vs in(out)-degree

does not guarantee that nodes trusted by target X also trust X, tends to provide meaningful results for nodes with a relevant number of social relationships; both for indegree and outdegree the threshold after which no nodes are missing is about 5.

4   Related Work

The recent work by Artz and Gil [1] provides a comprehensive discussion about trust within computer science and the semantic web; the notion of trust however is quite old, indeed it has been addressed in social sciences, business and psychology before it concerned computer science researchers, for instance in [7] trust is analyzed from a social perspective.

Considering the numerical representation of trustworthiness, we believe the range [-1,1] is the best choice, being simple, normalized and symmetric around the zero (representing the indifference). As claimed by [13] (the first work providing a formal model of trust), probably none of the extremes (i.e. full trust or distrust) is actually possible. In [17] however, the use of a range with negative values to model distrust is somehow criticized, mainly due to algorithmic-related issues (no more necessarily real values for trust matrix eigenvector). We use the [-1,1] range since negative values provide the right adjustment when evaluating reputation of a given node. Using interval [-1,1] is a comfortable way to study trustworthiness, but any other representation - a set of scalar values, different range of values, fuzzy sets - can be used without affecting the way those values will be searched and collected by the source agents going to calculates reputation on some target agent. In general, as reported in [18], the choice of continuous values is adopted by trust models based on aggregation mechanisms, as our proposal where reputation aggregates values from nodes that knows the node to be assigned a trustworthiness; the use of discrete values, as for instance the binary set {0,1}, is frequent in models that rely on probabilistic methods, e.g. [4].

The formula 3 we introduced to evaluate reputation is also used in other works, e.g. [4][22], being a simple approach that allows both to exploits the trustworthiness of people that directly know the target node X, and to mediate these values with trust relationships chains of requesting node A. [3] adopts the same approach, focusing on how to determine paths from A to X.

5 Conclusions

In this work metrics for reputation evaluation are introduced, exploiting them in assigning trustworthiness to an unknown person (target node). We showed that the enhanced metrics is effective in finding needed judgements. We also considered how to discover nodes with existing opinion about target node (judges), in order to exploit their experience. We applied such metrics on a real data set to test their assessment. This work suggests the investigation of several other promising issues, as friendship chains algorithm evaluation (discarding too long paths or with too low average), feedback (how trust changes over time) and robustness (resilience to empowering and other attacks).

References

1. Artz, D., Gil, Y.: A survey of trust in computer science and the semantic web. Web Semantics: Science, Services and Agents on the World Wide Web 5(2), 58–71 (2007)
2. Chervany, N.L., McKnight, D.H.: The meanings of trust. Technical report, Minneapolis, MN - USA (1996)
3. Dell'Amico, M.: Neighbourhood maps: Decentralised ranking in small-world p2p networks. In: 3rd International Workshop on Hot Topics in Peer-to-Peer Systems (Hot-P2P), Rhodes Island, Greece (April 2006)
4. Despotovic, Z., Aberer, K.: P2P reputation management: probabilistic estimation vs. social networks. Comput. Networks 50(4), 485–500 (2006)
5. Dunn, J.: The Concept of Trust in the Politics of John Locke. In: Philosophy in History. Cambridge University Press, Cambridge (1984)
6. Esfandiari, B., Chandrasekharan, S.: On how agents make friends: Mechanisms for trust acquisition. In: Fourth Workshop on Deception, Fraud and Trust in Agent Societies, Montreal, Canada, pp. 27–34 (2001)
7. Fagin, R., Halpern, J.Y.: I'm ok if you're ok: on the notion of trusting communication. Journal of Philosofical Logic 17, 329–354 (1988)
8. Gambetta, D.: Can we trust trust? Trust: Making and Breaking Cooperative Relations, 213–237 (1990)
9. Golbeck, J.: Trust and nuanced profile similarity in online social networks. ACM Transactions on the Web (to appear, 2008)
10. Gupta, M., Judge, P., Ammar, M.: A reputation system for peer-to-peer networks. In: NOSSDAV 2003: Proceedings of the 13th international workshop on Network and operating systems support for digital audio and video, pp. 144–152. ACM Press, New York (2003)
11. Levien, R.: Advogato data set (2004)
12. Luhmann, N.: Trust and Power. Wiley, Chichester (1979)

13. Marsh, S.: Formalising trust as a computational concept. Technical report, University of Stirling, PhD thesis (1994)
14. Marti, S., Garcia-Molina, H.: Limited reputation sharing in P2P systems. In: EC 2004: Proceedings of the 5th ACM conference on Electronic commerce, pp. 91–101. ACM Press, New York (2004)
15. Massa, P., Avesani, P.: Controversial users demand local trust metrics: An experimental study on epinions.com community. In: AAAI, pp. 121–126 (2005)
16. Misztal, B.: Trust in Modern Societies. Polity Press (1996)
17. Raghavan, P., Guha, R., Kumar, R., Tomkins, A.: Propagation of trust and distrust. In: Proc. of WWW 2004 conf. (2004)
18. Sabater, J., Sierra, C.: Review on computational trust and reputation models. Artificial Intelligence Review 24, 33–60 (2005)
19. Garcia-Molina, H., Kamvar, S.D., Schlosser, M.T.: The eigentrust algorithm for reputation management in P2P networks. In: Proceedings of the Twelfth International World Wide Web Conference 2003 (2003)
20. Travers, J., Milgram, S.: An experimental study of the small world problem. Sociometry 32(4), 425–443 (1969)
21. Wellman, B.: Computer networks as social networks. Science 293(5537), 2031–2034 (2001)
22. Xiong, L., Liu, L.: Peertrust: Supporting reputation-based trust for peer-to-peer electronic communities. IEEE Trans. Knowl. Data Eng. 16(7), 843–857 (2004)

Linguistic Extraction for Semantic Annotation

Jan Dědek[1] and Peter Vojtáš[2]

[1] Charles University in Prague, Department of Software Engineering
Malostranské nám. 25, 118 00 Prague 1, Czech Republic
jan.dedek@mff.cuni.cz
[2] Academy of Sciences of the Czech Republic, Institute of Computer Science
Pod Vodárenskou věží 2, 182 07 Prague 8, Czech Republic
vojtas@cs.cas.cz

Summary. Bottleneck for semantic web services is lack of semantically annotated information. We deal with linguistic information extraction from Czech texts from the Web for semantic annotation. The method described in the paper exploits existing linguistic tools created originally for a syntactically annotated corpus, Prague Dependency Treebank (PDT 2.0). We propose a system which captures text of web-pages, annotates it linguistically by PDT tools, extracts data and stores the data in an ontology. We focus on the third phase – data extraction – and present methods for learning queries over linguistically annotated data. Our experiments in the domain of reports of traffic accidents enable e.g. summarization of the number of injured people. This serves as a proof of concept of our solution. More experiments, for different queries and different domain are planned in the future. This will improve third party semantic annotation of web resources.

1 Introduction

For the Web to scale, tomorrow's programs must be able to share and process data even when these programs have been designed totally independently. Web services provide a standard means of interoperating between different software applications, running on a variety of platforms and/or frameworks. Web services are characterized by their great interoperability and extensibility, as well as their machine-processable descriptions thanks to the use of XML. They can be combined in a loosely coupled way in order to achieve complex operations. Programs providing simple services can interact with each other in order to deliver sophisticated added-value services [7].

Still, more work needs to be done before the Web service infrastructure can make this vision come true. Current technology around UDDI, WSDL, and SOAP provide limited support in mechanizing service recognition, service configuration and combination (i.e., realizing complex workflows and business logics with Web services), service comparison and automated negotiation. In a business environment, the vision of flexible and autonomous Web service translates into automatic cooperation between enterprise services. Any enterprise requiring a business interaction with another enterprise can automatically discover and select the appropriate optimal Web services relying on selection policies. Services can be invoked automatically and payment processes can be initiated. Any necessary mediation would be applied based on data and process ontologies and the automatic translation and semantic interoperation. An example would be

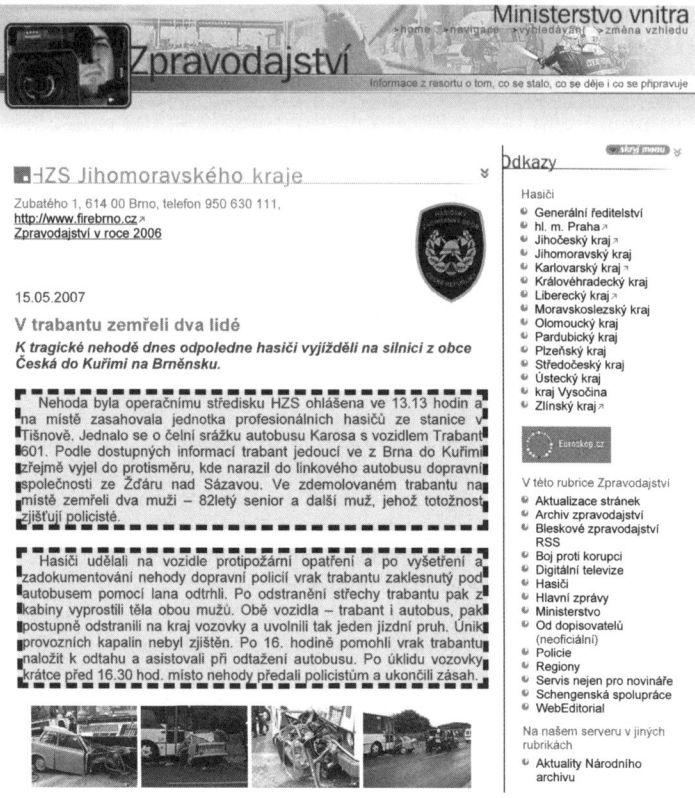

Fig. 1. Example of the web-page with a report of a fire department

supply chain relationships where an enterprise manufacturing short-lived goods must frequently seek suppliers as well as buyers dynamically. Instead of employees constantly searching for suppliers and buyers, the Web service infrastructure does it automatically within the defined constraints. Other applications areas for this technology are Enterprise-Application Integration (EAI), eWork, and Knowledge Management [6].

Bottleneck for semantic web services is lack of semantically annotated information. This is especially difficult for Web resources described in natural language, especially for IndoEuropean flexitive type languages like Czech Language. We deal with linguistic information extraction from Czech texts from the Web for semantic annotation.

In this paper we describe initial experiments with information extraction from traffic accident reports of fire departments in several regions of the Czech Republic. These reports are being published on the web[1] of the Ministry of Interior of the Czech Republic. An example of such report can be seen on the Figure 1. We would like to demonstrate the prospects of using linguistic tools from the Prague school of computational

---

[1] http://www.mvcr.cz/rss/regionhzs.html

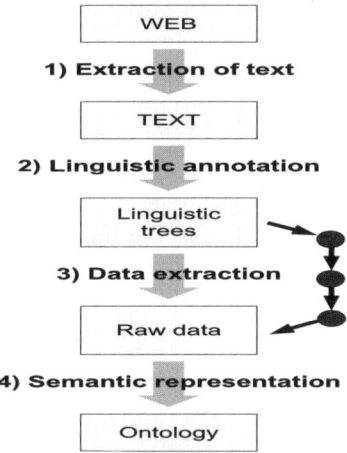

Fig. 2. Schema of the extraction process

linguistic (described in 3). Our experiments are promising, they e.g. enable the summarization of the number of injured people.

Main contributions of this paper are:

1. Experimental chain of tools which captures text of web-pages, annotates it linguistically by PDT tools, extracts data and stores the data in an ontology.
2. In the third phase – data extraction – methods for learning queries over linguistically annotated data.
3. Initial experiments verifying these methods and tools

2   Chain of Tools for Extraction and Annotation

Here we describe our chain of tools for the linguistic extraction of semantic information from text-based web-resources (containing grammatical sentences in a natural language). The chain covers a process that consists of four steps. The Figure 2 describes it. Notice, more detailed structure of the third pahase we focus in this paper.

1. *Extraction of text*
   The linguistic annotating tools process plain text only. In this phase we have to extract the text from the structure of a given web-resource. In this first phase we have used RSS feed of the fire department web-page. From this we have obtained URLs of particular articles and we have downloaded them. Finally we have extracted the desired text (see highlighted area in the Figure 1) by means of a regular expression. This text is an input for the second phase.
2. *Linguistic annotation*
   In this phase the linguistic annotators process the extracted text and produce corresponding set of dependency trees representing the deep syntactic structure of individual sentences. We have used the linguistic tools described in the section 3 for this task. Out put of this phase are tectogrammatical trees (for example see Figure 3) of sentences in document under investigation.

3. *Data extraction*

We use the structure of tectogrammatical (i.e. deep syntactic) dependency trees to extract relevant data. Refinement of this step is the main focus of this paper, see section 4 for more details.

4. *Semantic representation*

This phase consists of quite simple data transformation or conversion to the desired ontology format. But it is quite important to choose suitable ontology that will properly represent semantics of the data. Output are two fold. An ontology with instances. Annotation of a web resource (e.g. using API to an RDFa editor of html pages).

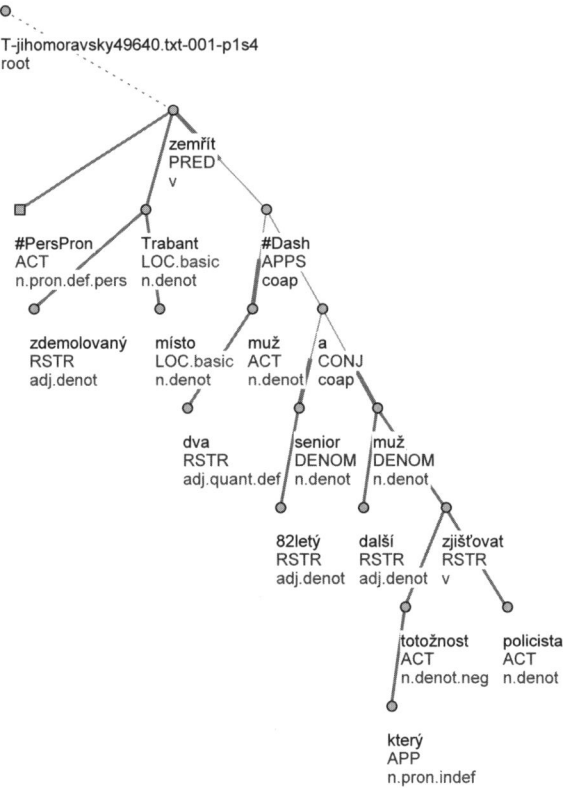

Fig. 3. Example of a tectogrammatical tree

3   PDT Linguistic Tools for Automatic Linguistic Annotation of Texts

In this section we will describe the linguistic tools that we have used to produce linguistic annotation of texts. These tools are being developed in the Institute of Formal

Table 1. Linguistic tools for machine annotation

Name of the tool	Results (proclaimed by authors)
Segmentation and tokenization	precision(p): 98,0%, recall(r): 91,4%
Morphological analysis	2,5% unrecognized words
Morphological tagging	93,0% of tags assigned correctly
Collins' parser (Czech adapt.)	precision: 81,6%
Analytical function assignment	precision: 92%
Tectogrammatical analysis [3]	dependencies p: 90,2%, r: 87,9%
	f-tags p: 86,5%, r: 84,3%

and Applied Linguistics[2] in Prague, Czech Republic. They are publicly available – they have been published on a CD-ROM under the title PDT 2.0 [2] (first five tools) and in [3] (Tectogrammatical analysis). These tools are used as a processing chain and at the end of the chain they produce tectogrammatical [4] dependency trees. The Table 1 shows some details about these tools.

1. **Segmentation and tokenization** consists of tokenization (dividing the input text into words and punctuation) and segmentation (dividing a sequences of tokens into sentences).
2. **Morphological analysis** assigns all possible lemmas and morphological tags to particular word forms (word occurrences) in the text.
3. **Morphological tagging** consists in selecting a single pair lemma-tag from all possible alternatives assigned by the morphological analyzer.
4. **Collins' parser – Czech adaptation** [1]
   Unlike the usual approaches to the description of English syntax, the Czech syntactic descriptions are dependency-based, which means, that every edge of a syntactic tree captures the relation of dependency between a governor and its dependent node. Collins' parser gives the most probable parse of a given input sentence.
5. **Analytical function assignment** assigns a description (*analytical function* – in linguistic sense) to every edge in the syntactic (dependency) tree.
6. **Tectogrammatical analysis** produces linguistic annotation at the tectogrammatical level, sometimes called "layer of deep syntax". Such a tree can be seen on the Figure 3. Annotation of a sentence at this layer is closer to meaning of the sentence than its syntactic annotation and thus information captured at the tectogrammatical layer is crucial for machine understanding of a natural language [3].

4   The Linguistic Extraction - Learning a Query

Extraction of information in this phase of our research and development is based on specific queries. Here for example, to get from web resources number of injured people in traffic accidents based on concrete traffic accidents reports (in certain time and region

---

[2] http://ufal.mff.cuni.cz

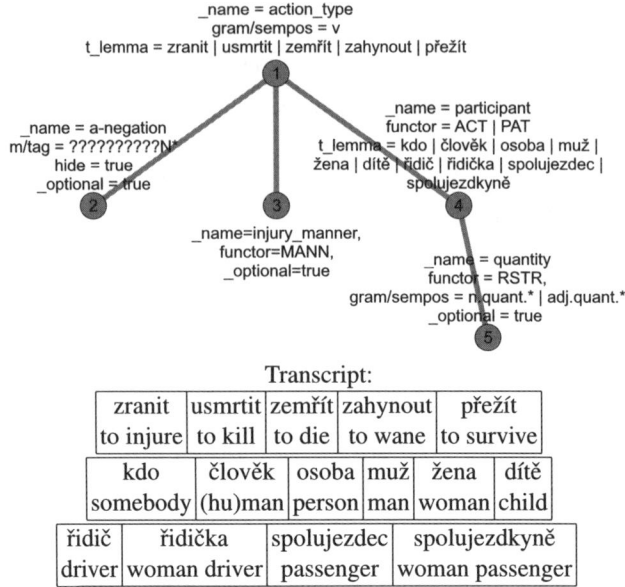

_name = action_type
gram/sempos = v
t_lemma = zranit | usmrtit | zemřít | zahynout | přežít

_name = a-negation
m/tag = ??????????N*
hide = true
_optional = true

_name = participant
functor = ACT | PAT
t_lemma = kdo | člověk | osoba | muž |
žena | dítě | řidič | řidička | spolujezdec |
spolujezdkyně

_name=injury_manner,
functor=MANN,
_optional=true

_name = quantity
functor = RSTR,
gram/sempos = n.quant.* | adj.quant.*
_optional = true

Transcript:

zranit	usmrtit	zemřít	zahynout	přežít
to injure	to kill	to die	to wane	to survive

kdo	člověk	osoba	muž	žena	dítě
somebody	(hu)man	person	man	woman	child

řidič	řidička	spolujezdec	spolujezdkyně
driver	woman driver	passenger	woman passenger

Fig. 4. Netgraph query – extract rule

- but these are "easy attributes"). Such an informal query will be translated, in order to be applied to the results of second phase of our process, namely to tectogrammatical trees of traffic accidents reports.

Our linguistic extraction method is based on extraction rules. These rules correspond to query requests of Netgraph application. The Netgraph application [5] is a linguistic tool used for searching through a syntactically annotated corpus of a natural language. It was originally developed for searching the analytical and tectogrammatical levels of the Prague Dependency Treebank, a richly syntactically annotated corpus of Czech [2]. Netgraph queries are written in a special query language. An example of such Netgraph query can be found in the Figure 4. The Netgraph is a general tool for searching trees, it is not limited only to the trees in the PDT format. In our application we use it for searching the tectogrammatical trees provided by a set of language processing tools described in the previous chapter. The tectogrammatical trees have a very convenient property of containing just the type of information we need for our purpose, namely the information about inner participants of verbs - actor, patient, addressee etc.

4.1 Extraction Method

The extraction works as follows: the extraction rule is in the first step evaluated by searching through a set of syntactic trees. Matching trees are returned and the desired information is taken from particular tree nodes.

Let us explain it in more detail by using the example of extraction rule from the Figure 4. This rule consists of five nodes. Each node of the rule will match some

```
<injured_result>
    <action type="zranit">
        <sentece>
            Při požáru byla jedna osoba lehce zraněna -- jednalo se
            o majitele domu, který si vykloubil rameno.
        </sentece>
        <sentece_id>T-vysocina63466.txt-001-p1s4</sentece_id>
        <negation>false</negation>
        <manner>lehký</manner>
        <participant type="osoba">
            <quantity>1</quantity>
            <full_string>jedna osoba</full_string>
        </participant>
    </action>
    <action type="zemřít">
        <sentece>
            Ve zdemolovaném trabantu na místě zemřeli dva muži -- 82letý
            senior a další muž, jehož totožnost zjišťují policisté.
        </sentece>
        <sentece_id>T-jihomoravsky49640.txt-001-p1s4</sentece_id>
        <negation>false</negation>
        <participant type="muž">
            <quantity>2</quantity>
            <full_string>dva muži</full_string>
        </participant>
    </action>
    <action type="zranit">
        <sentece>Čtyřiatřicetiletý řidič nebyl zraněn.</sentece>
        <sentece_id>T-jihomoravsky49736.txt-001-p4s3</sentece_id>
        <negation>true</negation>
        <participant type="řidič">
            <full_string>Čtyřiatřicetiletý řidič</full_string>
        </participant>
    </action>
</injured_result>
```

Fig. 5. Example of the result of the extraction procedure

node in each matching trce. So we can investigate the relevant information by reading values of tags of matching nodes. We can find out the number (node number 5) and kind (4) of people, which were or were not (2) killed or injured (1) by an accident that is presented in the given sentence. And we can also identify the manner of injury in the node number 3.

We have evaluated the extraction rule shown in the Figure 4 by using the set of 800 texts of news of several Czech fire departments. There were about 470 sentences matching the rule and we found about 200 numeric values contained in the node number 5. This extraction rule (from the Figure 4) is a result of a learning procedure described in the section 4.2.

Small part of the result of the extraction is shown in the Figure 5. This result contains three pieces of information extracted from three articles.

Each piece of information is closed in the <action> element and each deals with some kind of action that happened during some accident.

The attribute type specifies the type of the action. So in the first and in the third case there was somebody injured (*zranit* means to injure in Czech) and in the second case somebody died (*zemřít* means to die in Czech).

The element <negation> holds the information about negation of the clause. So we can see that the participant of the third action was **not** injured.

The element <participant> contains information about the participants of the action. The attribute type specifies the type of the participants and the element <quantity> holds the number of the participants. So in the first action only a single person (*osoba*) was injured. In the second action two men (*muž*) died and in the third action a driver (*řidič*) was not injured.

4.2   Query Learning Procedure

So far the process of building up the extraction rules is heavily dependent on skills and experience of a human designer. Fulfillment of this process is quite creative task. But we will try to pick it up as precisely as possible. We assume that a formal description of this process can help us in two ways. First – we can develop tools that will assist the designer of the extraction rules. Second – we can work on the automatization of the process. This process consists of two parts:

Learning the Netgraph Query

The procedure of learning the Netgraph query is demonstrated in the Figure 6. One obvious preposition of this learning procedure is that we have a collection of learning texts.

The procedure starts with frequency analysis of words (their lemmas) occurring in these texts. Especially frequency analysis of verbs is very useful — meaning of a clause is usually strongly dependent on the meaning of corresponding verb.

Frequency analysis helps the designer to choose some representative words (**key-words**) that will be further used for searching the learning text collection. Ideal choice of key-words would cover a majority of sentences that express the information we are looking for and it should cover minimal number of the not-intended sentences (maximization of relevance). An initial choice need not be always sufficient and the process could iterate.

Next step of the procedure consists in **investigating trees** of sentences covered by key-words. System responds with a set of **matching trees**. The designer examines corresponding syntactic trees — looks for the position of key-words and their matching **neighbors** in the trees.

After that the designer can formulate an initial (Netgraph) **tree query** and he or she can compare result of the Netgraph query with the coverage of key-words. Based on this he or she can reformulate the query and gradually **tune** the query and the **query coverage**.

There are two goals of the query tuning. The first goal is maximization of the relevance of the query. The second goal is to involve all important tree-nodes to the query. This second goal is important because the **complexity of the query** (number of involved nodes) makes it possible to extract more complex information. For example see the query on the Figure 4 — each node of it keeps different kind of information.

Semantic Interpretation of the Query

After the designer have successfully formulated the Netgraph query he or she have to supply semantic interpretation of the query. This interpretation expresses how to

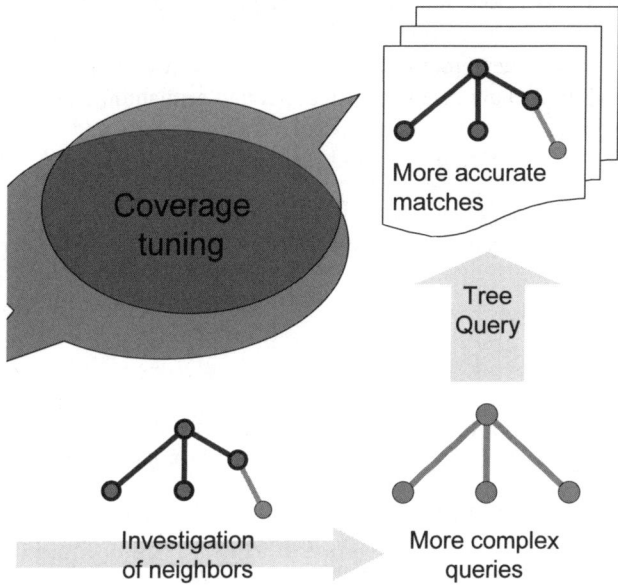

Fig. 6. Schema of the query learning procedure

transform matching nodes of the query (and the available linguistic information connected with the nodes) to the output data. The complexity of the transformation varies form simple (e.g. putting value of some linguistic attribute of the node to the output) to complex. For example a translation of a numeral to a number can be seen in the Figure 5 (element <quantity>). This is a candidate for our task to select number of killed and/or injured people in traffic accidents. In an inductive procedure (as an another ILP task) we have to learn rules which try to interpret results of extraction procedure in the sense of our task. One example of such rule, can be red as follows: if <negation> has value true, then number of injured people is 0 (e.g. nobody was injured). Another rule can from <negation>false</negation> and <quantity>2</quantity> deduce that number of injured people is two.

Our experiments have shown that the whole chain works and linguistic extraction and semantic annotation are realizable. Nevertheless, it is still a long way to go, especially in automating our process and improving learning on several steps of our procedure.

5   Conclusion

We have presented a proposal of and experiments with a system for linguistic extraction and semantic annotation of information from Czech text on Web pages. Our system relies on linguistic annotation tools from PDT [2] and the tree querying tool Netgraph [5]. Our contributions are an experimental chain of tools which captures text of web-pages, annotates it linguistically by PDT tools, extracts data and stores the data in an ontology. Especially in the third phase – data extraction – we have presented methods for learning queries over linguistically annotated data. Our initial experiments verified these methods and tools. In the near future we would like to extend this method by domain oriented lexical net and semiautomatic search for interesting extraction rules, more experiments with different queries and different domain. In a more distant future we plan to include our method in a semantic web service.

Acknowledgment. This work was partially supported by the Ministry of Education of the Czech Republic (grant MSM0021620838) and by Czech projects 1ET100300517 and 1ET100300419.

References

1. Collins, M., Hajič, J., Brill, E., Ramshaw, L., Tillmann, C.: A Statistical Parser of Czech. In: Proceedings of 37th ACL Conference, University of Maryland, College Park, USA, pp. 505–512 (1999)
2. Hajič, J., Hajičová, E., Hlaváčová, J., Klimeš, V., Mírovský, J., Pajas, P., Štěpánek, J., Vidová-Hladká, B., Žabokrtský, Z.: Prague dependency treebank 2.0 cd-rom. Linguistic Data Consortium LDC2006T01, Philadelphia 2006 (2006)
3. Klimeš, V.: Transformation-based tectogrammatical analysis of czech. In: Sojka, P., Kopeček, I., Pala, K. (eds.) TSD 2006. LNCS (LNAI), vol. 4188, pp. 135–142. Springer, Heidelberg (2006)
4. Mikulová, M., Bémová, A., Hajič, J., Hajičová, E., Havelka, J., Kolářová, V., Kučová, L., Lopatková, M., Pajas, P., Panevová, J., Razímová, M., Sgall, P., Štěpánek, J., Urešová, Z., Veselá, K., Žabokrtský, Z.: Annotation on the tectogrammatical level in the prague dependency treebank. annotation manual. Technical Report 30, ÚFAL MFF UK, Prague, Czech Rep. (2006)
5. Mírovský, J.: Netgraph: A tool for searching in prague dependency treebank 2.0. In: Hajič, J., Nivre, J. (eds.) Proceedings of the Fifth Workshop on Treebanks and Linguistic Theories (TLT), Prague, Czech rep., vol. 5, pp. 211–222 (2006)
6. SWSI. Semantic web services initiative
7. W3C. Web services activity statement (2008)

Xord: An Implementation Framework for Efficient XSLT Processing

Jana Dvořáková and Filip Zavoral

Department of Software Engineering
Faculty of Mathematics and Physics
Charles University in Prague, Czech republic
{Jana.Dvorakova,Filip.Zavoral}@mff.cuni.cz

Summary. We introduce an implementation of Xord - an XSLT processing framework which enables us to design and implement efficient algorithms for clearly characterized classes of XSLT transformations with known memory requirements. Within the framework, we design and implement a streaming algorithm using stack of the size proportional to the depth of the input document and associate it with the class of simple order-preserving and branch-disjoint transformations. The framework provides an unified interface to the underlying algorithms and acts as a standard XSLT processor.

1 Introduction

In this paper, we focus on streaming processing of XML transformations. Common processors of XML transformation languages XSLT and XQuery store the whole input in the memory and then perform the transformation itself. This kind of processing is called *tree-based processing*. In early days of XML, tree-based processing was sufficient since the existing XML documents were small and stored in files. However, nowadays it is quite common to see very large XML documents or XML data streams in practice. In both cases, the classical processing is apparently not suitable - in the former case, it is not acceptable or even possible to store the whole input document in the memory, while in the later one, the XML data become available stepwise and need to be processed "on the fly". It is thus natural to employ the *streaming processing*, i.e., to read the input document sequentially in the document order as well as to generate the output document sequentially. It is easy to see that, for certain classes of XML transformations, such streaming processor is less memory-consuming than the tree-based processors.

The main contributions of this paper are the following:

- We introduce an implementation framework called Xord framework for efficient XSLT transformations. It enables us to develop streaming algorithms, each of them capable to process certain XSLT transformation class. A contributing feature of our approach is exactly the association of streaming algorithms (differing in their complexity - memory, number of passes) with clearly characterized transformation classes. An algorithm is shown to be efficient

C. Badica et al. (Eds.): Intel. Distributed Comput., Systems & Appl., SCI 162, pp. 95–104, 2008.
springerlink.com

when processing the associated class - the efficiency is proved with mathematical rigor using the underlying formal models. The framework provides a unified interface to the streaming algorithms and thus can be viewed as a standard XSLT processor.

- Within the framework, we implement a stack-based streaming algorithm for processing simple order-preserving and branch-disjoint XSLT transformations. We exactly characterize this class of transformations. The algorithm uses a stack of the size proportional to the depth of the input XML document. Such memory usage is highly-efficient in practice, since real XML documents typically contain only few levels of elements.
- We support theoretical results for the stack-based algorithm by evaluation tests. As shown the algorithm indeed consumes constant amount of memory when processing transformations from associated class on large and shallow XML inputs.

Related work. Several streaming processors for XSLT and XQuery have been implemented. However, their efficiency was demonstrated only by experiments on a small number of XML transformations and input XML documents. It is thus not known how much memory is consumed on clearly characterized transformation classes.

XML Streaming Machine (XSM) [8] processes a subset of XQuery on XML streams without attributes and recursive structures. It is based on a model called XML streaming transducer. The processor have been tested on XML documents of various sizes against a simple query. Using XSM the processing time grows linearly with the document size, while in the case of standard XQuery processors the time grows superlinearly. However, more complex queries have not been tested.

BEA/XQRL [4] is a streaming processor that implements full XQuery. The processor was compared with Xalan-J XSLT processor on the set of 25 transformations and another test was carried on XMark Benchmarks. BEA processor was fast on small input documents, however, the processing of large documents was slower since the optimizations specially designed for XML streams are limited in this engine.

FluXQuery [7] is a streaming XQuery processor based on a new internal query language *FluX* which extends XQuery with constructs for streaming processing. XQuery query is converted into FluX and the memory size is optimized by examining the query as well as the input DTD. FluXQuery supports a subset of XQuery. The engine was benchmarked against XQuery processors Galax and AnonX on selected queries of the XMark benchmark. The results show that FluXQuery consumes less memory and runtime.

SPM (Streaming Processing Model) [5] is a simple one-pass streaming XSLT processor without an additional memory. Authors present a procedure that tries to converts a given XSLT stylesheet into SPM. However, no algorithm for testing the streamability of XSLT is introduced, and thus the class of XSLT transformations captured by SPM is not clearly characterized.

2 Formal Base

The Xord framework is based on the abstract framework introduced in [2, 3]. The abstract framework captures transformations of XML documents without data values. It consists of two groups of formal models. The basic *general XML transducer (GXT)* is used to model all (algorithmically computable) XML transformations and their tree-based processing. By imposing various restrictions on the GXT, different XML transformation classes can be defined. On the other hand, the basic *streaming XML transducer (SXT)* is used to model one-pass streaming processing without an additional memory. It can be extended by a memory to store temporary data or by allowing more passes over the input document. This way, we obtain models of several streaming processors.

When designing a new streaming algorithm within the framework, it is necessary to find a correspondence between a restricted GXT and an extended SXT. In this paper, we present one such algorithm in which we consider the simple order-preserving branch-disjoint GXT and the simple SXT (SSXT). The algorithm is called *SSXT algorithm* according to the streaming XML transducer employed. In this paper, we focus on XSLT transformations and therefore we directly consider XSLT transformation classes instead of GXTs.

At the design level, the Xord framework for XSLT transformations consists of two basic modules:

1. *Static analyzer* analyzes the given XSLT stylesheet *xsl* and determines to which of the known classes of XSLT transformations it belongs.
2. *Transformer* processes *xsl* using the algorithm associated with the determined transformation class.

3 SSXT Algorithm

In the SSXT algorithm we consider only simple XSLT transformations which are described below. A simple XSLT transformation must conform to two further conditions in order to be processable by the algorithm - it must be order-preserving and branch-disjoint. The conditions are input-dependent, i.e., in order to check their conformance, both the XSLT stylesheet and the structure of the input XML documents must be examined.

3.1 Simple XSLT Transformations

Simple XSLT stylesheet contains an *initializing template* and several *transforming templates*. The initializing template sets the mode to $m0$ and calls processing of the root element. A transforming template is called by an element name and a mode:

```
<xsl:template match="a" mode="m1"> ... body ...</xsl:template>
```

The template body consists of output elements (possibly nested) and template calls which call application of other templates by an XPath expression and a mode. The template calls are of the form:

```
<xsl:apply-templates select="child::a/descendant::b" mode="m2"/>
```

A subset of XPath expression is allowed in transforming templates - they may contain only child and descendant axis, and they select nodes by name:

$XPath := Step \mid Step/XPath$
$Step \quad := (\texttt{child} \mid \texttt{descendant})::name$

where *name* refers to an element name. The evaluation function for expression *exp* with respect to the XML document *d* and one of its nodes *u* is denoted by *eval(exp, d, u)*. The semantics of the *eval* function directly follows the semantics of the evaluation of XPath expressions - the only difference is that in our case it is sufficient to consider a single node as the current context set.

In order to determine whether a stylesheet is simple, it must be checked that it conforms to the structure described. Now we define order-preserving simple XSLT transformations and branch-disjoint simple XSLT transformations. In case a simple XSLT transformation does not conform to these conditions, the SSXT algorithm is not applicable - moreover, in majority of cases, additional memory buffers are needed in order to process the transformation. We first define an auxiliary function *eval-exp*: Let *tmp* be a transforming template, *d* be an XML document, and *u* be a node of *d*, then

$$eval\text{-}exp(tmp, d, u) = eval(exp_1, d, u) \dots eval(exp_n, d, u)$$

where $exp_1, \dots, exp_n$ is a sequence of XPath expressions appearing in the template calls of *tmp* (in this order). Thus, the *eval-exp* function returns the concatenation of the node sequences returned by individual expressions.

Order-preserving simple XSLT. A simple XSLT *xsl* is order-preserving on a set of XML documents $\mathcal{D}$ if and only if,

- for each transforming template *tmp* of *xsl*,
- for each XML document $d \in \mathcal{D}$,
- for each node *u* of XML document *d*,

it holds *eval-exp(tmp, d, u)* returns a sequence of nodes of *d* in document order.

Branch-disjoint simple XSLT. A simple XSLT *xsl* is branch-disjoint on a set of XML documents $\mathcal{D}$ if and only if it holds *eval-exp(tmp, d, u)* does not contain two nodes located within the same branch of *d* where *tmp, d, u* are as above.

3.2   Simple Streaming XML Transducer

Now we describe the transducer SSXT and the way how it processes the input XSLT stylesheet *xsl*. The SSXT is defined as a stream-to-stream tranducer since such specification is closer to the implementation than the tree-to-tree transduction used in the previous work [2, 3]. We denote by $\mathcal{D}_\Sigma$ a set of XML documents over an alphabet of element names Σ.

SSXT. The SSXT has a single input head that reads the input document sequentially, and a single output head that generates the output document sequentially.

The SSXT is equipped with a stack to store temporary data. Formally, the SSXT is a 7-tuple

$$T = (M, \Sigma, \Delta, \Gamma, m_0, Z_0, R)$$

where M is a set of states, Σ is an input alphabet, Δ is an output alphabet, Γ is a finite set of stack symbols, $m_0 \in Q$ is the designated initial state, $z_0 \in \Gamma$ is the initial stack symbol, and R is a set of rules of the form

$$(m, label, tag, la\text{-}tag, z) \rightarrow s(m', move, \gamma)$$

In the left-hand side, m is the current state, $label \in \Sigma$ is an input element name, $tag, la\text{-}tag \in \{start, end\}$ is a type of the current tag and the following tag (*lookahead tag*), respectively, and z is the current top stack symbol. In the right-hand side, s is a constant string that represents the part of the output to be generated, m' is a new state, γ is a new sequence of stack symbols, and *move* is an action of the input head: $\otimes$ - no move, $\rightsquigarrow$ - preorder move.

The configuration of SSXT T with respect to the input XML document d_{in} is of the form

$$s_{out}(m, e, tag, \gamma)$$

where s_{out} is the output XML stream generated so far, m is the current state, e is the current element of d_{in}[1], tag is the type of the current tag and $\gamma \in \Gamma^*$ is the current content of the stack. The transformation induced by T is the function $\mu_T : D_\Sigma \rightarrow D_\Delta$ such that $\mu_T(s_{in}) = s_{out}$ if and only if $s_{out} \in D_\Delta$ is generated by computation of T starting at the initial configuration (with respect to s_{in}) of the form $(q_0, 1, start, z_0)$ and terminating in the final configuration $s_{out}(q, 1, end, z_0)$.

The SSXT reads the input document d_{in} sequentially in one pass and apply the stylesheet *xsl* stepwise. First, the template matching the root element of d_{in} in the initial mode m_0 is set to be the currently processed template (*current template*). The processing proceeds in cycles. During a single cycle, a single template call of the current template is processed.

Processing cycle. All XPath expression within a template are evaluating concurrently. The evaluation is realized by deterministic finite automata (DFA)[2]. A single DFA is constructed for each expression. When the processing of a template starts, the sequence of the initial states of DFAs is pushed on the stack. The input head of SSXT reads the elements of d_{in} in document order. When a start-tag is encountered, new sequence of DFAs is computed. Three situations may occur:

a) new sequence contains no final state - the input head continues in evaluation,
b) new sequence contains a single final state which belongs to the DFA evaluating the lastly-matched expression or an expression located *after* the lastly-matched expression - the corresponding template call is processed,

---

[1] We consider dynamic-level numbering for unique identification of the elements within an XML documents, i.e., the root element has identifier 1, its children 1.1., 1.2, 1.3, etc.

[2] We refer the reader to [1] for a more detailed description of this evaluating method.

c) new sequence contains a final state which belongs to the DFA evaluating expression located *before* the lastly-matched expression, or it contains two or more final states - error.

In case b), the current cycle configuration *(template id, matched expression id)* is pushed on the stack and new cycle for processing the called template starts. The cycle configuration is popped after the whole called template has been processed and the control moves back to the current template. In case a), the evaluation continues. Here if an end-tag is encountered, the sequence of the DFA states located at the top of the stack is popped. Hence, the XPath expression of the current template are evaluated on "branches" of d_{in}.

3.3 Algorithm

We outline the overall streaming algorithm. It accepts a simple XSLT stylesheet xsl and an XML document d_{in} as the input. It tries to process the transformation specified in xsl on d_{in}. In case xsl is non-order-preserving or non-branch-disjoint on d_{in}, an error is reported. Otherwise, a proper output XML document is generated.

Stack items. The stack may contain two kinds of items (and the initial symbol z_0):

- *cycle configuration* is a pair of integers referring to the current template and to the matched expression (see variables below).
- *sequence of DFA states* used for evaluating XPath expression in the current template.

Variables. The algorithm uses the following variables:

- *top*: returns the symbol on the top of the stack,
- *label*: returns the label of the current tag,
- *tag*, *la-tag*: returns the type of the current tag and the lookahead tag.
- *current-template*: refers to the currently processed template of xsl,
- *matched-expression*: refers to the lastly-matched XPath expression of the current template. Initially, it contains null reference.

```
 1: set current-template to template with head <xsl:template match="label" mode="m0">;
 2: set matched-expression to 0;
 3: push initial stack symbol z_0;
 4: set transformed to false;
 5: while not transformed do {Iterates over elements of d_in in document order}
 6:   if top is a sequence S of DFA states then
 7:     if tag is start-tag then {Downwards evaluation}
 8:       let S' be a sequence of DFA states obtained after transition from S on
         symbol label;
 9:       if S' contains no final state then {No match}
10:         push sequence of DFA states S';
11:         advance;
```

12:     **else if** S' contains final state for single XPath expression $exp \geq$ *matched-expression* **then** {Match found}
13:         generate fragment of *current-template* between *matched-expression* and *exp*;
14:         let m be a mode associated with *exp*;
15:         let *tmp* be template with head `<xsl:template match="label" mode="m">`;
16:         **if** *tmp* is no-call template **then**
17:             generate content of *tmp*;
18:             set *matched-expression* to *exp*;
19:             **if** *la-tag* is start-tag **then**
20:                 push S';
21:             **end if**
22:         **else if** *tmp* contains some call **then**
23:             push cycle configuration *(current-template,exp)*;
24:             set *current-template* to *tmp*;
25:             set *matched-expression* to 0;
26:         **end if**
27:     **else**
28:         error;
29:     **end if**
30:     **else if** *tag* is end-tag **then** {Upwards evaluation}
31:         **if** *la-tag* is end tag **then**
32:             pop;
33:             advance;
34:         **end if**
35:     **end if**
36:     **else if** *top* is a cycle configuration (i, j) **then**
37:         **if** *tag* is start-tag **then** {Cycle start}
38:             **if** *la-tag* is start tag **then**
39:                 push sequence of initial states of DFAs for *current-template*;
40:             **end if**
41:             advance;
42:         **else if** *tag* is end-tag **then** {Cycle end}
43:             generate fragment of *current-template* between *matched-expression* and the end of the template;
44:             set *current-template* to i;
45:             set *matched-expression* to j;
46:             pop;
47:         **end if**
48:     **else if** *top* is initial stack symbol z_0 **then**
49:         **if** *tag* is start-tag **then** {Initial cycle start}
50:             push sequence of initial states of DFAs for *current template*;
51:             advance;
52:         **else if** *tag* is end tag **then** {Initial cycle end}
53:             generate fragment of *current-template* between *matched-expression* and the end of the template;
54:             set *transformed* to **true**;
55:         **end if**
56:     **end if**
57: **end while**

Memory usage. In the SSXT algorithm, a single sequence of DFA states is pushed on the stack when reading start-tags and no match is found (3.3), and a single sequence of DFA states is popped from the stack when reading end-tags and moving upwards in the element hierarchy (3.3). The sequences are obviously of constant length since the number of states in a sequence depends on the number of XPath expressions in the templates of xsl. A new processing cycle starts when a match is found for some template call (3.3, 3.3). Here an extra item - a cycle configuration - is pushed on the stack. However, when returning from processing the call, the cycle configuration is popped (3.3). Based on this observation, it is easy to see that the size of the stack never exceeds the number *(depth of $d_{in} * 2$)*.

We treat two boundary situations in a special way - processing templates without template calls (3.3) and processing matches at leaves of d_{in} (3.3).

Static analysis. The static analyzer for the SSXT algorithm currently checks the order-preservation and branch-disjointness of simple XSLT stylesheet in case the set of input XML documents is not restricted by a schema.

4   Implementation and Evaluation

We have designed and implemented the Xord framework based on .Net technologies for static analyzing of stylesheets and XML schemas and running different classes of sequential transformation algorithms depending on the analysis results. The overall structure of the Xord is depicted on Fig. 1.

The evaluation of the SSXT algorithm implementation shows that it requires a memory proportional to the depth of the input XML document. Since this depth is generally not depending on the document size and documents are relatively shallow (99% of documents have fewer than 8 levels whereas the average depth is 4 according to [6]), our memory requirements for most of the XML documents are constant, independent to the document size. On contrary, XSLT processing using standard processors like Xalan or Saxon constructs DOM for the whole document which implies memory requirements proportional to the document size. Our measurements confirmed this expectation. Fig. 2 (**a**) shows a comparison of transformation memory requirements between DOM-based and streaming processing. While the DOM-based processing requires a memory

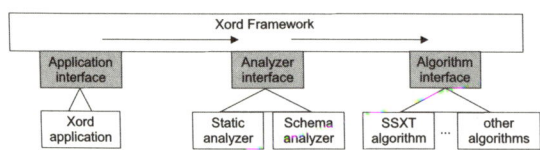

Fig. 1. A schema of the Xord Framework

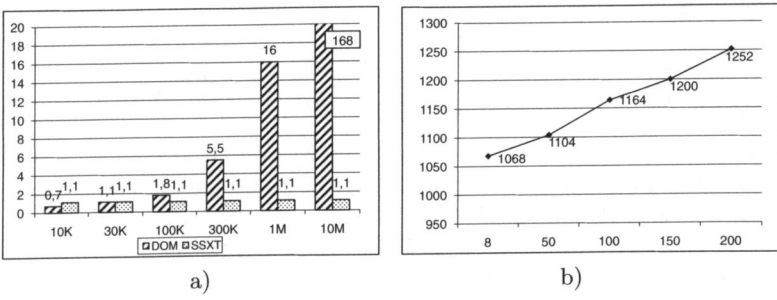

Fig. 2. Evaluation results

linear to the input size, the sequential processing memory requirements do not depend on the input size, it remains constant. On the other hand, the SSXT memory requirements depend on the document depth, but the differences are not significant (they would even not be observable using the scale as in Fig. 2 (**a**). The dependence on the document depth is depicted in Fig. 2 (**b**) in more detail, even relatively big differences in a document depth imply only minor increase of memory requirements.

We have not included comparison to the transformations written manually using event-based parsers since the effectiveness of a hand-written program depends solely on the programmer skills.

5   Conclusion

We introduced an implementation framework for efficient XSLT processing. Within this framework, we designed and implemented a streaming algorithm, called the SSXT algorithm, which can process simple order-preserving and branch-disjoint XSLT transformations using stack of the size proportional to the depth of the input document. In practice, such algorithm is highly efficient since real-world XML documents are shallow as shown by experiments.

In the future, we intend to extend the static analysis for XML documents constrained by a schema so that the analyzer will be able to determine the class of given XSLT transformation by examining both the stylesheet and the schema. Such static analysis will also help us to design more complex buffer-based streaming algorithms for processing transformations outside the class processable by the SSXT algorithm.

Acknowledgments. This work was partially supported by the Ministry of Education of the Czech Republic (grant MSM0021620838) and by the grant VEGA 1/3106/06. A part of the results presented comes from a PhD thesis of Comenius University in Bratislava, Slovakia.

References

1. Diao, Y., Altinel, M., Franklin, M.J., Zhang, H., Fischer, P.: Path sharing and predicate evaluation for high-performance XML filtering. ACM Trans. Database Syst. 28(4), 467–516 (2003)
2. Dvořáková, J.: Automatic Streaming Processing of XSLT Transformations Based on Tree Transducers. In: Proceedings of IDC 2007. Studies in Computational Intelligence. Springer, Heidelberg (2007)
3. Dvořáková, J., Rovan, B.: A Transducer-Based Framework for Streaming XML Transformations. In: Proceedings of SOFSEM (2), pp. 50–60 (2007)
4. Florescu, D., Hillery, C., Kossmann, D., Lucas, P., Riccardi, F., Westmann, T., Carey, M.J., Sundararajan, A., Agrawal, G.: The BEA/XQRL Streaming XQuery Processor. In: Proceedings of VLDB 2003, pp. 997–1008 (2003)
5. Guo, Z., Li, M., Wang, X., Zhou, A.: Scalable XSLT Evaluation. In: Yu, J.X., Lin, X., Lu, H., Zhang, Y. (eds.) APWeb 2004. LNCS, vol. 3007. Springer, Heidelberg (2004)
6. Mlýnková, I., Toman, K., Pokorný, J.: Statistical Analysis of Real XML Data Collections. In: COMAD 2006: Proc. of the 13th Int. Conf. on Management of Data, New Delhi, India, pp. 20–31 (2006)
7. Koch, C., Scherzinger, S., Schweikardt, N., Stegmaier, B.: FluXQuery: An optimizing XQuery processor for streaming XML data. In: Proceedings of VLDB 2004, pp. 1309–1312 (2004)
8. Ludäscher, B., et al.: A Transducer-Based XML Query Processor. In: Proceedings of VLDB 2002, pp. 227–238 (2002)

A Simple Trust model for On-Demand Routing in Mobile Ad-Hoc Networks

Nathan Griffiths*, Arshad Jhumka, Anthony Dawson, and Richard Myers

Department of Computer Science, University of Warwick, Coventry, CV4 7AL, UK
{nathan,arshad}@dcs.warwick.ac.uk

Summary. In a mobile ad-hoc network, nodes cannot rely on any fixed infrastructure for routing purposes. Rather, they have to cooperate to achieve this objective. However, the absence of any trusted third party in such networks may result in nodes deviating from the routing protocol for selfish or malicious reasons. The concept of trusted routing has been promoted to handle the problems selfish and malicious nodes cause to the network. In this paper, we focus on using trust in routing, and show how trust can mitigate against malicious behaviour.

1 Introduction

A mobile ad-hoc network (MANET) is a wireless network with no fixed infrastructure and no central administration. Nodes in the network usually have limited resources for computation, bandwidth, memory, and energy. Because nodes are mobile, the topology of the network varies. Message routing in MANETs is a significant problem. The lack of central administration means that nodes cannot be forced to cooperate for message routing. Nodes may deviate from the protocol for selfish or malicious reasons. For example a selfish user may wish to preserve energy resources, while a malicious user might attempt a denial of service attack. Routing protocols must cope with such selfish and malicious behaviours.

Recently, a new class of routing protocol has been proposed, namely trusted routing. Trusted routing protocols consist of two parts: a routing part and a trust model. Routing decisions are made according to the trust model. Trust and reputation have been used in many settings to cope with uncertainty in interactions. Trust is used to assess the risk associated with cooperating with others; it is an estimate of how likely another is to fulfil its commitments [2, 5]. Trust can be derived from direct interactions and from reputation.

Our work is inspired by Pirzada and McDonald's (hereafter referred to as P&M) trusted routing model [7, 8]. Based on Marsh's [5] work on computational trust, P&M use trust for routing in ad-hoc networks and obtain promising simulation results. Their approach (described below) is sophisticated and combines a range of situational trust assessments into an overall trust assessment for making decisions. Our view is that although such sophistication offers rich information on which to base decisions, similar levels of resistance to malicious behaviour can be achieved with a simpler approach.

---

* Contact author.

C. Badica et al. (Eds.): Intel. Distributed Comput., Systems & Appl., SCI 162, pp. 105–114, 2008.
springerlink.com

Although we accept P&M's results we also find some limitations. For example, they consider a range of mechanisms for malicious behaviour, and their results do not discern the effect of trust against specific types of behaviour. Aspects of P&M's results are counter-intuitive, e.g., network latency decreases as the number of malicious nodes is increased.

2 Background

In this section, we briefly introduce key work that relates to our approach. We begin by introducing the Ad-hoc On-demand Distance Vector (AODV) routing protocol, and then discuss selected trust models and how trust relates to routing.

2.1 Routing Protocols

There are two major classes of routing protocols for MANETs: proactive and reactive protocols. In proactive protocols nodes devote resources to tracking routes in a routing table, whereas in reactive protocols, routes are discovered when needed to preserve nodes' resources. In this paper, we focus on the AODV reactive protocol as it is an efficient low-overhead approach. There also exist hybrid protocols, that combine features of proactive and reactive protocols, but these are beyond the scope of this paper.

In AODV [6], when a source node wants to communicate with a destination node, but does not have a route to the destination, it initiates a route discovery. The source node broadcasts a RREQ (route request message) to all of its neighbours. Each neighbour that receives the RREQ will check in its own routing table to see if it has a route to the specified destination. If not, it will set up a reverse path towards the sender of the RREQ and then re-broadcast the RREQ. Any node receiving the RREQ will generate a RREP (route reply message) if it either has a fresh enough route to the destination, or is itself the destination. This RREP is then unicast to the next hop towards the originator of the RREQ. When a node receives a RREP, it updates the appropriate fields in its routing table and in the RREP, and then forwards the RREP to the next hop until it reaches the original sender. A sender node can have multiple routes to the destination. However, the chosen route is the shortest one between the sender and destination. This relies on the underlying assumption that all nodes are trustworthy and will never deviate from the protocol. In this paper we do not make this assumption, and use trust to mitigate against malicious or faulty behaviour.

2.2 Dependable Routing

The majority of routing mechanisms for MANETs rely on the assumption that nodes will never deviate, but in a real-world MANET this assumption is unrealistic. Because resources in a MANET are scarce, nodes may act selfishly such as not forwarding a message. In the worst case, nodes may act in an arbitrary fashion, i.e., display Byzantine behaviour [1]. Hence, to handle these problems, techniques such as secure routing [11] and trusted routing [7] have been proposed. In secure routing, cryptographic primitives are used to ensure properties such as confidentiality, integrity etc. However, secure routing requires a centralised trusted third party, making it impractical for MANETs. Trusted routing, on the other hand, can be used to handle both selfish and Byzantine nodes. In trusted routing, a trust model is embedded within the routing algorithm, and routing decisions are taken based not on shortest path but on trust values. Thus, in trusted routing the path with the highest trust is chosen.

2.3 Trust Models

Numerous models of trust and reputation exist to support cooperation in computational environments [4, 9]. One of the earliest approaches is Marsh's formalism [5]. Marsh uses the outcomes of direct interactions among entities to calculate situational and general trust. Situational trust is the level of trust in another for a specific type of situation, while general trust refers to overall trustworthiness irrespective of the situation. After each interaction an entity considers whether the other entity fulfilled its obligations. If so, then trust increases, but trust decreases if commitments are broken. To minimise the risk of failure entities will interact with the most trusted of the potential interaction partners.

Marsh's formalism is the base of many subsequent models, which supplement trust based on direct interactions with other information sources to inform decision making. For example, sophisticated approaches such ReGreT [10] and FIRE [3] add reputation information provided by third parties and knowledge of social structures to arrive at overall trust assessments. However, whilst powerful, such sophisticated models are not appropriate for routing in MANETs where resources are scarce and knowledge of social relationships between nodes is unlikely to be available.

Several trust models have been developed for peer-to-peer systems [12, 13, 14], based on sharing recommendation information to establish reputation. Although in principle these could be applied to routing in MANETs, there are two important problems. First, there is significant network overhead due to the additional information exchanged. Second, addressing the potential for malicious recommendations requires a trusted third party (or a computationally expensive public-key infrastructure), which goes against the nature of MANETs.

There are few trust mechanisms for ad-hoc networks. Zhou and Haas [15] describe a cryptographic scheme to ensure node integrity. However, their approach requires complex pre-configuration of servers to provide a distributed certification authority and relies on cryptographic operations which are costly in computation and power. P&M propose arguably the most appropriate mechanism, where nodes calculate situational trust according to observed events and then use an aggregated general trust for routing decisions. Nodes record information about others for various event types: acknowledgements, packet precision, gratuitous route replies, blacklists, HELLO packets, destination unreachable messages and authentication objects. For each type, the proportion of positive events is taken to correspond to the situational trust. Situational trust values are then aggregated using a weighted product to give overall trust. When routing, nodes will forward packets to maximise trust (rather than minimising cost in standard AODV). P&M have obtained promising simulation results, but we argue that similar positive effects can be obtained with a greatly simplified trust model.

3   The Proposed Model: Simple Trusted AODV

3.1   Network Model

The setting for our approach is a simple MANET in which we assume that nodes are situated in a bounded 2-dimensional space, within which they are free to move. For simplicity we assume they move randomly around the space. Each node has individual characteristics that define its speed of movement and the range over which it can transmit messages. The positions and transmission ranges define the network neighbourhood, since nodes can only transmit to others within their transmission range, and can only receive messages from others when they are within their range. Thus, if two nodes are within each others' transmission range they are free to communicate, but otherwise intermediate nodes are needed to forward packets. We assume that nodes use AODV as described above, and we describe below our approach for incorporating trust into AODV.

3.2   Attack Model

The standard AODV protocol assumes that nodes are fully functional and benevolent, and does not cope well if this is not the case. This has led to the development of trusted routing protocols such as that proposed by P&M. In developing their protocol, P&M describe several possible attacks, and their simulations allow malicious nodes to use any of these. Consequently, it is impossible to evaluate their trust model against *specific* attack types. In this paper, therefore, we concentrate on a small number of specific attacks and test our model against each type individually.

We consider two varieties of blackhole and a greyhole attack. A blackhole is a malicious node that attempts to drop all packets, typically by forging route replies to create fake routes with it as an intermediate node. This allows the blackhole to divert and intercept traffic from across the network, and subsequently drop all packets that it receives. A greyhole can be viewed as a faulty node, rather than explicitly malicious. Greyholes do not falsify route replies, but instead will periodically drop packets. This might be due to a fault or due to malicious intentions. Regardless of the reason, greyholes appear as intermittently faulty nodes to the rest of the network. There are several possible mechanisms to implement these attacks within AODV, and we use the following definitions.

Blackhole on route (`Blackhole-OnRoute`)

This is our simplest blackhole definition, and operates by replying that it has a fresh enough route to the destination whenever it receives a RREQ, regardless of whether it actually knows a route. AODV uses sequence numbers to track the freshness of routes. When nodes issue a new RREQ or the destination responds the sequence number is increased. A `Blackhole-OnRoute` node claims to have an existing fresh route to the destination and so the generated RREP has the same sequence number as the RREQ, causing it to be accepted by the original sender, which subsequently creates a route with the blackhole as an intermediate node. This kind of a blackhole is partially guarded against within AODV, since if the original RREQ eventually reaches the intended destination a

RREP will be generated. The reply from the destination itself has an increased sequence number over the RREQ and so will overwrite the malicious route setup by the blackhole. Despite this, in our simulations `Blackhole-OnRoute` was able to cause significant packet loss, as the routes it created intercept the first packets sent across any new route until the destination's RREP was received.

Blackhole fake destination reply (`Blackhole-FakeDestReply`)

This blackhole is more malicious than `Blackhole-OnRoute`, since in addition to claiming to have a recent enough route to the destination it also increases the sequence number in the RREP and so appears to offer a new route. The effect is that `Blackhole-Fake-DestReply`'s route is not overwritten by any reply subsequently returning from the destination itself. Thus, a route to the actual destination will only be established when the destination's RREP is received before that generated by the `Blackhole-Fake-DestReply` node.

Greyhole (`Greyhole`)

The `Greyhole` does not falsify route replies in order to intercept packets, but instead simulates a node having intermittent faults. We characterise a `Greyhole` using two time periods:

- `MAX_TIME_TO_BURST_FAULT`: maximum time to the next burst fault (seconds)
- `MAX_TIME_BURST_FAULT_LASTS`: maximum burst fault duration (seconds)

Using these time periods a node will start a burst fault at a random time between 0 and `MAX_TIME_TO_BURST_FAULT`. The burst fault lasts for a random period between 0 and `MAX_TIME_BURST_FAULT_LASTS`. These parameters can be modified to alter the nature of the faults.

3.3   Trust Model — Simple Trusted AODV (ST-AODV)

There are many potential mechanisms for determining whether a node can be trusted, based on observing the nodes' activities and behaviours. The influence of these observations can be combined to determine a trust level. P&M use several aspects of node behaviour including acknowledgements, packet precision, gratuitous route replies etc., as described in Section 2. Our view is that the effect of malicious nodes can be significantly reduced using a much simpler scheme. We build our trust models using acknowledgements as the single observable factor for assessing trust. We believe that acknowledgements offer an effective indication of a node's trustworthiness.

An acknowledgement is a means of ensuring that packets which have been sent for forwarding have actually been forwarded. There are a number of ways that this is possible, but *passive acknowledgement* is the simplest. Passive acknowledgement uses promiscuous mode to monitor the channel, which allows a node to detect any transmitted packets, irrelevant of the actual destination that they are intended for. Using this method a node can ensure that packets it has sent to a neighbouring node for forwarding are indeed forwarded.

To record trust information about a node, we introduce a `TrustNode` data store, which comprises a `nodeID`, a `packetBuffer`, and an integer `trustValue` for the node. Each node maintains a `TrustNode` for each of the nodes that it has sent packets to for forwarding. To detect whether a packet is successfully forwarded, the packets that have been recently sent for forwarding are stored in the `packetBuffer`. This is a circular buffer, meaning that if packets are not removed frequently enough the buffer will cycle, erasing the oldest elements. Thus, if a node is dropping packets or is being unacceptably slow at forwarding packets then the buffer will cycle. Otherwise, if the node is performing acceptably then when the promiscuous mode detects a forwarded packet, it can be found and removed from the buffer.

In ST-AODV we use a simple trust model, where the `trustValue` for each node is initialised to 0. With each observation, the value is incremented for nodes that are detected to forward packets and decremented for nodes that do not appear to forward packets. To check whether a node is sufficiently trusted we introduce a `minTrust` threshold such that nodes with `trustValue <= minTrust` are considered untrusted. If a node is untrusted then it is not sent packets for forwarding, and any replies it gives to route requests are ignored. Once a node becomes untrusted it is barred from consideration for packet forwarding by dropping it from the set of neighbours, removing all routes that use it, and sending out a new RREQ to re-establish the removed routes. Similarly, when receiving a RREP the first hop node is checked and if it is untrusted then the reply is disregarded. Thus, only routes where the first hop is trusted are established. Nodes make routing choices based on trust as well as the number of hops, such that the selected next hop gives the shortest trusted path.

4   Simulation and Results

To evaluate the effectiveness of ST-AODV we have performed simulations using the ns-2 network simulator[1]. Nodes are situated in a bounded 2-dimensional world about which they wander randomly. We use a network of 50 nodes in the simulations discussed below. The network contains benevolent nodes that use ST-AODV to make routing decisions, and malicious nodes that use one of the attacks defined in Section 3. The `minTrust` threshold used for barring nodes is set at -10. We obtain the following metrics from our results (which are averaged over a number of runs):

- **Packet throughout:** ratio of packets received by the destination to the number of packets sent (%)
- **Average latency:** average time for packets to reach their destination (seconds)
- **Packet overhead:** ratio of control packets generated to the total number of data packets sent (%)
- **Byte overhead:** ratio of control bytes generated to the total number of data bytes sent (%)

We record these metrics using both standard AODV and ST-AODV for each attack type under various proportions of malicious nodes. Figures 1, 2 and 3 show the results for `Blackhole-FakeDestReply`, `Blackhole-OnRoute` and `Greyhole` attacks

---

[1] http://www.isi.edu/nsnam/ns/

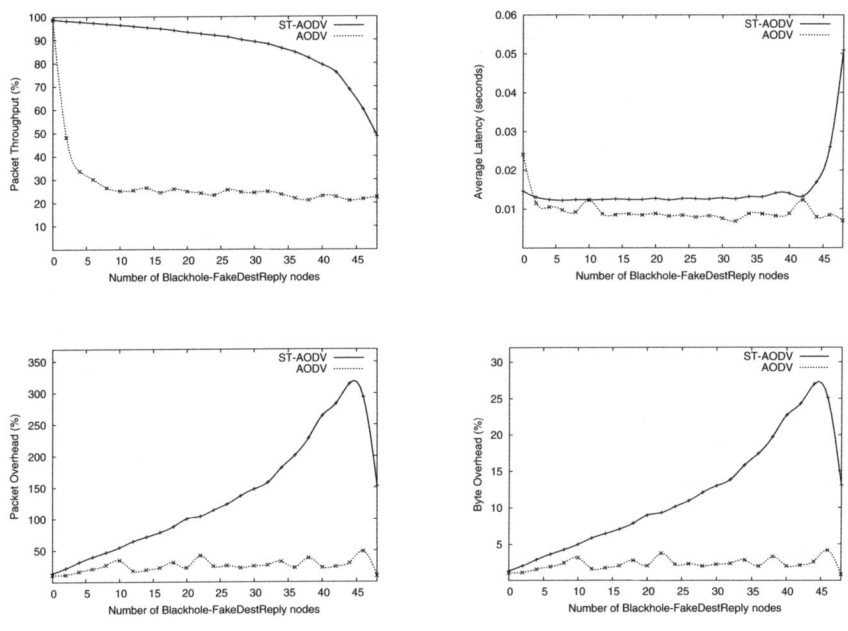

Fig. 1. Results for the `Blackhole-FakeDestReply` attack

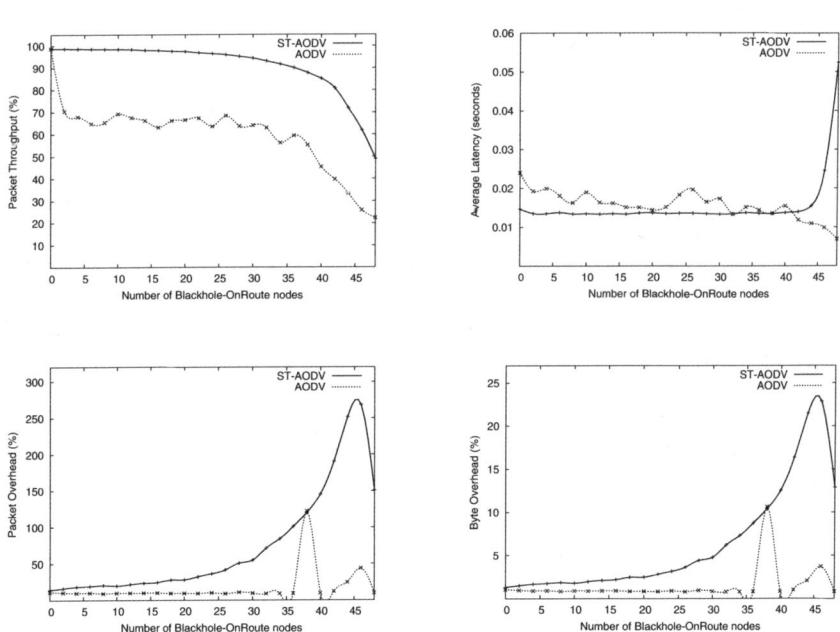

Fig. 2. Results for the `Blackhole-OnRoute` attack

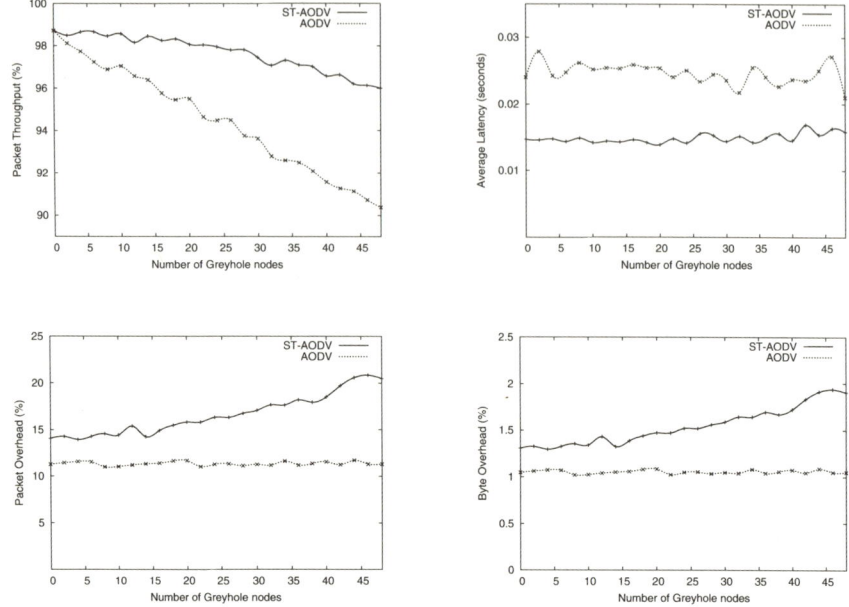

Fig. 3. Results for the Greyhole attack

respectively. The results show that ST-AODV significantly improves packet throughput under all attack types. As the number of malicious nodes is increased each attack type reduces throughput, but ST-AODV mitigates against this.

In standard AODV a small number of blackhole nodes dramatically reduces throughput, the effect stabilises for moderate numbers, and for Blackhole-OnRoute falls off for high numbers (Blackhole-FakeDestReply does not fall off further since throughput has already fallen significantly). The Greyhole attack results in a fairly linear throughput reduction as the number of malicious nodes increases. As predicted, Blackhole-FakeDestReply has the most effect. For AODV, increasing the number of Blackhole-FakeDestReply nodes very soon reduces throughput to around 25% with 10 malicious nodes, while a similar number of Blackhole-OnRoute nodes gives around 65% throughput. Regardless of attack type, ST-AODV achieves a good and fairly consistent throughput. For both blackhole attacks a throughput of over 90% is maintained if less than half the nodes are malicious. With standard AODV just 2 malicious nodes reduces throughput to below 70%. Under a Greyhole attack the throughput using ST-AODV reduces linearly with the number of malicious nodes (as for AODV), but the rate of reduction is reduced meaning trust is more beneficial with higher numbers of malicious nodes.

For blackhole attacks there is relatively little effect on latency using ST-AODV. Performance is slightly improved for Blackhole-OnRoute attacks (by < 0.005 seconds) while it is slightly worse for Blackhole-FakeDestReply (again by < 0.005 seconds). Under Greyhole attacks latency is reduced by approximately 0.01 seconds using

ST-AODV, regardless of the number of malicious nodes. As expected, the packet overhead and the byte overhead are increased by using ST-AODV under all attack types. As the number of malicious nodes is increased the overhead also increases, and more significantly so with higher numbers of malicious nodes. For the `Greyhole` attack the packet overhead is increased by approximately 5% where under half the nodes are malicious, rising to around 10% with more malicious nodes. In the `Blackhole-On-Route` attack the overhead is below 25% with below 25 malicious nodes, but this rises rapidly for higher numbers, peaking at over 200% overhead for 45 malicious nodes. The `Blackhole-FakeDestReply` attack causes the overhead to rise more rapidly, to nearly 100% where half the nodes are malicious. This is as expected, since the `Blackhole-FakeDestReply` attack is more malicious.

5   Conclusions and Summary

We have described a simple trust model that extends AODV to cope with malicious nodes. Our simulations show significant improvements in throughput, at the expense of packet and byte overhead. For low proportions of malicious nodes in the population the increase in overhead is relatively small given the improvement in throughput. Our results also show how different attacks affect a network. In particular, using standard AODV a `Blackhole-FakeDestReply` attack significantly reduces throughput compared to `Blackhole-OnRoute` and `Greyhole` attacks. Using ST-AODV we are able to minimise this difference and to protect the network effectively against all three attacks.

The results presented above are preliminary findings and there are many areas of ongoing investigation. Our results compare favourably to those obtained by P&M in terms of the improvement in throughput. We find a higher packet and byte overhead than P&M and this requires further investigation. However, P&M's results are unintuitive in that the overhead and latency decrease as more malicious nodes are added. These differences require further investigation.

We are considering several extensions to ST-AODV, including a more flexible (non-linear) trust update function and improved monitoring using promiscuous mode to monitor all traffic, rather than only a node's own packet forwarding requests. We are also investigating more flexible sanctions against untrusted nodes, such as temporary blacklisting. Finally we aim to explore how different trust models perform against different attacks and combinations of attack.

References

1. Awerbuch, B., Holmer, D., Nita-Rotaru, C., Rubens, H.: An on-demand secure routing protocol resilient to byzantine failures. In: Proceedings of the 1st ACM workshop on Wireless security, pp. 21–30 (2002)
2. Gambetta, D.: Can we trust trust? In: Gambetta, D. (ed.) Trust: Making and Breaking Cooperative Relations, pp. 213–237. Basil Blackwell, Malden (1988)
3. Huynh, T.D., Jennings, N.R., Shadbolt, N.R.: An integrated trust and reputation model for open multi-agent systems. Autonomous Agents and Multi-Agent Systems 13(2), 119–154 (2006)

4. Jøsang, A., Ismail, R., Boyd, C.: A survey of trust and reputation systems for online service provision. Decision Support Systems 43(2), 618–644 (2007)
5. Marsh, S.: Formalising Trust as a Computational Concept. PhD thesis, University of Stirling (1994)
6. Perkins, C., Royer, E.M., Das, S.: Ad hoc on-demand distance vector (AODV) routing. IETF RFC 3561 (2003)
7. Pirzada, A.A., McDonald, C.: Trust establishment in pure ad-hoc networks. Wireless Personal Communications 37(1–2), 139–168 (2006)
8. Pirzada, A.A., McDonald, C., Datta, A.: Performance comparison of trust-based reactive routing protocols. IEEE Trans. on Mobile Computing 5(6), 695–710 (2006)
9. Ramchurn, S.D., Huynh, D., Jennings, N.R.: Trust in multi-agent systems. Knowledge Engineering Review 19(1), 1–25 (2004)
10. Sabater, J., Sierra, C.: Reputation and social network analysis in multi-agent systems. In: Proceedings of the 1st Int. Conf. on Autonomous Agents in Multi-Agent Systems, pp. 475–482 (2002)
11. Sanzgiri, K., Dahill, B., Levine, B.N., Shields, C., Belding-Royer, E.M.: A secure routing protocol for ad hoc networks. In: Proceedings of the 10th IEEE Int. Conf. on Network Protocols, pp. 78–89 (2002)
12. Selçuk, A.A., Uzun, E., Pariente, M.R.: A reputation-based trust management system for P2P networks. In: IEEE/ACM Int. Symposium on Cluster Computing and the Grid, pp. 251–258 (2004)
13. Song, S., Hwang, K., Zhou, R., Kwok, Y.-K.: Trusted P2P transactions with fuzzy reputation aggregation. IEEE Internet Computing 9(6), 24–34 (2005)
14. Xiong, L., Liu, L.: PeerTrust: Supporting reputation-based trust in peer-to-peer communities. IEEE Trans. on Knowledge and Data Engineering 16(7), 843–857 (2004)
15. Zhou, L., Haas, Z.J.: Securing ad-hoc networks. IEEE Network Magazine 13(6), 24–30 (1999)

A Platform for Collaborative Management of Semantic Grid Metadata

Michael Hartung[1], Frank Loebe[2], Heinrich Herre[3], and Erhard Rahm[2]

[1] Interdisciplinary Center for Bioinformatics, University of Leipzig, Germany
hartung@izbi.uni-leipzig.de
[2] Department of Computer Science, University of Leipzig, Germany
(loebe,rahm)@informatik.uni-leipzig.de
[3] Institute for Medical Informatics, Statistics and Epidemiology,
University of Leipzig, Germany
hherre@imise.uni-leipzig.de

Summary. Grid environments, providing distributed infrastructures, computing resources and data storage, usually show a high degree of heterogeneity in their metadata. We propose a platform for collaborative management and maintenance of common metadata for grids. As the conceptual foundation of this platform, a meta model is presented which distinguishes structured descriptions and classification structures. On this basis, the system allows for the user-friendly creation and editing of grid relevant metadata and provides various search and navigation facilities for grid participants. We applied the platform to the German D-Grid initiative by establishing the D-Grid Ontology (DGO).

1 Introduction

Grid computing offers scientists a distributed infrastructure for collaboration and provides massive amounts of computing, storage, and data resources. Such grid initiatives, e.g., the German D-Grid[1], are highly complex and involve many heterogeneous components. They offer resources of different types (e.g., hardware or software resources). Furthermore, these resources belong to many participating organizations, e.g., universities, research centers or enterprises, which themselves have affiliated persons or take part in different grid sub projects representing individual communities such as medicine or physics.

Metadata at varying levels of detail is needed to describe all these grid resources as well as the participating organizations, projects, and persons. Frequently, grid metadata is managed independently in each participating project, i.e., a project is responsible for its specific metadata. This may be appropriate for the management of project-specific or domain-specific metadata, for example, biomedical grid projects typically use life science ontologies for data annotation. On the other hand, there are common types of metadata which apply to all grid projects. Information about projects, grid resources and organizations

---

[1] http://www.d-grid.de

C. Badica et al. (Eds.): Intel. Distributed Comput., Systems & Appl., SCI 162, pp. 115–125, 2008.
springerlink.com   © Springer-Verlag Berlin Heidelberg 2008

can be managed in an integrated form, and should be accessible on-line and directly editable for all authorized participating persons and projects. Furthermore, metadata especially about resources should be offered to grid applications and services, e.g., through metadata service interfaces. Providing an integrated access to grid metadata permits projects to better exchange information about their ongoing work. For example, grid participants can more easily notice related work in other projects, so that cooperation can be improved and duplicate efforts be reduced. It is important that a metadata management system offers simple user interfaces for the extension and change of the metadata (usability aspect), since persons of different domains with diverse technical backgrounds (e.g., computer scientists, physicians, or librarians) meet in a grid's virtual organization. We make the following contributions in this paper:

- We propose a simple yet flexible meta model suitable for management of semantic grid metadata including content types for structured information and ontological categorization for content classification.
- We describe a web-based and wiki-like platform using the defined meta model and supporting the collaborative creation and editing of grid metadata. The platform also addresses usability issues such as powerful search, navigation and visualization capabilities.
- An application of our platform is presented, namely the D-Grid Ontology (DGO) of the German D-Grid initiative available under *http://buell.izbi.uni-leipzig.de/dgo*. In particular, we outline the current organization of the semantic metadata.

The remainder of the paper is organized as follows. In Section 2 we describe models for the collaborative management of grid metadata, with a focus on the meta model level. Section 3 presents the model of DGO, while usability features of the platform are illustrated in Section 4. Implementation details are provided in Section 5. Section 6 discusses related work. We conclude with a summary and an outlook on future work.

2   Models of the Platform

We build on a three-layered representation of metadata and data (see Fig. 1) differentiating between the following layers: *meta model, models* and *instance data*. The model (or schema) is specific to a particular grid or virtual organization, e.g., D-Grid, and prescribes the structure of possible instances and their semantic annotations. The meta model defines the constructs which can be used for defining the models, in particular for describing the structure of instances (content) and the use of ontologies for semantic annotation of instances. In this section we describe the meta model, whereas Section 3 focuses on the D-Grid Ontology (DGO) with its model and instances.

The meta model consists of two main parts, *content types* and *categories*. Content types are used to define the meta information (structure) for instantiable information or content. Categories, on the other hand, are not directly instantiable but serve for a semantic annotation of content, in particular *content items*.

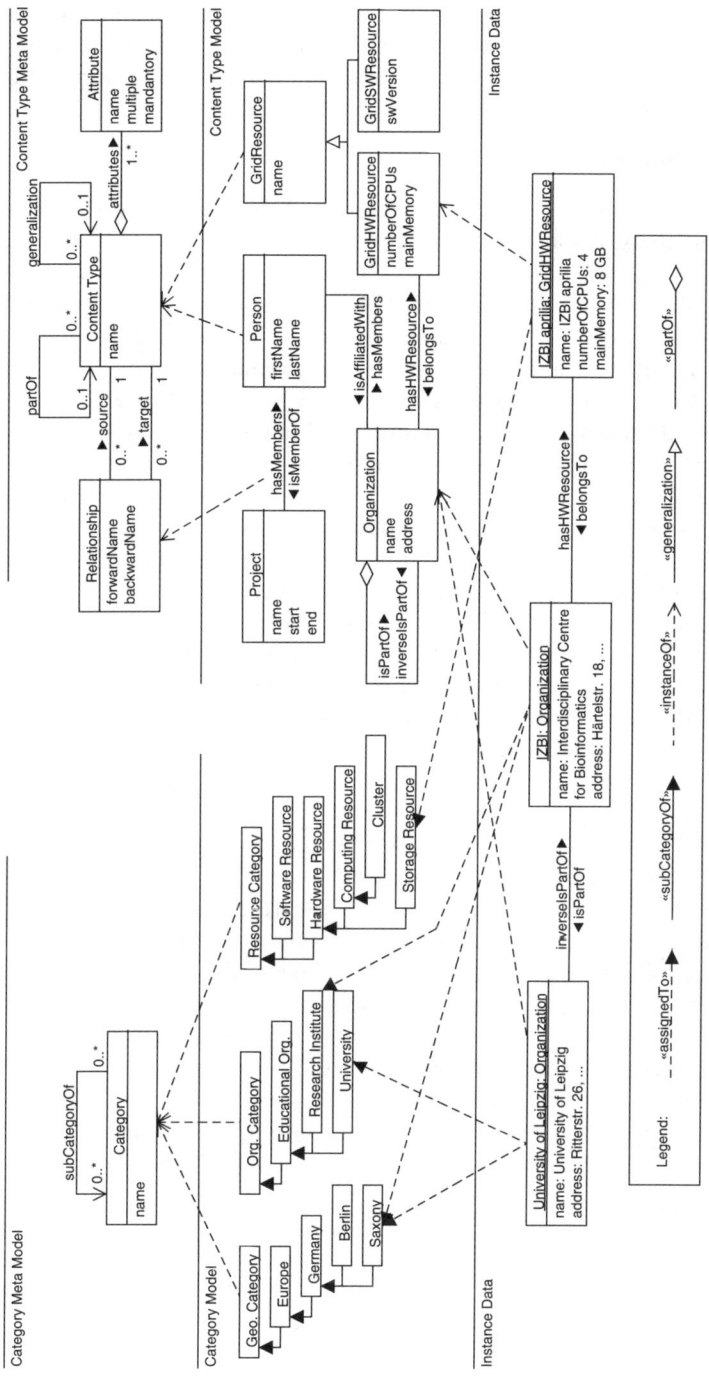

Fig. 1. Three-layered representation of metadata

Each content item is associated to a particular content type, i.e., a content item instantiates a specific content type of the model. In the following subsections, we describe content types, categories and related aspects in more detail.

2.1 Content Types

A content type has a *name* and a set of *attributes* describing simple properties for content items. An attribute has a *name*, a data type and a cardinality of one or many. The latter allows for arbitrarily many values of that attribute within a content item. Attributes may also be defined as *mandatory*, i.e., they must be specified during content instantiation (e.g., the first and last name of a person). The attribute's data type restricts the permissible values, e.g., date, URL or string. Furthermore, allowed values can be restricted to a controlled vocabulary to guarantee well-defined terms. We further distinguish between *generic* and *specific* attributes. Generic attributes are predefined and exist for all content types, e.g., the 'ID' and 'Synonym' attributes. Specific attributes describe application-specific properties of content types.

Content types can be interrelated by binary *relationships* of a specified cardinality. Relationships are managed bidirectionally and thus consist of a forward and backward relationship. Hence content items participating in a relationship are accessible from both directions. For instance, assume a content type Person has a relationship with a second content type Organization. When a content item A of Person 'isAssociatedWith' a content item B of Organization (forward relation), we also maintain that B is connected to A through a 'hasMembers' relationship (backward). In order to keep our model simple and flexible, we currently do not use relationship attributes.

In addition to such application-specific relationships we support two general kinds of relationships with predefined semantics: *generalization* and *partOf*. Firstly, content types can be part of generalization hierarchies supporting inheritance. Hence, derived content types reuse the metadata of their predecessors in the generalization hierarchy and may define additional attributes or relationships. The topmost (root) nodes of the generalization relation are called *base content types*. For instance, a base content type 'GridResource' may inherit its attributes and relationships to more specific content types such as 'GridHardwareResource' or 'GridSoftwareResource'. Secondly, the partOf relationship interrelates content types to construct aggregation hierarchies. For example, we use a recursive partOf relationship between organizations. Such partOf hierarchies are used in our platform to support navigation and to specify the context of content items. For instance, we may have several items called 'Department of Computer Science'. Their meaning only becomes clear by considering their predecessors within the organizational partOf hierarchy, e.g., to differentiate between 'University of Leipzig' / 'Department of Computer Science' and 'TU Munich' / 'Department of Computer Science'.

2.2   Categories

Categories have a *name* and are hierarchically organized within *subCategoryOf* relationships. These relationships are assumed to form directed acyclic graphs (DAGs) of categories. Moreover the subCategoryOf relationship involves different semantics depending on what categories are interrelated, e.g., 'Germany' is part of 'Europe' or a 'University' is an 'Educational Organization'. *Roots* are special categories without predecessor for the subCategoryOf relationship and therefore act as entry points of a category structure.

We build on this simple yet flexible category model to broadly support semantic annotations, i.e., the ontological structuring and classification of content items (instance data). Categories can be used to manage content items of different content types *independently* of the content structure. In particular, content items can be categorized along multiple categories. Notably, the associations between content items and categories exhibit the character of annotations (see 'assignedTo' associations in Fig. 1). Such associations may be used in many cases, e.g., to instantiate categories or to associate objects to a geographical category. For example, the content item 'University of Leipzig' may be associated to a 'University' category and a 'Saxony' category.

Categories can be used to improve the navigation within the platform (along the lines of faceted classification) and to support semantic queries. For instance, if somebody is interested in all universities participating in a grid, one navigates through the organization category structure to the university category to see all associated university organizations.

3   Sample Application – The D-Grid Ontology

D-Grid started in 2005 as a Germany-wide grid initiative. Its aim is to provide a common grid infrastructure for e-Science projects in Germany and to prove the viability and advantages of grid usage in different scientific domains. D-Grid entails many community projects, e.g., for medical and physics applications, and a common integration project (DGI).

Currently, metadata about D-Grid and its structures is highly heterogeneous and distributed across many websites and project-specific repositories, e.g., information about projects, persons, or available hardware and software resources. Furthermore, there are almost no relations or explicit semantic links between these independently maintained information objects. The goal of our metadata platform is to integrate and semantically categorize this heterogeneous information in a common system and to offer it to all D-Grid participants, applications and interested users. New participants in D-Grid can thus quickly inform themselves about ongoing work in D-Grid projects and the organizations and persons involved. Further, resource providers, i.e., institutes providing hardware or software to the grid, can specify parameters about their resources which may be useful for scheduling and distribution of grid applications. Our platform semantically categorizes its content within a so-called D-Grid Ontology (DGO). It simplifies the manual creation and maintenance of metadata using a collaborative, wiki-like platform. Through the use of the meta model including content types and ontological annotations a high data consistency and quality is pursued.

On the basis of our meta model described in Sec. 2, we use four basic grid content types in the DGO model, namely *Person*, *Project*, *Organization* and *GridResource* (see content type model in Fig. 1). As an example, the content type Person uses attributes such as first name, last name, email or phone number for the registration of personal information. Furthermore, relationships to content items of other content types show a person's semantic neighborhood, e.g., the projects a person is working in ('isMemberOf') or the organization to which a person is affiliated ('isAffiliatedWith'). Furthermore, DGO exploits recursive partOf relationships for projects and organizations. In particular, 'D-Grid' is the topmost project of DGO and contains a number of sub projects such as 'MediGRID', 'HEP-Grid' or the 'Integration Project (DGI)', which themselves include further sub projects. Furthermore, DGO uses several category hierarchies for ontological classification of content items (see category model in Fig. 1). Every content item of DGO is assigned to a minimum of one category. For instance, a community project such as 'MediGRID' is assigned to the category 'Community Project' (in terms of project type) and 'D-Grid I' (funding aspect) since it was funded as one of the starting projects of the D-Grid initiative.

The current version of DGO (as of April 2008) categorizes and interrelates about 40 projects, 150 organizations, 300 persons, and 75 grid resources. There are about 950 bidirectional relationships between content items.

4   Usability Features

In the following, we describe some of the features of our platform to illustrate its usability. In particular, we firstly illustrate how semantic metadata is displayed within the platform. Furthermore, we present navigation and search capabilities as well as options for creation, classification and editing of content. For a hands-on experience the interested reader may directly use the system (after registration) under *http://buell.izbi.uni-leipzig.de/dgo*.

4.1   Content Visualization

Each content item is shown on its own article page, providing information about its name, basic attributes, relationships, category classifications, explanations (free text), images and versioning. Relationships to other content items are presented as hyperlinks allowing the user to traverse to the content page of the referenced item. Specific tabs allow the direct change of content pages, in particular editing, renaming or category assignment.

Our platform exploits Web 2.0 techniques, such as maps and navigable trees, to display semantic metadata in different forms. In particular, we use Google Maps[2] to geographically locate content items such as organizations or D-Grid

---

[2] http://maps.google.com

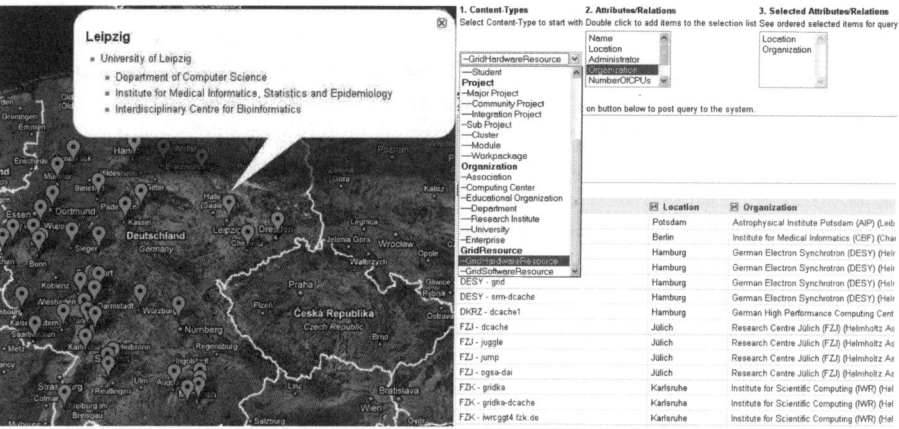

Fig. 2. Organizations of D-Grid on a map (left) and query generator (right)

hardware resources on a map. For example, users are able to notice what organizations in their local environment also participate in the same grid project and hence regional cooperation is improved or duplicate work can be reduced. Furthermore, we employ partOf relationships between content types to generate trees representing hierarchical structures such as organization or project structures.

The sample map in Fig. 2 (left) includes all organizations currently participating in D-Grid. When selecting a location, e.g., Leipzig, all organizations in this place participating in D-Grid are listed and may be further explored. In order to generate these maps, we utilize location attributes of a content type as well as partOf relationships between content items. Currently, the location attributes represent the city, e.g., of an organization. The geographical coordinates (latitude/longitude) of a city needed for the map visualization is obtained from a publicly available web service[3]. For each location on the map, we use the partOf structure among content items to aggregate all corresponding items for display.

4.2 Search and Navigation Facilities

The platform provides different search and navigation facilities. A simple text search supports keyword-based search over all attributes of content items. Furthermore, semantic query capabilities on content types and categories are provided. In particular, a query generator (Fig. 2 right) for interactive specification of semantic queries is available so that users can pose powerful queries without having to learn a complex query syntax or query rules. Users choose a specific content type and their attributes or relationships they are interested in. For

---

[3] http://www.geonames.org

instance, a query to determine the email and names of all persons working in D-Grid can be generated within a few seconds. The results are presented in tables which can be interactively sorted on different attributes or relationships, e.g., person name or the affiliated organization.

Besides search, the platform provides extensive navigation capabilities for content retrieval. A category browser (Fig. 3 left) enables simple and fast navigation to content of interest. It dynamically generates a navigation tree representing categories and content items in an integrated form, by attaching content items as leaves to their most specific categories. For instance, with some clicks a user can navigate from the top category 'Person' to 'Researcher' or 'Professor' to see all associated content items. All nodes of the tree are linked, i.e., a click on a category displays the corresponding category page with all assigned content items, and a click on a content item shows the article of the content item, respectively.

4.3   Creation and Editing of Content

For every content type the system provides an interactive input form to create new content items. These forms are dynamically created from the current meta information (attributes, relationships, category associations) of a content type. To change existing content items, the current content attribute values, relationships and category associations are presented for editing within UI forms analogous to the ones for creating new content items (Fig. 3 right).

A UI form for creation or editing of content consists of different kinds of form fields, in particular mandatory fields, autocomplete-aware fields, single- / multivalued fields, category association fields and free text. Mandatory fields reflect mandatory attributes, i.e., they need to be filled out in order to create a new content item, e.g., a person's name. In order to simplify user input and to avoid duplicate entries, autocompletion is utilized in the following way. As soon as a user clicks on an autocomplete field or types some letters into it, value suggestions are offered

Fig. 3. Category browser (left) and editing of content (right)

for selection. For example, an input field capturing a relationship to the content type 'Organization' (e.g., a person's affiliation) suggests organization items matching the input. Furthermore, if an attribute is restricted to a controlled vocabulary, we suggest values matching current entries of such a vocabulary. In order to enter multiple values for an attribute or relationship we utilize multivalued fields with a common separator to separate multiple values. The category association field provides the possibility to assign the current content item to different categories. Here, we again make use of autocompletion to simplify categorization and to guarantee correct category associations. Finally, a free text field allows for entering content not covered by attributes, relationships or category association. The different fields just described are marked with different background colors and labels to improve user interaction and the input dialog.

5 Implementation

The presented platform builds upon a widely used semantic wiki implementation, the Semantic Media Wiki (SMW) [5]. SMW, in turn, extends the MediaWiki[4] implementation, which is also used by Wikipedia. MediaWiki provides a powerful infrastructure for collaborative management of text-based articles. It is also aware of categories and sub categories, but links between articles in MediaWiki are un-typed (have no semantics) and search capabilities are limited to simple text searches. SMW introduces semantic properties for wiki articles and thus supports a semantic annotation and enhanced querying of wiki contents.

We extended MediaWiki and SMW in several directions. Firstly, we introduce content types (based on the template feature of MediaWiki) to capture semantic metadata in the form of structured content. Secondly, we introduce bidirectional relationships (on the basis of SMW semantic properties) between content types to automatically maintain referential integrity and to provide better navigation capabilities. Thirdly, we support the use of controlled vocabularies and user-friendly UIs for content creation and change, e.g., autocompletion to avoid duplicates. Finally, we utilize Web 2.0 techniques for novel visualization and interaction options, e.g., dynamic generation of maps for content items and interactive specification of semantic queries.

6 Related Work

Our approach builds upon established wiki technology [6] and its combination with semantic technology, cf. [8, 10]. The initially visible distinction between semantic wikis originating from 'classical' wikis, e.g., the Semantic MediaWiki [5], and editors for knowledge bases or ontologies with wiki-like, collaborative features, e.g., IkeWiki [9] or OntoWiki [1], is currently diminishing [3].

In general, the platform presented herein aims at the collaborative and user-friendly collection and maintenance of structured data. A major difference to

---

[4] http://www.mediawiki.org

other systems concerns our meta model. The meta models of many semantic wikis are based on Semantic Web standards, most often RDF (e.g., WikSAR [2], SweetWiki [3], etc.) and sometimes OWL [7] (e.g., IkeWiki, OntoWiki). In contrast, our meta model supports both a database-oriented and an ontological part. The first comprises multiple content types, relationships and attributes for expressing structured contents. The ontological part provides multiple hierarchies of categories for the classification of content items. These aspects result in a clearly structured system configuration and facilitate a user-friendly access and maintenance of grid metadata. In contrast, the sole use of RDF and OWL models often result in complex graph structures and reduced user friendliness. Another feature of our platform is the bidirectionality of the relationships. This can be considered as a simple form of reasoning which still allows for efficient system behavior. Many semantic wikis avoid the use of Semantic Web reasoning for efficiency reasons (cf. [3, p. 87]; exceptions are e.g. IkeWiki and BOWiki [4]).

As already discussed in the previous section, the presented system utilizes the features of the meta model (content types, bidirectional relationships, categories, controlled vocabularies) for improved consistency and usability, e.g., semantic queries and powerful navigation, visualization and editing (e.g., autocompletion). This is a clear improvement over approaches in which editing of information is only possible in terms of wiki syntax as used for free text editing and markup.

7 Summary and Future Work

We presented a meta model and a platform for the collaborative management of semantic metadata in grids. The platform provides grid participants of large-scale grid initiatives such as D-Grid with a collaborative, web-based and user-friendly way of creating, editing and using grid metadata, e.g., on grid resources, projects, and participating organizations and persons. We applied the platform within the German D-Grid initiative in order to build a semantic metadata repository for D-Grid and to improve the collaboration between participating projects. The platform is currently running under *http://buell.izbi.uni-leipzig.de/dgo* and is actively used by D-Grid members.

In the future, we will extend the platform based on new requirements from the D-Grid communities. We further investigate automatic support of the evolution of the domain model, i.e., changes in the content types and categories (instances with respect to the meta model level).

Acknowledgement. This work is supported by BMBF grant 01AK803E "Medi-GRID – Networked Computing Resources For Biomedical Research".

References

1. Auer, S., Dietzold, S., Riechert, T.: Ontowiki – a tool for social, semantic collaboration. In: Cruz, I., Decker, S., Allemang, D., Preist, C., Schwabe, D., Mika, P., Uschold, M., Aroyo, L.M. (eds.) ISWC 2006. LNCS, vol. 4273, pp. 736–749. Springer, Heidelberg (2006)

2. Aumüller, D., Auer, S.: Towards a semantic wiki experience – desktop integration and interactivity in WikSAR. In: ISWC 2005, vol. 175, pp. 212–217. CEUR-WS.org, Aachen (2005)
3. Buffa, M., Gandon, F.L., Ereteo, G., Sander, P., Faron, C.: SweetWiki: A semantic wiki. Journal of Web Semantics 6(1), 84–97 (2008)
4. Hoehndorf, R., Prüfer, K., Backhaus, M., Herre, H., Kelso, J., Loebe, F., Visagie, J.: A proposal for a gene functions wiki. In: Meersman, R., Tari, Z., Herrero, P. (eds.) OTM 2006 Workshops. LNCS, vol. 4277, pp. 669–678. Springer, Heidelberg (2006)
5. Krötzsch, M., Vrandečić, D., Völkel, M., Haller, H., Studer, R.: Semantic Wikipedia. Journal of Web Semantics 5(4), 251–261 (2007)
6. Leuf, B., Cunningham, W.: The Wiki Way: Collaboration and Sharing on the Internet. Addison-Wesley Professional, Reading (2001)
7. McGuinness, D.L., van Harmelen, F.: OWL Web Ontology Language overview. W3C Recommendation, World Wide Web Consortium (W3C), Cambridge, Massachusetts (2004)
8. Riehle, D., Noble, J. (eds.): Proc. of the 2006 International Symposium on Wikis. ACM, New York (2006)
9. Schaffert, S.: IkeWiki: A semantic wiki for collaborative knowledge management. In: Proc. of the 15th IEEE International Workshops on Enabling Technologies: Infrastructures for Collaborative Enterprises, WETICE 2006, Manchester, UK, June 26-28, pp. 388–396. IEEE Computer Society Press, Los Alamitos (2006)
10. Völkel, M., Schaffert, S. (eds.): SemWiki 2006 – From Wiki to Semantics: Proc. of the First Workshop on Semantic Wikis, Budva, Montenegro, June 12, vol. 206, CEUR-WS.org, Aachen (2006)

Distributed Approach for Genetic Test Generation in the Field of Digital Electronics

Eero Ivask, Jaan Raik, and Raimund Ubar

Tallinn University of Technology, Raja 15, 12618 Tallinn, Estonia
{ieero,jaan,raiub}@pld.ttu.ee

Summary. Distributed computing attempts to aggregate different computing resources available in enterprises and in the Internet for computation intensive applications in a transparent and scalable way. Digital test generation aims to find minimal set of test vectors to obtain maximum fault coverage for digital electronic circuits. In this paper we focus on distributed environment and parallelization of the computationally intensive genetic algorithm based test generation for sequential circuits. We discuss the concept and implementation of our system infrastructure, task partitioning, allocation, test generation algorithm and results.

1 Introduction

As the complexity of modern digital devices is increasing dramatically, the demand on test quality and reliability is getting higher for most products. On the other hand, as the sizes of circuits grow, so do the test costs [1]. Test costs include not only the time and resources spent for testing a circuit, but also time and resources spent to generate appropriate test vectors (input values) for circuit under test.

Despite of their efficiency, GA-based techniques may still require large amount of CPU time due to simulation costs. Speeding up current test generation tools is challenge nowadays, as we see. One way to gain practical speedup is to parallelize the task execution. In our case, most time consuming part of the algorithm is certainly fault simulation- verification of usefulness of produced test vectors. There are several methods to parallelize the fault simulation: algorithm can be parallelized, circuit model can be partitioned into separate components and simulated in parallel, partitioning the fault set data and simulating faults in parallel (fault parallelism). In this paper, we present distributed test generation approach, which relies on fault parallelism. Fault set is divided and faults are simulated in parallel on different computers in wide or local area network.

Current distributed solution was initially inspired from MOSCITO system [2], which had a goal to provide the functionality of the existing local work tools to potential users in LAN mainly. Major obstacle for Internet based use was TCP/IP socket based communication, which conflicted with firewalls. More flexible web-based solution for remote tool usage was proposed in [3]. Socket communication was replaced with HTTP. Java Servlets and Applets were used along with Java applications, also database was introduced to store user information and intermediate results. In current paper this concept is revised and improved to support adaptive distributed computing.

C. Badica et al. (Eds.): Intel. Distributed Comput., Systems & Appl., SCI 162, pp. 127–136, 2008.
springerlink.com

There are, of course other, general purpose frameworks for distributed computing, like BOINC (Open Infra-structure for Network Computing), which is a non-commercial middleware system for volunteer computing, originally developed to support the SETI@home project, but intended to be useful for other applications in areas as diverse as mathematics, medicine, molecular biology, climatology, and astrophysics [4].

Major drawback of such infrastructure however is the use of remote procedure call (RPC) mechanisms, which is considered security risk (even if it is configured for connections from the same computer). Use of PHP instead of Java cannot be considered as an advantage.

Internet based grid-computing middleware AliCE using Java based Jini technology is described in [5]. Jini technology is based on RMI (Remote Method Invocation)- although elegant programming solution for distributed computing where one program can remotely invoke methods physically residing in other machine, however, firewall traversal can be problematic again, as dedicated communication ports are needed. Strict security policy might not allow that.

The paper is organized as follows: overall concept of web-based infrastructure and communication is described in section 2. In section 3 implementation details are outlined including communication, data management and user interface. Section 4 describes workflow with distributed computing. Section 5 describes digital test tool. Sections 6 and 7 describe task partitioning and task allocation. Experimental results are presented in Section 8 and conclusions are given in Section 9.

2   General Concept

A web-based solution to support parallel distributed test generation for digital sequential circuits is described below. User will be able to use test tool remotely over the Internet.

System core has client-server concept. There is one master server, several application servers and arbitrary number of clients (see Fig. 1). Master server maintains the information about application servers, which provide service. On application server so called agents can be invoked. Agents encapsulate actual test tool.

There is no need to install tools on the user's local computer. Therefore, user's effort for installation, configuration and maintenance of software will be drastically reduced.

The system is implemented in Java and can therefore run on different computing platforms. Actual work tools must run on their native platform of course.

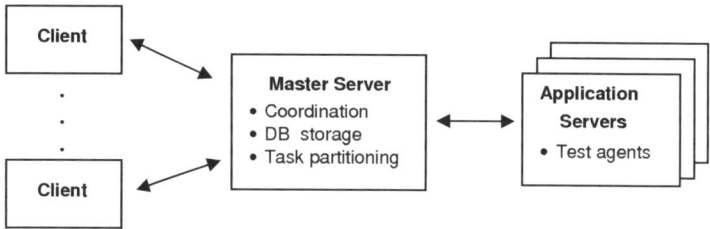

Fig. 1. Components in the system

Each tool must be provided with network communication abilities i.e. it must be wrapped with additional software an Agent is created. Agents work on Application servers. One Master server will serve many clients in parallel. There is task queue management. Results reside initially on the server computer. Each user has its own server-side workspace in the database.

Client and Agents work in polling mode, Master server is working in answering mode. Master server and Agents on Application servers must be started by administrators first. Client then initiates a task which is first passed to Master server where the task is stored until a free Agent is asking for a new task. When task is complete, Agent passes results back to Master server where results are stored again until user will ask for results. Subsequently, implementation specific details of general concept are given.

3 Implementation

The WEB based infrastructure is built according to the client-server three-tier concept using Java applet/servlet technology. MySQL database was chosen as backend DB for data persistency, for user tracking and to support management tasks. Principal solution in details is given in Fig. 2. Tomcat is the servlet container that is used in the official Reference Implementation for the Java Servlet and JavaServer Pages technologies. Tomcat and servlets running on it play important role in order to gain access to intranet resources on application servers and to MySql database (platform independent open source DB). Test tool is implemented partially in Java, partially in C language, i.e. simulation functions are written in C and invoked via Java Native Interface (JNI). Java language has excellent support for network programming.

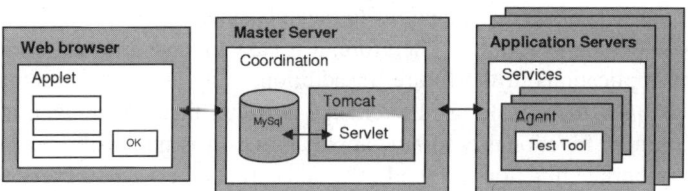

Fig. 2. Three layer architecture and communication details

Communication is based on HTTP protocol. The tools on different computers and on different computing platforms (UNIX, Linux, Windows) can easily exchange data as serialized Java objects (Transfer Object design best practice). HTTP protocol allows us also easy firewall traversal as we can use default web server port and Java servlet extensions on web servers as sort of proxies in order to reach intranet resources. There is no need for opening extra ports in the firewall on the user side as it is the case in TCP/IP socket based communication (which would be major restriction). Communication is secured via SSL encryption.

Data management module used in coordinating servlet running on the web server (i.e. Master server) is described below in this section. As we know, web-based http communication is by design stateless. This means that we have to keep track about all necessary information. As work tools tend to run long, then normal user's http session is not valid for such time period and data will be lost. As we want to provide our user with a possibility to come back later to receive his results, we have to identify (track) users and store all their relevant data. Using so called cookies could be one solution, but database approach used here offers many advantages like powerful SQL query mechanism, speed, reliability, consistency of data and ease of use. User tracking system allows us to monitor and control the usage of service. Without proper user management, anybody in the Internet could possibly use valuable computer resources. Therefore, better practice would be to allow registered users access the resources. It may allow also billing the business customers. Main goal was to provide sufficient set of basic functions to support user registration, tracking, and management of user tasks. Solution is based on relational database. Tool execution and data base access over Internet is carried out via Java servlet technology. Below the implementation details are given.

The data management module has open architecture, general API (application programming interface). With slight modifications, it is also reusable for similar web-based systems, where for example user tracking is needed. It has three layers: presentation layer (user tier), business logic tier (data base queries, etc.), physical database (MySQL-platform independent open source DB).

First two layers are implemented in Java programming language. User is accessing database via presentation layer, not directly. This makes architecture open. Database access is implemented using Data Access Object (DAO) design practice in mind. Data access is using also connection pooling to speed up DB transactions. Data passing between user and Master server and between Application server is implemented following Transfer Object (TO) design practice. Relevant information is not sent string by string but it is passed once as data bundle (datagram).

User tier consists of several functions to make business layer queries. User tier and business logic tier are decoupled. Therefore, it is easy to have different user interfaces for different applications in the future. In addition, if the database structure or business logic changes, there is no need to change user interfaces. Moreover, it is easy to introduce common functionality to new applications - it is much simpler for instance to invoke appropriate function (method), than construct a new query every time a new application needs one.

Graphical User interface. (GUI) is based on Java Applet, which can be integrated into HTML page when needed. GUI has fields to gather test tool's parameters, allows browsing for circuit model file, has button to start the tool, a console window to display all the messages from the running tool. When the task is complete, results download button is enabled. User can browse and select the folder where to save results.

Since accessing local hard drive for Java applet is restricted for security reasons, then GUI applet has been signed digitally, with so called self-signed certificate for simplicity. Certificate shows owner specific information. Only difference for end user is that when signed Applet is first time downloaded into user's computer, informative dialog box is displayed. It is user's responsibility to trust or untrust the origin and contents of the Applet. User can contact Applet owner about autenticy of certificate, when question arises.

4   Workflow with Distributed Computing

First, user specifies parameters and design file location for certain test tool. Thereafter user GUI contacts with coordinating web server and described parameters along the model are passed automatically. Task coordinator service process (Java servlet) on Web server records all requests from user(s) and divides the task into subtasks. Java based test agents poll constantly web server and if any subtask is scheduled by coordinator process, then test agents receive the appropriate parameters and design file and will start actual native test tool. Test agent waits then until his subtask will be completed and reports results back to coordinating web server, which in turn assembles sub results into final result and forwards it to user when requested.

Test agent will accept one task at time, but if the host computer has several processors then of course it is reasonable to have one agent for each processor. Operating system itself will assign running tasks to available processors. Test agent and native test tool must reside on the same computer as we assume that native test tool has no network support built in. Web server resides usually separately from test agents, they must not reside in the same local area network. Each agent can reside on different local area network.

This solution is relatively flexible and effectively works across the internet and through the firewalls as long as dedicated communication port for task coordinator service process on Master server is opened by administrator.

5   Genetic Test Generation

This section presents a genetic algorithm based approach to test generation for digital gate level sequential circuits (circuits with memory and feedback to inputs). This approach differs from most of the previous works by specifically targeting single faults, also some structural knowledge about the circuit is used. The priority was to improve the fault coverage, to detect additional faults.

In order to solve the problem, the following components are must in genetic algorithm [6]:

- Representation of solution to the problem,
- Way to create an initial population of solutions,
- An evaluation (fitness) function in order to estimate the quality of the solution
- Genetic operators that alter the structure of "children" during reproduction
- Fine-tuned parameters

Representation. In context of test generation for sequential circuits, sequence of test vectors will be the individual. Several concurrent sequences form the population.

Initialization takes place by generating a random set of test sequences. Such an initial test sequence set is subsequently given to a simulator tool for evaluation. Following steps of algorithm are carried out repeatedly.

Evaluation of test vectors measures fitness of the individuals, i.e. the quality of solutions in a population. Better solutions will get higher score. Evaluation function directs population towards progress because good solutions (with high score) will be selected for crossover and poor solutions will be rejected. We use fault simulation in order to evaluate test sequences. Simulation is carried out only for particular fault under consideration. Fault simulator was improved to keep track the number of fault effects activated and propagated onto flip-flops and primary outputs. Simulation procedure uses Structurally Synthesized BDD (SSBDD) description [7] as its internal model. It is a special case of Binary Decision Diagrams (BDD), which are now commonly used for representing Boolean functions because of their efficiency in terms of time and space [8]. They have become the state-of-the-art data structure in many VLSI CAD systems. In essence, a decision diagram is a directed acyclic graph, consisting nodes, edges and special terminal nodes. Traditional BDDs can be used only for representing functions and not for the faults in gate networks.

Fitness of the test sequence is calculated as follows: $C_a \cdot activated + C_p \cdot propagated$, where *activated* is number of clock cycles when particular fault effect was activated in the circuit and *propagated* is the number of clock cycles when fault effect was propagated onto some flip-flop. C_a and C_p are constants, which show how much stress is given to parameters. We selected 0.1 for C_a and 1 for C_p.

Genetic operators. *Selection* is needed for finding two candidates for crossover. Based on fault simulation results better test sequences are selected. Roulette wheel selection mechanism was used here. Number of slots on the roulette wheel will be equal to population size. Size of the roulette wheel slots is proportional to the fitness value of the test vector sequence. This means that better sequences have a greater possibility to be selected. If our population size is N, and N is an even number, we have N/2 pairs for reproduction. Candidates in pair will be determined by running roulette wheel twice. One run will determine one candidate. With such a selection scheme, it can happen that same candidate is selected two times. Reproduction with itself does not interfere. This means the selected test sequence is good and it carries its good genetic potential into new generation.

Crossover. Swapping genetic material of the two parents allows useful genes (relevant bits) to be combined in their offspring (new test sequence). Most successful parents reproduce more often. Beneficial properties of two parents combine. Crossover and selection (fitness function) are the keys to genetic algorithm's power. Here, one-point vertical and one-point horizontal crossover were implemented.

Mutation. Random mutation provides background variation and occasionally introduces beneficial genetic material [6]. Without the mutation, all the individuals in population will eventually be the same (because of the exchange of genetic material) and there will be no progress. We change randomly some bits in test vectors. Bit position corresponding to reset input is not altered during mutation. Using such knowledge based technique helps to reduce search space.

Working algorithm. GA works in two stages: In the *first stage*, fault activation sequence for the particular fault is generated: at first, short random test sequence is simulated with fault simulator. If fault was not activated then test sequence length is automatically doubled and fault simulation is repeated. This happens until fault is activated or test sequence length limit is exceeded. In latter case fault is aborted and next fault from list is taken. Activation process starts again with short sequence. Fault is considered activated when we could set up the necessary logic value in the particular schematic node. Important is that in such initialization sequence bit position corresponding to reset signal is filled with zeros. Only in second vector there is 'one' i.e. we describe behavior of the reset signal based on a priori knowledge. User can supply reset index.

Second stage of GA begins if fault activation was successful. At first, activation sequence is distributed into all individuals - the beginning of all test sequences in our population is filled with fault activation vectors (keeping reset 0). The rest of the sequences are filled with random patterns. Sequence length in this stage is twice as long as final fault activation sequence was. It is possible to select also vector sequence length dynamic increase - it takes into account if fault effect was not propagated onto primary outputs, but still progress was made compared to previous iteration. In such cases sequence length is doubled. This can happen until fault is detected (propagated to primary outputs) or until sequence length limit is reached. Such a technique has proved to be effective in terms of fault coverage increase and shorter test sequences, but requires longer runtimes.

After the population of test sequences is initialized, *GA main cycle* begins. At first, all sequences are evaluated subsequently by simulation procedure again. For each sequence, numeric fitness value is calculated. Then candidates for crossover are selected. Crossover type can be selected by user. New population will be filled only with newly constructed sequences, however optionally it is possible to conserve the best individual from the last generation- this is called elitist selection. Finally, before new cycle begins, some mutation is introduced into newly engineered test sequences. Mutation probability is increased dynamically when several subsequent generations did not improve fault propagation. When there was success finally, then mutation rate is lowered again down to initial value. After new population of test sequences is ready, GA main cycle is repeated. This will last until current fault is detected or number of predetermined (by user) number of generations is exceeded. Thereafter new fault is considered if any is left. More details of this test generator tool are given in [9].

6   Task Partitioning

In current solution, we exploited the fault parallelism- the fact that test generation can be done independently for subsets of faults using the same circuit model and resulting test vectors can be later assembled into final test set.

Two types of fault set partitioning were tried: adjacent fault selection where faults were selected one after another in a sequence and random fault selection. Experiments showed that the latter is able to ensure more equal execution times for subtasks and therefore contributes to overall shorter parallel simulation time.

Task size, i.e. how many faults will fault set contain varies and depends on number of users on the system and processors available.

7   Task Allocation

There could be possibly two goals while allocating tasks to different computers: 1) maximize the speed of particular test generation task execution for particular user 2) maximize system overall throughput from the perspective of all users (the level of the service quality- nobody should starve).

Generally, task allocation can depend on several factors that may change in time:

- Number of application users
- Number of computers in resource pool
- Computer's workload
- Speed of the communication links

In current solution we assume that communication delay is taken into account already with execution time for particular computer - i.e. communication link speed is not addressed separately here.

Resource allocation goes as following: all the computing resource is divided between particular application users. When number of users changes, computing resource is reallocated. Current users get less resource from pool and when number of users increases and vice versa. In extreme case when there is only one user working in the system, then his task is solved on all processing units. When number of users is equal to number of processors then user task is solved on single computer. Task allocation starts with small amount of data and will be increased by half after each iteration, assuring at the same time that all users get equally attention. Task sizing stops when there is no free computing resource available. Initial small amount of data also helps us to estimate different computers computational power in the resource pool. The time it takes to complete the initial subtasks is measured and speed coefficients are determined for each machine. These coefficients can be applied while next time allocating subtasks- faster machines get larger tasks respectively.

8   Experimental Results

In experiments we wanted to determine how well is the distributed solution with current task partitioning scalable when number of processing units is increasing. We measured fault simulation times as simulation is by far most ressource intensive task here. Experiments were carried out on SUN UltraSPARC-IIIi processors running at 1280 MHz. For each circuit equal number of test patterns was applied. In Table 1, the maximum time of all subtask simulation times is presented. Two types of fault set partitioning were tried: adjacent fault selection (subsequent) and random fault selection. The latter is able to ensure more equal simulation times for subtasks and therefore contributes to overall shorter parallel simulation time (up to 3 times faster in case of 8 processors for DIFFEQ circuit in comparison with adjacent selection, see Table 1).

Adjacent fault selection performs badly, we can see subtle differences between subtask results in Table 2 maximum difference is 20 times for DIFFEQ circuit in case of 8 processors. In case of 4 processors difference was 14 times (not sown in table). Task is not considered finished until all the subtasks are not completed.

Experiments show, that partitioning can be still improved, especially considering more 'equal' fault sets. *Max deviation* in table 2 shows the difference in percentage between the ideal (mean) time and maximum time (the time user has to actually wait) for fault simulation.

Table 1. Simulation results for example circuits (1-8 processors)

circuit	Subsequent fault selection (Time, s)				Random fault selection (Time, s)			
	1 proc.	2 proc.	4 proc.	8 proc.	1 proc.	2 proc.	4 proc.	8 proc.
mult8x8	156	92	49	29	156	79	41	21
diffeq	212	138	123	118	212	125	75	40
risc	437	279	151	78	437	221	115	60
ellipf	652	346	181	109	652	329	166	84

Table 2. Simulation results for subtasks (8 processors)

circuit	Subsequent fault selection				Random fault selection			
	mult8x8	diffeq	risc	ellipf	mult8x8	diffeq	risc	ellipf
Time for subtasks, s	**29**	**118**	55	**109**	20	33	54	79
	21	7	76	73	19	22	55	83
	20	6	75	80	20	37	56	82
	17	12	**78**	86	**21**	**40**	**60**	**84**
	21	6	48	84	19	22	56	83
	23	5	36	79	21	25	55	83
	20	22	41	81	19	24	51	81
	10	53	35	67	21	27	54	81
Total time, s	161	229	444	659	160	230	441	656
Mean of total time, s	20	29	56	82	20	29	55	82
Max deviation, %	45	307	39	33	5	37	9	2

Overall simulation speedup gained with current random fault selection solution varies. In case of 8 processors lowest speedup 5.3 times for DIFFEQ circuit was obtained and highest speedup 7.8 times for ELLIPF circuit was obtained. The same proportion remains also for other number of processors.

9   Conclusions

WEB-based distributed environment described here in the paper will allow to speed up the genetic test generation significantly and allows to work over the internet thereby extending the lifecycle and value of this tool. Concept of remote tool usage was improved to support adaptive parallel distributed computing, in order to overcome the difficult problem that genetic algorithm based tools tend to run too long on a single computer to obtain satisfactory test coverage in everyday tasks.

Current solution allows to run partitioned fault simulation task on several computing stations in parallel. Adaptive task partitioning and task allocation was introduced to the distributed system. Algorithm is adaptive to changes in number of computers available and their workload in the resource pool, changes in number of users are also taken account. Presented solution is flexible and effectively works across the Internet and through the firewalls.

Two types of fault set partitioning were tried in experiments: adjacent fault selection (subsequent) and random fault selection. The latter is able to ensure more equal fault distribution and therefore contributes to overall shorter execution time.

References

1. ITRS roadmap (2006), http://public.itrs.net/
2. Schneider, A., et al.: Internet-based Collaborative Test Generation with MOSCITO. In: Proc. DATE 2002, Paris, France, pp. 221–226 (2002)
3. Ivask, E., Raik, J., Ubar, R., Schneider, A.: WEB-Based Environment: Remote Use of Digital Electronics Test Tools. In: Proc. IFIP 18th World Computer Congress: Virtual Enterprises and Collaborative Networks, Toulouse, France. Kluwer Academic Publishers, Dordrecht (2004)
4. BOINC, http://boinc.berkeley.edu/
5. Teo, Y.M., Low, S.C., Tay, S.C., Gozali, J.P.: Distributed Geo-rectification of Satellite Images using Grid Computing. In: Proc. International Parallel and Distributed Processing Symposium IPDPS 2003 (2003)
6. Goldberg: Genetic algorithms. Addison-Wesley, USA (1991)
7. McGeer, P., McMillan, K., Saldanha, A., Sangiovanni-Vincetelli, A., Scaglia, P.: Fast discrete function evaluation using decision diagrams. In: Proc. ICCAD 1995, pp. 402–407 (1995)
8. Minato, S.: Binary Decision Diagrams and Applications for VLSI CAD. Kluwer Academic Publishers, Dordrecht (1996)
9. Ivask, E., Raik, J., Ubar, R.: Fault Oriented Test Pattern Generation for Sequential Circuits Using Genetic Algorithms. In: Proc. IEEE European Test Workshop, Cascais, Portugal, pp. 319–320 (2000)

A Planning-Based Approach for Enacting World Wide Argument Web

Ioan Alfred Leţia[1] and Adrian Groza[1]

Technical University of Cluj-Napoca
Department of Computer Science
Baritiu 28, RO-400391 Cluj-Napoca, Romania
{letia,adrian}@cs-gw.utcluj.ro

Summary. The goal of this research was to identify the suitable technologies for enacting the World Wide Argument Web (WWAW) in the context of newly arisen Pragmatic Web paradigm. The vision is to develop the WWAW based on the Argument Interchange Format (AIF) ontology. On the one hand, we propose concept maps for presenting AIF-based arguments to the human agents. On the other hand, the argumentation schemes are formalized as planning operators in order to provide software agents with the ability to build argumentation plans.

1 Introduction

We are in the age when we can imagine an infrastructure (World Wide Argument Web - WWAW), native to the Internet, which enhances software agents with the ability to debate, rise argumentation, or analyze ideas, in order to provide an effective dissemination of the information to the more and more knowledge driven, but lost, human agents. WWAW [7] is a large scale network of interconnected arguments created by human agents in a structured manner. Even if the idea of integrating argumentation within WWW backs its roots in 1997 [8], the current vision is to create an infrastructure for mass-collaborative editing of structured arguments in the style of Semantic Wikipedia.

During the past years, the research on argumentation theory has focused on identifying and formalizing the most adequate technical instrumentation for modeling argumentation. Defeasible logic seems to be one answer to these attempts. Recently, in the context of large scale argumentation, the interest has been shifted towards frameworks where all different inference mechanisms are able to co-exist under one umbrella. At the moment, the standard is given by the Argument Interchange Format (AIF) ontology.

From a complementary perspective, the Pragmatic Web [10] aims to give control to the information consumers so that they can customize how to use information. As stated in [9], the original syntactic web has given most of the control to the information producers, whilst the semantic web separates the content from presentation, without having received yet general acceptance. The Pragmatic Web intends to provide information consumers with the technical instrumentation for specifying how to turn existing data into context relevant information

C. Badica et al. (Eds.): Intel. Distributed Comput., Systems & Appl., SCI 162, pp. 137–146, 2008.

[9]. Within this three level architecture, our proposal regards: i) for the *pragmatic web level* to extend the WWAW with context; ii) at the *semantic level*, to use the AIF ontology, and iii) at the *syntactic level*, to build a computational model based on planning (PDDL).

2   WWAW as a Pragmatic Web Component

2.1   Extending AIF Ontology with Context

One desiderata is for the WWAW to employ a unified extendable argumentation ontology [7]. At present, there exist two extensions of the AIF ontology: one in which Argument Schemes (AS) are introduced [7], and one in which Protocol Interaction Application Nodes are attached [6]. The first one enhances agents with both reasoning capabilities: logic-based and scheme-based argumentation, and it also focuses on representing the *form of an argument*. The second one allows agents to represent dialectical part of arguments.

We introduce a new node type, namely *context node* ($CO - node$). We argue the necessity of this node, due to the fact that context exists independent of any object in the system. Thus, one context may be used to evaluate different arguments, whilst the same argument can be evaluated in different contexts. Arguments conveyed in a debate base their degree of acceptance on a complex background of beliefs (or epistemic state) of the audience.

Definition 1. *The extended-AIF ontology has five disjoints sets of nodes:*

- *An information node $I - node \in N_I$ represents passive information of an argument such as: claim, premise, data, locution, etc.*
- *A scheme node $S - node \in N_S$ captures active information or domain-independent patterns of reasoning. The schemes are split in three disjoint sets, whose elements are: rule of inference schemes ($RA - node$), conflict application node ($CA - node$), preference application node ($PA - node$).*
- *Forms of arguments $F - node \in N_F$ model argumentation schemes, by defining premises and conclusion descriptors, presumptions, and exceptions.*
- *Protocol interaction nodes ($PIA-node$) are used to constrain the dialog moves within an argumentation process.*
- *Context application nodes ($CO - nodes \in N_{CO}$) are used to capture the context of the above node types in order to increase the re-usability of arguments in WWAW.*

$RA - nodes$ are used to represent logical rules of inference such as modus ponens, defeasible modus ponens, modus tollens. Specific pragmatic inference schemes such as: entailment, implicature, presupposition, deixis can be encapsulated within these nodes. Because of the separation of the argument structure, modeled with *I-nodes* and *Scheme-nodes*, from contexts, more power to re-use arguments, and flexibility in representation and acceptance is provided. $CA - nodes$ represent declarative specifications of possible conflicts (such as *negation*). $PA-nodes$ allow to declaratively specify preferences among evaluated nodes (such as *legis*

Argument from expert opinion $\doteq AS\_EO$ ────────────

A_1 : E asserts that A is known to be true.

A_2 : E is an expert in domain D.

C : A may (plausibly) be taken to be true.

CQ_1 : Is A within D?

CQ_2 : Is E a genuine expert in D?

CQ_3 : Is A relevant to domain D?

CQ_4 : Is A consistent with what other experts in D say?

CQ_5 : Has the expert E a good reputation?

Fig. 1. Critical questions block the derivation of the conclusion

posterior, legis superior, or *legis specialis*). Allowing the application of $CA-nodes$ or $PA-nodes$ over $RA-nodes$ results in a very expressive formalism to model different types of arguments (meta-argumentation for instance). $PIA-node$ encodes the range of possible speech acts as reply to an *I-node* of type locution, and their preconditions and effects [6]. In our approach, PIA nodes always occur in a richly defined context. In WWAW, a mediator deploys $PIA-nodes$ for dialog representation that can be accessed by the participating agents that can take two actions: i) to use this node, by providing $I-nodes$ encapsulating the speech acts specified into the $PIA-node$, or ii) to attack the node by instantiating a scheme node having the $PIA-node$ as conclusion. $F-nodes$ focuses on the form aspect of arguments by allowing introduction of ASs in the AIF ontology. The next section addresses argumentation schemes related issues.

2.2 Argumentation Schemes as Protocol

From the practical viewpoint of enacting argumentation based applications, there is a gap between logic-based agents and human reasoning. The model of argumentation schemes aims to fill this gap by providing schemes capturing stereotypical patterns of human reasoning. One example is the pattern *Argument from expert opinion*, depicted in figure 1. Formally, an argumentation scheme is composed of a set of premises A_i, a conclusion C, and a set of critical questions CQ_i, aimed to defeat the derivation of the consequent.

One desiderata of the argumentation schemes is to simplify the argumentation process. This is done by hiding secondary premises and encapsulating them as critical questions. Based on the main premises A_1 and A_2, the consequent is defeasibly inferred. During the process of gradually revealing information in a dispute, when a counter-argument arises, the conclusion might be defeated. Each argumentation line sustaining a claim provides the correspondent critical questions that the opponent may use to challenge the pleading. When a critical question is conveyed, the conclusion of the argumentation scheme to which the respective CQ belongs is suspended, until the subject of the dispute is clarified.

Whoever is responsible for this clarification, in other words who has the burden of proof, depends on the type of the *CQ*.

Definition. *An undercutting CQ attacks the link between the premises and the conclusion. The burden of proof is shifted to the proponent of the argument. A rebuttal CQ challenges an argument by instantiating an AS sustaining the opposite conclusion. The burden of proof remains to the opponent.*

An undercutting *CQ* cannot be used to draw any conclusion, its only use is to prevent the derivation of some conclusions. Having the burden of proof, the proponent of the claim has to provide more justifications in favor of that conclusion. A rebuttal *CQ* is used to derive the opposite claim. Having the burden of proof, the opponent must instantiate a scheme sustaining the opposite conclusion. In current practice of law the burden of proof can be itself the subject of the dispute.

2.3 Enacting AIF Ontology as Concept Maps

The simplification of the argumentation process is done in two steps. First, one can start by splitting the premises in ordinary premises, presumptions and exceptions [3]. In the next step, the presumptions are translated to undercutting critical questions and the exceptions to rebuttal critical questions (see figure 2). Here, the same *Argument from expert opinion scheme* is represented as a $F-node$ in the AIF ontology.

We use the concept maps to provide intuitive visualizations of argument networks in WWAW. Concept maps have their origin in the learning movement called constructivism. Following the constructivist vision, in our approach the argumentation agents actively construct knowledge based on the available information and the current context. Regarding technical instrumentation, one possible candidate is Cmap servers (http://cmap.ihmc.us/). When an argument map is saved to a CmapServer, a web page version is also stored. Thus, a WWW browser is sufficient to browse the argumentation chains, opening the road to enact the WWAW architecture.

The functionalities provided by the Cmap servers that we exploit are: i) *Deploying arguments in WWAW* - the system allows users to save their arguments on the available public servers; ii) *Searching Arguments* - Cmaps tool provides searching capabilities for identifying arguments within both public argument maps and WWW; iii) *Validating and fixing links* - Due to the dynamics of WWW resources, web pages having the role of supporting or attacking arguments might be no longer available. The tool can check if any chain of arguments is available at a certain time; iv) *Public character of the arguments* - some debates, such as Online Dispute Resolution, need to maintain arguments as private. Even if they are posted on the WWW, only the arbitrator might have the right to read them; v) *Providing evidence* - An argument is stronger if evidence is provided for its premises. The system enhances parties with the ability to point towards relevant evidence in different formats.

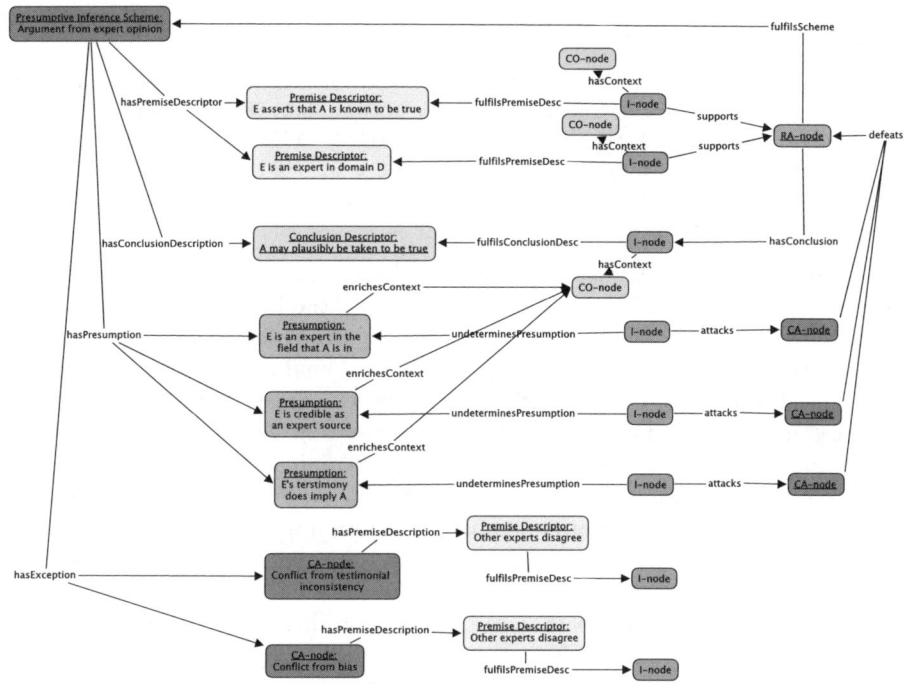

Fig. 2. F-node: *Argument from expert opinion* enriched with context, in CMaps

3 Formal Model of Argumentation Schemes

3.1 Interleaving Planning with Arguing

Consider the simple scenario in figure 3. The goal of the producer P is to obtain 20\$ profit. It can achieve its input item from two suppliers: S_1 at the price of 20\$, or S_2 at the price of 15\$, but the consumer C promises to pay only 30\$. In a planning domain, this can be formalized as follows:

```
(:Init (sell S1 20$)(sell S2 15)(buy C 30$))        (:Goal (profit P 20$))
```

In a classical planning approach, no plan can be generated given the initial facts. The idea is that agent P can slightly adjust the initial world state by

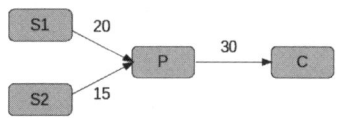

Fig. 3. Planning in supply chain

negotiating with its business partners. Given the current situation, the agent P can generate several plans to reach its objective:

```
p1: (sell S2 15$) (argue increase C from 30$ to 35$) (buy C 35$)
p2: (sell S1 20$) (argue increase C from 30$ to 40$) (buy C 40$)
p3: (argue decrease S2 from 15$ to 10$) (sell S2 10$)(buy C 30$)
p4: (argue decrease S1 from 20$ to 10$) (sell S1 10$)(buy C 30$)
p5: (argue decrease S2 from 15$ to 12$) (argue increase C from 30 to 32)
    (sell S2 12$)(buy C 32$)
```

Here, the communicative act *argue* is used to start an argumentation process with the client C, aiming to convince him to buy the item for 34$. Two issues are of utmost importance here: The first one relates to how the exact outcome of the action *argue* can be defined. In our view, a kind of contingency planning can be used: the expected outcome is defeasible considered, but the effect will be retracted in the light of new contradictory information. A re-planning phase will take place and the agent should consider the next plan according to a preference relation or persuasion strategy. This leads to the second issue: when does the agent pursue with the current communicative action in an attempt to persuade its partner or when does it decide to change the plan? We advocate that a machine learning approach can provide technical solutions to this issue.

Two threads of argumentation can run in parallel as in the plan p_5. The preference between plans can be encapsulated within $PA - nodes$, depending on past experience or on the current context extracted from $CO - nodes$. In order to be successful, the producer P has to provide arguments supporting the sentence *increase C from 30$ to 32$*. The arguments are built employing the available ASs, formalized as planning operators in the next section.

3.2   Implementation Issues

The classical solution for proving a sentence is to use an inference engine. We approach the problem from a different perspective: each AS is implemented as an action within a planning domain, whilst the sentence we want to prove represents the goal of that planning problem. Using PDDL, the following advantages arise by applying it to dispute resolution systems:

- It is highly expressive having different levels of richness of domain descriptions: types, probabilities, time constraints.
- It is supported by a wide range of planning engines. Having the schemes in PDDL, they may be delegated to the most suitable planner for the current issue. If explanations of the outcome are required, the schemes' chain will be computed by an abductive planner. In case of anticipating a tough debate, a defeasible planner may be adequate. When the legal complexity is high and domain knowledge is available, a hierarchical task decomposition planner would be more appropriate. In case of a typical debate, one can use case-based planners, whilst when there are hard time constraints, a genetic programming planner can be used.

- The *PIA − nodes* can be encapsulated in preconditions and effects, while the consequents are modeled with domain axioms and conditional effects, which models the fact that the derivation of a conclusion may depend on the proof standard required by the stage of the dispute.
- Several metrics can be attached to each argument in order to decide upon the most suitable one for a specific context: (: *maximise degree_of_support*) or (: *minimize number_of_critical_questions*). For instance, in a formal dispute one prefers arguments chains comprising preponderantly of *legal argumentation schemes*, or in a long run business relationship one seeks argumentation plans composed by *economical argumentation schemes*. Soft constraints, available in PDDL 3.0 [2], might be used, by applying them both to the claim and the preconditions.
- The core ontology can be easily extended by using PDDL domains. Online dispute resolution mediators are not necessarily lawyers or judges, or they can manifest different levels of experience. As they get experience, the mediators might extend and refine the basic ASs and CQs.
- Through WEB-PDDL or OWL2PDDL, the framework is compatible with semantic web. Therefore, structured argumentation systems can benefit from the existing translation tools or legal ontologies.

In our approach, the most suitable argumentation line for pleading is computing as a planning problem (figure 4). The domain starts by defining types for ASs, CQs, or the agents implied: suppliers, producers, consumers, and eventually the experts involved. Then, the needed predicates are defined.

In the case of claiming the argumentation scheme from expert opinion, the required parameters are: the agent ?*agent* who utters the scheme, the expert ?*e* cited by the agent, the fact ?*a* on which the expert has given his or her expertise, and the domain ?*d* to which the fact ?*a* belongs. The preconditions assure that

```
(define (domain SupplyChain)            (:action claim_AS_EXPERT
(:requirements :typing :adl)             :parameters (?agent agent ?e - expert
                                                      ?d - domain ?a - i_node)
(:types f_node, i_node, CQ, domain,
        agent - object                   :precondition (and (legal AS_EXPERT)
        as - f_node                                         (burdenofprof ?agent))
        presumption, exception - CQ
        cq1, cq2, cq3 - presumption      :effect (when (and (expert ?e ?d)
        cq4, cq5 - exception                               (claim ?e ?a))
        supplier producer expert - agent)      (and (plausibly ?a)
                                                   (legal cq1 belongs ?a ?d)
(:constants AS_EXPERT AS_COMMITMENT - as)           (legal cq2 genuine ?e ?d)
                                                   (legal cq3 relevant ?a ?d)
(:predicates (expert ?e - expert ?d - domain)       (legal cq4 consistent ?e)
             (claim ?e - expert ?a - i_node)        (legal cq5 reputation ?e))))
             (belongs ?a - i_node ?d - domain)
             (plausibly ?a - i_node)
             (evidence ?a - i_node)
             (legal ?as - f_node))
```

Fig. 4. Argument schemes as planning operators

the scheme can be conveyed, in our case: i) it is legal to be uttered in the current context and ii) the agent has the burden of proof. Thus, the mediation protocol is encapsulated as PDDL preconditions. Then, the conditional preconditions model the premises A_1 and A_2 of the AS from figure 1. The effect consists of asserting the conclusion $?a$ sustained by the expert, but it also introduces the legality to utter the associated CQs of the current scheme, where $cq1$, $cq2$, and $cq3$ are of type *presumption*, whilst $cq4$ and $cq5$ of type *exception*.

4 Estimated Impact

The long term goals of this research address two issues: context-aware argumentation and large scale argumentation. The objectives are related to: i) augmenting human collaboration and argumentation by appropriate technologies; ii) extending the WWAW towards a Pragmatic Web infrastructure for collaborative human-computer argument networks; iii) enhancing an individual's reasoning capabilities by increasing visibility, handle information overload, and providing users with re-usable patterns of argumentation. The current trend of forums, blogging, on-line debates is a positive social factor in the spirit of the current research. The technology has already started to be seen as an important part in the argumentation process [5], with an exciting impact on several domains such as:

Law. Structured argumentation will enhance the practice of conducting on-line consultation in domains such as Online Dispute Resolution. There are specific domains adequate for on-line dispute resolution services: e-commerce where the parties are not able to physically meet, or divorce disputes where parties, in some situations, don't want to directly interact. The impact of structured argumentation applied to law is related to: i) increasing transparency and trust; ii) enriching the ability to construct a judicial case; iii) costs and time saving due to on-line support.

Medical. A lot of research centered on applications of argumentation in medicine (such as risk assessment or treatment planning) has led to a comprehensive view of argumentation as a form of evidential reasoning. Our approach based on contextual-aware argumentation aims to identify methods of re-usable arguments for medical decisions based on multi-criteria factors: i) helping patient to understand him or her health state; ii) assisting medical staff to take decisions in case of contradictory information; iii) helping medical staff to apply forms of case based reasoning.

Education. Concept maps is a technique that allows the student to (1) see the connections between ideas they already have, (2) connect new ideas to knowledge that they already have, and (3) organize ideas in a logical but not rigid structure that allows future information or viewpoints to be included. In this context, concept mapping could be an effective teaching method for promoting learning, but also a mean to evaluate students' critical thinking. The method might be applied both to distant and traditionally learning.

Deliberative democracy (e-Government, e-Administration). It involves dialog with the public and it requires many feedbacks, which must present themselves in a structured manner in order to be effectively processed. The system helps when building multiple views of problems and resources among the following key actors: government and institutions, planners and technical experts, community. Services supporting structured argumentation impacts e-government in: i) increasing transparency; ii) collecting relevant and motivated ideas from citizens; iii) supporting multiple views representation. We anticipate the emergence of clusters of related debates, where arguments about general ideas and ones about specific issues interleave.

5   Related Work and Conclusions

Debatepedia[1] is a new wiki encyclopedia of arguments and debate related materials, including domains such as critical thinking, education, deliberative democracy. It provides a search-able repository of debates and the corresponding arguments supporting them, but without any formalization. We address the issue of large scale argumentation from a more structured viewpoint and in a more context dependent approach, in the spirit of the Pragmatic Web.

Araucaria[2] is a tool for analyzing arguments based on diagrammatic reasoning, which also deploys a repository of debates. It provides a user-customizable set of schemes with which the human agent can analyze arguments and save them in the Argument Markup Language format. We make use of the AIF ontology, which represents the state of the art standard at the moment when formalizing arguments. We also provide a computational model based on planning for building argument chains based on ASs.

Interleaving planning with arguing appears also in [1], but at a more abstract level, based on defeasible logic. By using ASs, we aim to provide a framework in which the interaction with the human agent is more flexible. From a different perspective, defeasible logic has already been proved to be the most suitable technical instrumentation for modeling argumentation. Quite the opposite, in the context of large scale argumentation, the interest is to provide a framework where all different inference mechanisms are able to co-exist under one umbrella, given for the moment by the AIF ontology. By introducing context, our research is closer to the idea of social dependent argumentation [4].

The ongoing work regards dealing with the context in PDDL, concerning the following: the extension of the WWAW with context in the spirit of the *pragmatic web level*; usage of the AIF ontology for the *semantic level*, and building a computational model of argumentation schemes based on planning and PDDL at the *syntactic level*. In this paper, we bring concept maps and AIF ontology

---

[1] http://wiki.idebate.org, launched on October 2007.

[2] http://araucaria.computing.dundee.ac.uk/

together with interesting advantages and challenges: concept maps are a semantically weak notation, which brings benefits in usability, but they have weaknesses in interpretation and automated manipulation, while AIF ontology reverses this balance.

Acknowledgment

We are grateful to the anonymous reviewers for useful comments. Part of this work was supported by the grant TD7 CNCSIS 534 from the National Research Council of the Romanian Ministry for Education and Research.

References

1. García, D.R., García, A.J., Simari, G.R.: Planning and defeasible reasoning. In: Durfee, E.H., Yokoo, M., Huhns, M.N., Shehory, O. (eds.) AAMAS, pp. 856–858 (2007)
2. Gerevini, A., Long, D.: Plan constraints and preferences for PDDL3. Technical report, University of Brescia, Italy (2005)
3. Gordon, T., Prakken, H., Walton, D.: The Carneades model of argument and burden of proof. Artificial Intelligence 171(10-15), 875–896 (2007)
4. Kalofonos, D., Karunatillake, N., Jennings, N.R., Norman, T.J., Reed, C., Wells, S.: Building agents that plan and argue in a social context. In: 1st International Conference on Computational Models of Argument, pp. 15–26 (2006)
5. Katsch, E., Rifkin, J.: Online Dispute Resolution: Resolving Conflicts in Cyberspace. John Wiley, Chichester (2001)
6. Modgil, S., McGinnis, J.: Towards characterising argumentation based dialogue in the argument interchange format. In: Rahwan, I., Parsons, S., Reed, C. (eds.) ArgMAS 2007. LNCS (LNAI), vol. 4946, Springer, Heidelberg (2008)
7. Rahwan, I., Zablith, F., Reed, C.: Laying the foundations for a world wide argument web. Artificial Intelligence 171(10-15), 897–921 (2007)
8. Reed, C.: Representing and applying knowledge for argumentation in a social context. AI and Society 11(1-2), 138–154 (1997)
9. Repenning, A., Sullivan, J.: The pragmatic web: Agent-based multimodal web interaction with no browser in sight. In: Rauterberg, M., Menozzi, M., Wesson, J. (eds.) INTERACT. IOS Press, Amsterdam (2003)
10. Schoop, M., de Moor, A., Dietz, J.L.G.: The pragmatic web: a manifesto. Commun. ACM 49(5), 75–76 (2006)

A Distributed Immune Algorithm for Solving Optimization Problems

Mariusz Oszust[1] and Marian Wysocki[2]

[1] Rzeszow University of Technology, Department of Computer and Control Engineering
W. Pola 2, 35-959 Rzeszow, Poland
moszust@prz-rzeszow.pl
[2] Rzeszow University of Technology, Department of Computer and Control Engineering
W. Pola 2, 35-959 Rzeszow, Poland
mwysocki@prz-rzeszow.pl

Summary. The mammal immune system is a distributed multiagent system. Its properties of distributive control and self organization have created interest in using immune principles to solve complex engineering tasks such as decentralized robot control, pattern recognition, multimodal and combinatorial optimization. In this paper a new immunity-based algorithm for solving optimization problems is proposed. The algorithm differs from the representative immune algorithm CLONALG. The agents participating in distributed problem solving enrich their knowledge about the solution via communication with other agents. Moreover they are decomposed into groups of specialists that can modify only some decision variables and/or use their own method of local improvement of the solution. The empirical results confirming usability of the algorithm and its advantage over CLONALG are presented. Obtained estimates of the global optima of multimodal test functions and traveling salesperson problem (TSP) are closer to the theoretical solutions and require fewer tentative computations.

1 Introduction

The mammal immune system contains a set of tissues and cells protecting the body from foreign structures. The foreign structures that activate cells of the immune system (the *lymphocytes*) are called *antigens*. The lymphocytes cooperate to defense the organism against antigens. Part of them called *B lymphocytes* proliferate after the contact with the antigen and differentiate themselves during the *clonal selection* process. In the differentiation stage the B lymphocyte provides accelerated somatic mutations (*hypermutation*) on its *antibodies* in order to acquire better binding (*affinity*) to the antigen. After successful elimination of the antigen the best lymphocytes are maintained as memory cells. According to Jerne's idiotypic network (immune network) hypothesis [5, 3] antibodies are stimulated not only by antigens but also by other antibodies. The basic idea of this hypothesis is that the immune system is constructed of a lymphocyte network. These lymphocytes interact with each other. In this way, the immune system is a parallel distributed system.

There are many algorithms based on immune metaphors. The authors of [2] propose an immune algorithm CLONALG based on the clonal selection principle. The algorithm operates on a population of lymphocytes. The lymphocytes with best antibody-antigen

C. Badica et al. (Eds.): Intel. Distributed Comput., Systems & Appl., SCI 162, pp. 147–155, 2008.
springerlink.com
© Springer-Verlag Berlin Heidelberg 2008

affinity (best stimulated by the antigen) are cloned and the clones are hypermutated. Next the best clones replace the worse ones. The process ends after predefined number of generations. Wierzchon in [10, 11] proposed a modified clonal selection algorithm extending CLONALG by preselection and crowding mechanisms.

Due to the similarities observed between the immune system and the multi-agent system, (e.g.: distributed or decentralized structure, autonomous entities with individual and global goals, ability of communication and coordination, adaptability, knowledge with which they make intelligent decisions) various artificial immune systems are created using agents. The paper [8] provides an agent system AISIMAM solving a mine detection problem. The proposed model of the system defines two types of agents: antigens and lymphocytes. The lymphocytes (robots) cooperate trying to detect and diffuse the antigens (mines). More complex example of a mutlitagent system is described in [6]. Its main goal is to provide an integrated solution to control and coordinate distributed systems with large number of autonomous agents. The lymphocytes in this system identify the antigen, use communication with other lymphocytes to generate new capabilities, gain knowledge and make distributed decisions.

This paper presents an algorithm inspired by the concept of agents participating in distributed problem solving. The proposed distributed immune algorithm (DIA) does not directly use clonal selection but it rather resembles the Jerne's idiotypic network. The lymphocytes in DIA enrich their knowledge about the solution of the problem via communication with other lymphocytes. Moreover, they are decomposed into groups of specialists that can modify (mutate) only some decision variables and/or use their own method of mutation. The paper is organized as follows. Section 2 presents the new algorithm. Section 3 presents the results provided by the algorithm using a set of benchmark functions and compares them with the results obtained by other immune algorithms. It also contains discussion about the parameters of the algorithm and introduces the results of implementation 100 cities TSP. Finally, section 4 concludes this paper and suggests future research directions.

2   The New Algorithm

The aim of this algorithm is to solve optimization problems. Without loss of generality we consider maximizing a scalar objective function $J(x)$ with respect to the vector x of decision variables from a feasible region X. Like in other immune algorithms, two types of agents appear in the proposed method, i.e. the antigen and a set of lymphocytes. The antigen represents unknown global solution of the problem. The lymphocytes, identified here with antibodies, represent candidates of the solution. They are decomposed into groups of specialists. The specialist can modify (mutate) only some decision variables in x and/or it uses its own method of mutation, both defined at the initialization stage. At the beginning of the algorithm the antigen sends its estimation of the solution (the starting point) to a randomly chosen lymphocyte. This lymphocyte tries to improve the obtained solution through a mutation (hypermutation) process to get higher value of the objective function J. Next it sends the result to some set of randomly chosen lymphocytes. Each receiver repeats the activities of the sender and the process is continued. Each transferred solution is provided with a token, which says how many transfers have been performed or, equivalently, how many agents have participated in building this solution. When the token related to a solution reaches the given threshold, this solution is sent to the antigen which determines the final result. The algorithm is presented below. The SEQ (sequential) construct causes all of the following processes indented by "-" to execute in the listed order, the PAR (parallel) statement defines a set of processes which execute concurrently (or in parallel).

Algorithm. DIA(N, Ns, n, T, M, S)

N - population size

Ns - number of specializations

n - number of lymphocytes a tentative solution is directly transferred to, $n \leq N$ (optionally, the number of transfers is randomly selected from the set 1, 2, ... , n in each iteration)

T - threshold value of the token (it determines the number of main iterations)

M - number of iterations in the mutation process

S - number of solutions used by the antigen in determining the final estimation of the global optimum, $S \leq n^{T-1}$

Initialization

- Create a population of N lymphocytes. Each lymphocyte has its individual, randomly generated initial solution x_{mem}.
- Decompose the population into Ns sub-populations, each containing specialists of the same type.
- Create the antigen with an initial solution (the starting point of the algorithm), provided with the token = 0.

PAR

- *Antigen*
    SEQ
    - **send**(starting point, token, value of the objective function, to a randomly chosen lymphocyte)
    - **receive**(S solutions with related values of the objective function, from lymphocytes), determine the best solution
    - **broadcast**(end)
- *Lymphocytes*
    PAR
    - **if received**(end, from the antigen) **then** terminate the computation
    - **if received**(solution, token, value of the objective function, from a lymphocyte or from the antigen) **then** place the triple [solution, token, value of the objective function] into FIFO
    - **while** FIFO not empty **than do**
    SEQ
        - take [solution, token, value of the objective function], token ← token +1
        - x_{mut_a} ← solution, perform the mutation process on the received solution:

> **for** $m = 1$ **until** M **do**
>> $x_{mut\_old} \leftarrow x_{mut\_a}$, modify x_{mut_old} accordingly to the specialization of the lymphocyte, if the result is better, retain it as x_{mut_a}
>
> **end**
>
> - $x_{mut\_b} \leftarrow x_{mem}$, perform the mutation process on x_{mem}:
>
> **for** $m = 1$ **until** M **do**
>> $x_{mut\_old} \leftarrow x_{mut\_b}$, modify x_{mut_old} accordingly to the specialization of the lymphocyte, if the result is better retain it as x_{mut_b}
>
> **end**
>
> - set $x_{mem} \leftarrow$ better of two solutions x_{mut_a}, x_{mut_b}
> - **if** token $< T$ **then**
>> **send**(x_{mem}, token, value of the objective function, to n randomly chosen lymphocytes)
>
> **else**
>> **send**(x_{mem}, value of the objective function, to the antigen).

The lymphocytes can be identified with distributed autonomous agents that try to cooperatively solve a decision problem. The antigen can be defined as the separate entity or its function can be assigned to a lymphocyte. Looking for similarities of the proposed algorithm with the clonal selection, we can consider the receivers as the clones of the sender. The process of improving the solution by successive groups of lymphocytes can be compared with the creation of the generations of lymphocytes. Furthermore, we can speak about mutual stimulation, because the lymphocyte that sends its result to other lymphocytes, influences (stimulates) them. The lymphocyte sends and stores its best solution. To find this solution it uses the stored result from the previous iteration. This is comparable with storage and cloning of the best lymphocytes, important for the convergence of the algorithm based on clonal selection [9]. Furthermore, such a procedure extends the search region and makes it possible to escape from local optima. Mutation performed on x_{mem} allows to explore a local area around it. Because mutations with lower values of the objective function J are lost, the x_{mut_b} tends to go up the hill, leading to a local optimum. Occasionally, the solution received from other lymphocyte will be on the side of the hill where the climbing region (thus the mutation process performed on x_{mut_a}) is more promising, which means that it will lead to the global solution. The loops concerning x_{mut_a} and x_{mut_b} are independent and they do not have to be performed sequentially. The parameter S of the algorithm requires a comment. If the number n of transfers remains constant in each iteration the user can set S equal to or less than n^{T-1}. The second alternative may be useful when communication failures are considered. Similar situation arises if the number of transfers is randomly selected from the set $1, 2, ..., n$ in each iteration. Alternatively, one can modify the algorithm introducing a parameter representing the maximal allowed computation time. Independently, one can allow to stop computations after the antigen receives satisfactory solution.

3   Evaluation of the Algorithm

In this chapter we show the results of experiments. First we compare the algorithm DIA with CLONALG [2] and with the modified CLONALG proposed in [11], with respect to finding the global optima of some typical multimodal test functions. Next we discuss the influence of the parameters of the DIA. At the end we apply the algorithm to the 100 European cities TSP. The experiments have been performed using an agent-based system and the MadKit platform [4].

3.1   Finding Global Optima of Multimodal Functions

We consider three multimodal functions which turn out to be difficult for any search algorithm because they have numerous peaks [10].

$$F_A(x) = x_1 sin(4\pi x_1) - x_2 sin(4\pi x_2 + \pi) + 1, x \in [-1, 2] \tag{1}$$

$$F_B(x) = An + \sum_{i=1}^{n}(x_i^2 - Acos(2\pi x_i)), A = 10, n = 20, x \in [-5.12, 5.12] \tag{2}$$

$$F_C(x) = \alpha \sum_{i=1}^{n} \frac{x_i^2}{4000} + 1 - \prod_{i=1}^{n} cos(\frac{x_i}{\sqrt{i}}), \alpha = 0.1, n = 10, x \in [-600, 600] \tag{3}$$

Function F_A of two variables has the unique global maximum equal to 4.254. Function F_B is the Rastrigin function of 20 variables, F_C is the Grievank function of 10 variables. Both functions take their unique global minima equal to zero at the origin of the system of coordinates.

The mutation process in the DIA is performed only on decision variables determined by the specialization of the lymphocyte. In our experiments the specialists in the i-th of Ns groups modify the i-th set of p decision variables, where p = (number of decision variables)/Ns. First, a decision variable is randomly selected from the set assigned to the specialist. Then one of two mutation variants is selected with the probability 0.5. In the first variant the result of the mutation is drawn from the domain $[lb, ub]$ of the decision variable, in the second variant the result is taken as $min(ub, lb + (v - lb) * l)$ where v is the actual value of the decision variable and l is randomly selected from [0, 2]. Other parameters used in the experiments are given in table 1. Table 2 compares the results obtained with three algorithms: (i) DIA, (ii) CLONALG [2], and (iii) modified CLONALG [11] denoted here as MCLONALG.

As we can see, the results obtained using the DIA outperform the remaining results. They are closer to the theoretical solutions in the sense of the best value, as well as in the sense of the mean and dispersion of the best results obtained in repeated executions

Table 1. Parameters of the DIA used in the experiments

Function	N	Ns	n	T	M	S
F_A	4	2	2	4	40	8
F_B	5	5	2	8	100	128
F_C	5	5	2	10	100	512

Table 2. Comparison of the results obtained with various immune algorithms

Function	Algorithm	Number of executions	Average number of generated tentative solutions	The maximal value of the objective function	The minimal value of the objective function	The mean of the best values of the objective function	Standard deviation of the best values of the objective function
F_A	**DIA**	**30**	**1200**	**4.25379**	**3.98556**	**4.22618**	**0.05114**
	CLONALG	30	50100	4.13	3.209	3.711	0.281
F_B	**DIA**	**30**	**51000**	**0.00226**	**1e-15**	**0.00018**	**0.00055**
	CLONALG	10	800100	35.0729	12.67206	22.72131	6.30522
	MCLONALG	10	not reported	not reported	0.00646	0.066525	not reported
F_C	**DIA**	**30**	**204600**	**0.00420**	**0.00009**	**0.00130**	**0.00120**
	CLONALG	10	800100	0.07838	0.01205	0.0408	0.01752
	MCLONALG	10	not reported	not reported	0.01667	0.020942	not reported

of the algorithm. Moreover the DIA uses significantly fewer tentative solutions to find the final result.

Influence of the parameters of the algorithm

In the DIA the best solution is determined by the antigen on the basis of the solutions received from lymphocytes. These solutions arise as the effect of the process in which many tentative solutions are generated. The number of solutions generated by the DIA is equal to $2 * M * n^{T-1}$. So it is proportional to the number of mutations executed by the lymphocyte during the main iteration and to the $(T - 1)$-th power of the number of lymphocytes a tentative result is transferred to. Increasing the values of the parameters M, n, T, we expect that the final solution will be closer to the global optimum. Exemplary relations concerning the function F_C are shown in fig. 1.

Fig. 2. illustrates how the performance of the algorithm is influenced by the parameter S.

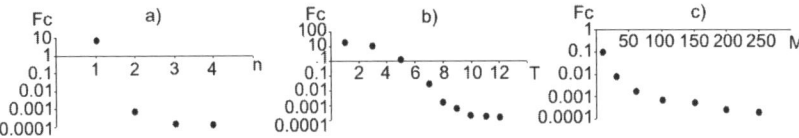

Fig. 1. The minimal value of the objective function F_C (averaged on the base of 10 executions of the algorithm) vs.: a) n, for $M = 100$, $T = 9$, b) T, for $M = 100$, $n = 2$, c) M, for $n = 2$, $T = 9$

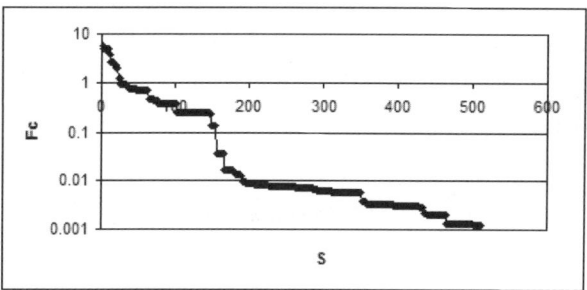

Fig. 2. The minimal value of the objective function F_C (averaged on the base of 30 executions of the algorithm) vs. the number S of successive solutions received by the antigen and used by it in determining the actual estimation of the optimum ($n = 2$, $M = 100$, $T = 10$)

3.2   TSP

In this known NP-hard problem a collection of cities resides on a plane, and we have to find the shortest tour starting in some city, visiting every other city exactly once, and returning to the starting point. In our experiments the tour of the salesperson is represented as the integer valued vector of numbers representing the cities. The population of the lymphocytes is divided into groups of specialists. All specialists in the group use the same method of mutation. The following known variants have been used [1, 7]: (1) *switching* - randomly selecting two cities and switching them in the tour, (2) *translation* - removing a section of the tour and then replacing it in between two randomly selected consecutive cities, (3) *inversion* - removing a section of the tour and then replacing it with the same cities running in the opposite order, (4) combination of the methods (2) and (3).

   The DIA algorithm has been tested on the 100 cities TSP [1]. The following values of the parameters have been used: $N = 15$, $Ns = 5$ (each group with 3 specialists, translation of single city is considered as the fifth specialization), $n =$ randomly selected from the set 1, 2, ... , 15 in each iteration, $T = 16$. The length of the shortest route is equal to 21134 km. After 15 executions of the DIA the following results have been obtained: the best result = 21224 km (i.e. 0.43% worse than the global optimum), the mean and the standard deviation of the best solutions from each execution: 21691 km (i.e. 2.64% worse than the global optimum), and 398 km, respectively. Fig. 3. illustrates how the performance of the algorithm is influenced by the parameter S.

   The same problem has been solved with the slightly modified CLONALG. As opposed to the original version of CLONALG only the best lymphocytes are cloned. The lymphocytes representing tours that are more than α percent longer than the best solution in the population are removed. The number of clones of the lymphocyte remains constant, but the population size changes from generation to generation, because it depends on the numbers of the best and of the worst lymphocytes in the previous generation. All lymphocytes use identical method of mutation, determined at the start of the algorithm (no specialization). The tests have been performed with the following parameters: population size in the first generation = 2, number of generations = 2000, number

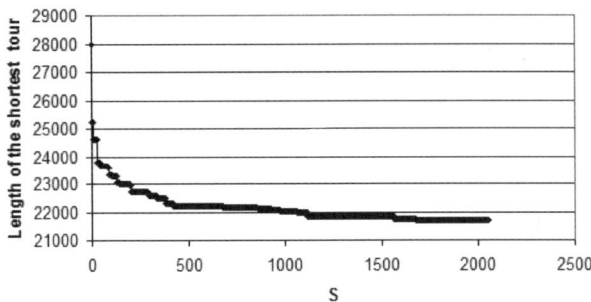

Fig. 3. The minimal length of the tour (averaged on the base of 15 executions of the algorithm) vs. the number S of successive solutions received by the antigen and used by it in determining the actual estimation of the optimum

of clones produced by the lymphocyte = 220, $\alpha = 0.2$. The computations have been executed 10 times for each mutation method. The best results are as follows: (1) for translation 22249 (the best result in the experiment, 5.27% worse than the global optimum), (2) for combination of inversion and translation 22424, (3) for inversion 22924, (4) for switching 28976, (5) the mean of the 10 results 23244 (obtained for translation, 10% worse than the global optimum) - with the standard deviation 796.

The results obtained with the DIA are significantly better although this algorithm generated fewer tentative solutions. The DIA uses specialization. So the result obtained by a lymphocyte is improved by agents with other specializations. Moreover, in CLONALG the best solution is not improved until the next generation. In the DIA it is directly sent for improvement to other lymphocytes. The additional important factor influencing the quality of the final result is that the best tentative solution generated by the lymphocyte is stored in its memory. After receiving a solution from other agent the lymphocyte tries to improve two solutions, i.e. the stored and the received. The better of two solutions is retained in the local memory and transferred to other lymphocytes.

4   Conclusion and Further Work

A new immune-based algorithm DIA for solving optimization problems has been proposed. The algorithm has been compared with two known immune-based algorithms CLONALG [2] and modified CLONALG [11]. The results concerning optimization of the typical multimodal test functions as well as the 100 cities TSP outperform the related solutions obtained with the remaining methods. They are closer to the exact solutions in the sense of the best value, as well as in the sense of the mean and dispersion of the best results obtained in repeated executions of the algorithm. Moreover, these results required fewer tentative solutions.

The DIA uses specialization. So the result obtained by a lymphocyte is improved by agents with other specializations. Moreover, in CLONALG the best solution is not improved until the following next generation. In the DIA it is directly sent for improvement to other lymphocytes. The additional important factor influencing the quality of the result is that the lymphocyte retains its tentative best solution. After receiving new solution, it tries to improve both results. Then the better solution is retained in the local memory and transferred to other lymphocytes. This procedure helps to escape from local optima.

The algorithm is faster than the remaining two algorithms, even if comparable numbers of tentative solutions are generated. This is because the lymphocytes in the DIA are focused on local improvement of its solutions and thus the elements of the population are not ranked. The lymphocytes in the DIA are not replaced with new individuals. They enrich their knowledge by the information from other lymphocytes. Thus they can generate good solutions based on communication with other lymphocytes, without direct contact with the antigen. This resembles the idiotypic network of Jerne [5].

The lymphocytes in the DIA can be identified with distributed autonomous agents (e.g. robots, controllers) that cooperatively solve a decision problem in a decentralized manner. In further research we will use the algorithm for energy management via negotiation among networked intelligent appliances.

References

1. Aarts, E., Korst, J.: Simulated Annealing and Boltzmann Machines. John Willey & Sons (1989)
2. De Castro, L.N., Von Zuben, F.J.: Learning and Optimization Using the Clonal Selection Principle. IEEE Transactions on Evolutionary Computation 3, 239–251 (2002)
3. Carter, J.H.: The Immune System as a Model for Pattern Recognition and Classification. Journal of the American Medical Informatics Association 1, 28–41 (2000)
4. Gutknecht, O., Ferber, J., Michel, F.: The MadKit Agent Platform Architecture. Rapport De Recherche, LIRM, Universite Montpellier, France (2000)
5. Jerne, N.K.: The Immune System. Scientific American 229(1), 52–60 (1973)
6. Lau, H.Y.K., Wong, V.W.K.: An Immunity-Based Distributed Multiagent-Control Framework. IEEE Transactions on Systems, Man, and Cybernetics - part A: Systems and Humans 1, 91–108 (2006)
7. Michalewicz, Z.: Genetic Algorithms + Data Structures = Evolutionary Programs. Springer, Heidelberg (1996)
8. Sathyanath, S., Sahin, F.: AISIMAM An Artificial Immune System Based Intelligent Multi Agent Model and its Application to a Mine Detection Problem. In: Proc. ICARIS (2002)
9. Villalobos-Arias, M., Coello Cello, C.A., Hernandez Lerma, O.: Convergence Analysis of Multiobjective Artificial Immune Algorithm. In: Proc. ICARIS, pp. 226–235 (2004)
10. Wierzchon, S.T.: Artificial Immune Systems. Theory and Applications (in Polish). AOW EXIT, Warszawa (2001)
11. Wierzchon, S.T.: Multimodal optimization with artificial immune system. In: Klopotek, M.A., Michalewicz, Z., Wierzchon, S.T. (eds.) Intelligent Information Systems. Physica-Verlag (2001)

Evaluation of Selective Distributed Discovery within Distributed Bio-active Agent Community

Ognen Paunovski[1], George Eleftherakis[2], Konstantinos Dimopoulos[2], and Tony Cowling[3]

[1] South East European Research Centre (SEERC), 17, Mitropoleos str., 54624 Thessaloniki, Greece
ogpaunovski@seerc.org
[2] City College, 13, Tsimiski str., 54624 Thessaloniki, Greece
{eleftherakis,k.dimopoulos}@city.academic.gr
[3] University of Sheffield, Regent Court, 211 Portobello Str., Sheffield, S1 4DP. UK
a.cowling@dcs.shef.ac.uk

Summary. The increased demand and complexity of the services operating within open distributed environments has emphasized the need for more robust, adaptive and self-organizing solutions. To address these problems some agent oriented approaches, like the Bio-Networking architecture, have adopted ideas from large scale biological collectives as a solution. However the introduction of biological properties, like birth and death events, generates an extremely dynamic system, making it difficult to maintain the overall connectivity of the agent network and discovery of resources within the system. Towards this end, in this paper the performance of a selective discovery mechanism is evaluated through multi-agent simulation studies. The primary focus of this study is on the impacts which death and (sexual/asexual) reproduction events have on the effectiveness of the discovery process in different overlay networks.

1 Introduction

Today, services operating in distributed environments are faced with growing number of users demanding more advanced, more efficient and secure services. In addition, developers aim to design more robust, decentralized, self-optimizing applications which will require minimal configuration and management. All these requirements impose significant constraints on the services, which are becoming complex, difficult to design, develop and maintain.

In recent years considerable effort has been put on the advancements of agent oriented approaches which show promise for dealing with these problems. The basic idea of multi-agent systems is to deploy autonomous entities with limited knowledge and capabilities which will cooperate in order to achieve a common goal [1]. In this context some of the approaches, like the Bio-Networking architecture [5], incorporate ideas from biology in order to overcome various obstacles as well as move towards self-organizing adaptive behaviour. The rationale for introducing ideas from biology is based on the notion that large scale biological collectives, as they exist in nature, are able to produce a variety of

C. Badica et al. (Eds.): Intel. Distributed Comput., Systems & Appl., SCI 162, pp. 157–166, 2008.
springerlink.com

complex behaviours like: self-organization, adaptation, scalability, availability etc. These properties can be very useful for solving problems common in distributed environments.

Nevertheless incorporating biological properties, especially death and reproduction, makes the entire multi-agent system extremely dynamic. Even a small fluctuation in the agent population can render the agent communication and interaction to be inefficient or impossible. In order to resolve this problem there is a need for intelligent discovery mechanisms, which will facilitate the location of agents and other resources within a distributed environment. The discovery algorithms used are heavily influenced by the solutions proposed in the area of peer-to-peer systems (see [3] for more details). However up to now there are no studies which aim to evaluate how bio-related properties like death and reproduction can influence the performance of the discovery mechanism within the agent population.

This paper presents the results of an investigation aimed to evaluate the impacts of death and (sexual/asexual) reproduction events on the performance of a discovery mechanism relying on selective query forwarding. The evaluation is done through an agent oriented simulation model based on the Bio-Networking architecture [5]. The work presented in this paper is a continuation of the work presented in [3] which concentrated on the selective distributed discovery in a bio-static environment.

2   Bio-Networking and Distributed Discovery

Bio-networking [5] is an attempt to create a framework for the development of bio-inspired agent oriented services operating within open, decentralized environments. It is both a paradigm as well as a middleware composed out of two major components, the Bionet platform and Cyber Entities (CEs) [5]. The Bionet platform corresponds to an adaptation of the biological environment, where the main application components called Cyber Entities operate. A Cyber Entity (CE) is an autonomous mobile agent, able to perform functions which are part of the specific Bio-Networking application design. In this context the major distinction between a CE and a mobile agent is the introduction of bio-properties (birth,death, natural selection) as part of the CE's life cycle.

Since each CE is completely autonomous there is no central entity or repository that manages the CE population. Consequently a vital element in the operation of the entire application is the ability of the entities to interact and communicate between each other. One of the approaches, proposed by Moore and Suda [2], for facilitating discovery in Bio-networking is based on selective query forwarding. The discovery is performed in a relationship overlay network formed by links between the individual CEs. During the discovery process, each CE autonomously performs the query processing and query routing procedure. When a query is received by an entity, it decides where to forward the query in the case it cannot satisfy it. However since query forwarding consumes bandwidth as well as processing power, the aim is to devise a strategy which will guide the forwarding process by indicating which entities are more likely to satisfy the query. One of the strategies elaborated in [2] is based on the use of keywords. Each entity is assigned keywords which describe its capabilities and knowledge. Consequently the query contains the keywords required by an entity to complete its task.

3 Simulation Model

The simulation model developed for the particular study is based on the Bio-Networking Architecture [5, 4] and partially on the discovery algorithm proposed by Moore and Suda [2]. The model has two main layers, the lower layer is composed out of a set of nodes and communication links between them. It abstracts a real network infrastructure with servers and network connections. The upper layer is composed of agent like entities (resembling CEs) and an overlay relationship network formed with the links between the entities. The data and the functionality of a particular entity are described by a specific set of keywords. Using these keywords other entities (agents or processes) are able to determine whether a specific agent can provide the required service or information. Furthermore the keywords are used to define a Keyword Similarity Value (KSV), which is the ratio of common keywords between two agents or an agent and a discovery query.

3.1 Types of Relationships in the Overlay Network

Each agent in the simulation has a limited number of relationships with other agents, forming an overlay network in which discovery is performed. A relationship between a pair of entities can be classified in two ways: firstly as **simplex** (a one-way relationship) or **duplex** (a two-way relationship), and secondly as **evaluated** or **random** relationships.

In a simplex relationship only one of the two entities can initiate communication (A knows B but B does not know A). In a duplex relationship either entity can initiate the communication (A knows B and B knows A). In an evaluated relationship, meta-data (KSV) about the relationship are kept. On the other hand in a random relationship no meta-data are kept. One entity can contain both evaluated as well as random relationships at the same time, while having either simplex or duplex relationships.

Due to the dynamic nature of the entity population, a certain relationship could be rendered invalid at a particular point of operation. This raises the need for relationship update, which is performed in two steps. First, discard the invalid relationship and second, acquire a new relationship (for details see [3]). The relationship update process is designed to acquire relationships with high KSV, while discarding the ones with low KSV. This process leads to organization of the relationship graph by grouping the agents with high number of similar keywords closer together, thus forming similarity clusters. Random relationships ensure the connectivity between the clusters.

3.2   Distributed Discovery in the Simulation Model

The discovery strategy is a query forwarding algorithm that decides to which entity the query will be forwarded next. The distributed discovery in the simulation model is based on "selective" query forwarding. In this scheme, the decision about where to forward the query depends on the KSV meta-data about the relationship partners and the keywords contained in the query. In the case when the keyword similarity between the entity and the query is high enough, the query is forwarded to the highest KSV partner. Otherwise the query is forwarded to a random or low KSV relationship. Each entity knows which relationship partner has sent the query and to which partners it has forwarded the query. In the case where an entity has forwarded the query to all of its relationship partners without successful match, the entity will return the query to the relationship partner it received it from, signalling that there are no more paths to explore.

The discovery process can end either with successful query match or with a failure. There are two ways in which the discovery can fail. First, by exploring all possible paths and second, by reaching the search limit. The search limit acts as a time-to-live (TTL) value which is decreased by one each time the query is processed by an entity. Using the described discovery strategy the relationship network is explored in a depth first search manner. At a particular time instance (time in the simulation is discrete) only one entity performs query matching and forwarding.

4   Simulation Conditions and Evaluation

The main aim of the simulation study was to evaluate the performance of the selective discovery strategy in simplex and duplex relationship networks during death and reproduction events in an agent population. This aim was the main factor which influenced the development of the simulation methodology and evaluation criteria.

The simulation model and the simulation environment were implemented as a standalone Java application. In order to address the study objectives a total of 12 different simulation scenarios were created. Each scenario started with different initial conditions varying in: relationship type (simplex and duplex), death/birth probability (0.01, 0.05 and 0.2) or a reproduction strategy (sexual, asexual). Since birth and death events are mutually dependent the same probability value was used for both events in a particular simulation run, thus maintaining constant search space. To address continuity of operation, one hundred consecutive simulation runs were executed for each simulation scenario, each for a different query with different initial entity. A population of 1500 entities was used in all scenarios. In order to ensure statistically sound results every scenario was repeated multiple times.

Each query contained eight keywords, whereas each entity contained fifteen keywords that described its data and functions. The query was designed to be successfully matched by at least one entity in the initial population. Each agent contained 10 relationships to other entities, 8 of which were evaluated while 2 were random.

In order to evaluate the performance of the distributed discovery mechanism, a number of parameters were used. These parameters are categorized as: discovery success, search cycles measurements and bio-life telemetry.

The **discovery success** gives an indication of successful/failed query matches in respect to all simulation runs executed with the same setup specification. It contains data indicating the success rate value as well as the failure rate value due to exploration of all paths or reaching the search limit.

The **search cycles** give an indication of the average cycles required to perform the discovery in a particular simulation run. One cycle is the time required for any agent to receive, perform query matching and forward the query to one other agent. Two measurements were used in this category. The total number of cycles corresponding to the total time required to perform the search. A unique search cycle which denotes the number of cycles that a query has been processed by an entity for the first time. The unique over total ratio expresses the useful search cycles in a single simulation run.

The **bio-life telemetry** contains bio-data gathered during the simulation. These are kept as measurements about the number of death and reproduction events which have occurred in a particular simulation run. In addition the bio-data also contain the population uniformity. The uniformity is an indication of the average number of entities with identical capabilities (keywords) in the population.

5   Results, Analysis and Discussion

A number of experiments have been performed in order to investigate how asexual/sexual reproduction and death events in the agent population affect the discovery performance in simplex and duplex relationships networks. To this end the analysis of the findings is presented and according to the reproduction strategy used.

5.1   Asexual Reproduction in Duplex and Simplex Networks

The asexual reproduction strategy implies creation of a child entity which is basically an identical copy of its parent. The child inherits all the keywords possessed by its parent. As the child entity is formed, a relationship between the child and the parent is introduced. As it can be seen from figure 1(A), the discovery success rates decrease as the probability of birth and death events increase. The success rate decrease is much higher in a simplex compared to a duplex network. For example the average decrease in simplex networks goes from 92% for birth/death probability of 0.01, to 7% when the probability is 0.2. The major reason for the decrease is that all possible paths have been explored (91% for p=0.2). Furthermore, data in figure 1(B) show that when the birth and death probability is 0.01, the number of average total search cycles is much higher than

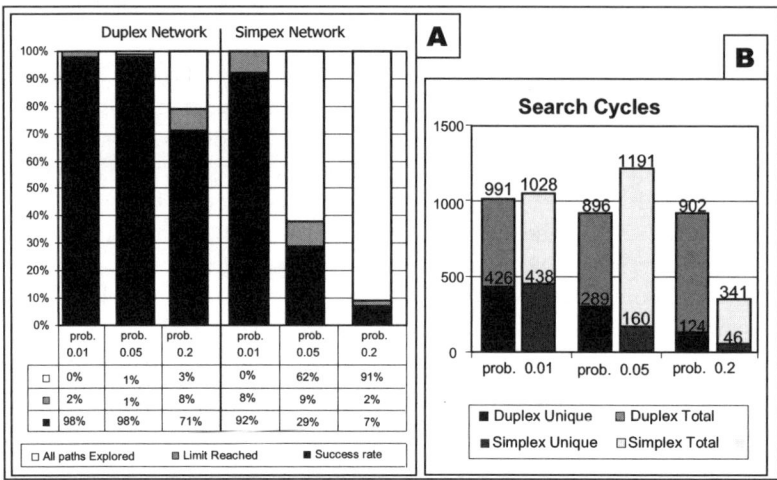

Fig. 1. A(left): Discovery success rates for asexual reproduction with different probabilities for birth/death events. B(right): Average total/unique search cycles for different probabilities for birth/death events during asexual reproduction.

in case of the 0.2 (1028 for p=0.01 to 341 for p=0.2). This implies fragmentation and clustering of the overlay relationship network. Figure 2 presents a view of the simplex network degradation process in a typical simulation scenario. During the initial runs, the fragmentation causes formation of big clusters. The discovery process needs many cycles to explore a cluster. In some cases, where the target entity is in a different cluster, the discovery reaches the search limit. However in later discovery runs, the network degradation process divides the clusters even further, creating smaller and smaller clusters. This leads to a situation where only a few discovery cycles are needed to explore all possible paths within the cluster.

The main reason for the fragmentation of the network in the case of simplex relationships is the inability to update invalid relationships. The crucial point in the relationship update is the detection of the invalid (dead) relationship partners, which in simplex networks can only be performed when the entity tries to contact its relationship partner. However since entities communicate only during the discovery process (spooling consumes too much bandwidth), the detection of invalid relationships cannot cope with the increase of the birth/death events. Thus entities perceive invalid relationships as valid.

On the other hand in duplex networks the detection of invalid relationships is handled more smoothly. Before an entity dies it informs its relationship partners (through the return relationship) that it is about to die, so the relationship partners can immediately render this relationship as invalid. Afterword they can replace the invalid relationship with a valid one. This is the main reason why the success rate in duplex networks is much higher (98% for p—0.01 and 71% for

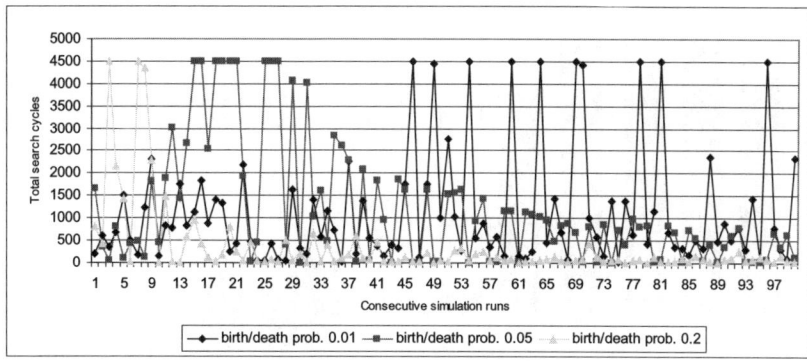

Fig. 2. Discovery performance during asexual reproduction (100 consecutive runs)

p=0.2). Nevertheless the success rate still decreases as the number of birth/death events increases due to fragmentation of the network.

The main reason for fragmentation in a duplex network is the formation of uniform fragments (clusters). A uniform cluster is an environment with large number of entities belonging to one (or very few) agent families with no outside relationships. The formation of family clusters occurs because a family with more members has greater chances of producing offspring. Additionally the child and parent introduce a relationship between them which has the highest possible KSV (having identical keywords), making the relationship permanent. Moreover, the relationship update gives preference to relationships with other members of the family over other entities, since family members have the highest KSV.

5.2 Sexual Reproduction in Duplex and Simplex Networks

Unlike asexual reproduction, in the case of sexual reproduction two entities play the role of the parents. The child entity is a result of a random combination of the keywords and relationships from both parents. In addition, during sexual reproduction there is a possibility of mutation (variation in the inherited keywords). This increases the probability that a newborn entity will be a unique instance in the entire population.

Figure 3(A) presents the success rates for both simplex and duplex networks during sexual reproduction. The general trend is that the success rate decreases as the birth and death probability increases. However if we compare the success in the sexual and asexual reproduction strategies (fig. 4(B)), then we clearly observe that the success rate decrease is much lower in the case of sexual reproduction. For example in the simplex network case when the birth/death probability is 0.05, the success rate is 29% during asexual reproduction. In comparison, the success rate for the same probability during sexual reproduction is 64%. An examination of figures 3(B) and 4(A) reveals the reason for this trend. In the case of birth/death probability of 0.05 we can observe in figure 3(B) that the

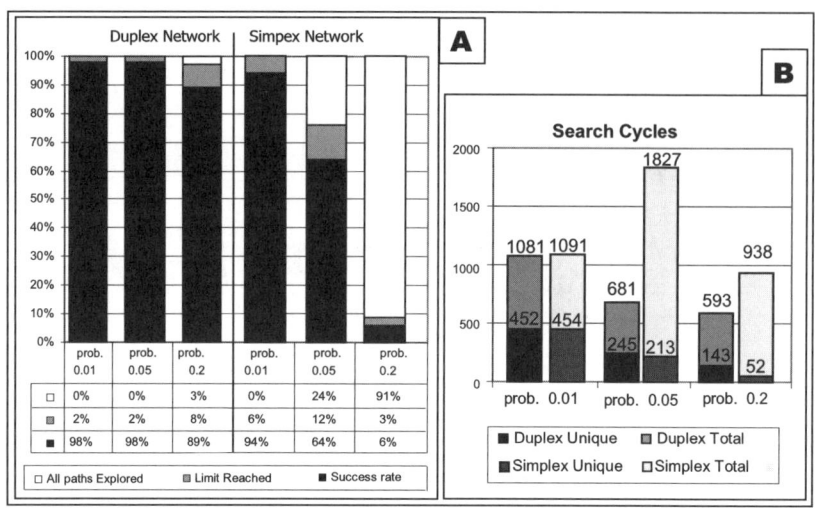

Fig. 3. A(left): Discovery success rates during sexual reproduction with different birth/death event probabilities. B(right): Average unique/total search cycles ratio during sexual reproduction with different birth/death event probabilities.

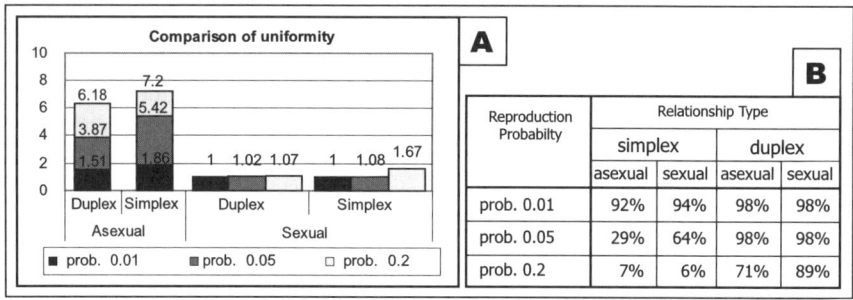

Fig. 4. A(left): Comparison of entity uniformity during sexual and asexual reproduction. B(right):Comparison of success rates for asexual and sexual reproduction.

unique/total cycles ratio is very high, while figure 4(A) shows that the uniformity of the population is very low. This means that the network is not completely fragmented and there are still a few relationships which connect the fragments. The discovery process manages to find an exit from the cluster after it has exhausted the high KSV relationships within the cluster. This is further supported by the simulation logs which indicated that usually an exit relationship has low KSV. These relationships exist because the sexual reproduction introduces diversity in the offspring's relationships, which in turn results in slower degradation of the network.

The case of sexual reproduction with probability 0.2 for simplex network is a special case (see fig. 4(B)). In this case there is no visible improvement in the success rate (6%) of sexual over asexual reproduction strategy. This is because in both situations the network is completely fragmented. Thus the success rate is strongly influenced by the initial point (starting entity) of the discovery process. If the query is initiated in the "correct" cluster then the query is matched in very few search cycles. However the significantly higher number of search cycles and better total/unique search cycle ratio suggest that fragments are relatively bigger during sexual reproduction.

Distributed discovery in duplex networks during sexual reproduction shows significant improvement compared to asexual reproduction for probability of 0.2 (89% compared to 71%). This is attributed to the fact that during sexual reproduction the population uniformity is much lower (figure 4(A)). Thus while the entities maintain relationships with their parents and children, the KSV is not as high as during asexual reproduction and therefore these relationships are not permanent. This slows down the fragmentation of the network and blurs the cluster boundaries.

6   Conclusions and Future Work

This paper is a continuation of the work presented in [3]. The paper examines the performance of a selective distributed discovery algorithm in a bio-active agent population through a multi-agent simulation model based on the Bio-Networking architecture. The focus of the study is on discovery performance in simplex and duplex relationships networks when there are death and (sexual/asexual) reproduction events in the agent population. In addition the paper has elaborated the impacts of sexual and asexual reproduction strategies on the fragmentation of the overlay relationship graph.

The major finding was that the degradation level in a duplex network was significantly lower (compared to simplex network) even under extreme population flux. This is due to the improved detection of invalid relationships which ensured better connectivity in the overlay network. Another important finding was that asexual reproduction by reducing the population diversity created uniform clusters of identical entities. This resulted in a reduction of the success rate and unique/total search cycles ratio. On the other hand the sexual reproduction increased the diversity, allowing better discovery performance by decreasing the success rate drop for high population flux.

This study can be of benefit to the design of intelligent search and retrieval processes for various content delivery systems, peer-to-peer systems, web services, and other systems operating within distributed environment. Furthermore the findings presented may prove vary valuable in the case of distributed web services where it is useful to incorporate death and reproduction as means of self-optimization based on the user demand for particular web service.

Therefore future work will aim to explore the selective discovery performance in different user request/demand scenarios. Furthermore it would be interesting to examine how migration and various migration heuristics can impact the discovery performance.

References

1. Jennings, N.R., Sycara, K., Wooldridge, M.: A roadmap of agent research and development. Journal of Autonomous Agents and Multi-Agent Systems 1(1), 7–38 (1998)
2. Moore, M., Suda, T.: A decentralized and self-organizing discovery mechanism (2001)
3. Paunovski, O., Dimopoulos, K., Eleftherakis, G.: Impacts of the relationship overlay network on the distributed discovery performance in a decentralized agent community. In: Proceedings of the 3rd Balkan Conference in Informatics (BCI 2007), Sofia, Bulgaria, pp. 277–288 (2007)
4. Suzuki, J., Suda, T.: An overview of the bio-networking architecture. In: The Super Distributed Objects Forum, number sdo/02-10-05. Object Management Group TC meeting at Helsinki (October 2002)
5. Wang, M., Suda, T.: The bio-networking architecture: A biologically inspired approach to the design of scalable, adaptive, and survivable/available network applications. In: Proceedings of the 1st IEEE Symposium on Applications and the Internet (SAINT) (2001)

VPOET: Using a Distributed Collaborative Platform for Semantic Web Applications*

Mariano Rico[1], David Camacho[1], and Óscar Corcho[2]

[1] Escuela Politécnica Superior, UAM
{Mariano.Rico,David.Camacho}@uam.es
[2] Ontology Engineering Group, Departamento de Inteligencia Artificial, UPM
ocorcho@fi.upm.es

Summary. This paper describes a distributed collaborative wiki-based platform that has been designed to facilitate the development of Semantic Web applications. The applications designed using this platform are able to build semantic data through the cooperation of different developers and to exploit that semantic data. The paper shows a practical case study on the application VPOET, and how an application based on Google Gadgets has been designed to test VPOET and let human users exploit the semantic data created. This practical example can be used to show how different Semantic Web technologies can be integrated into a particular Web application, and how the knowledge can be cooperatively improved.

Keywords: Distributed collaborative systems, Semantic Web, Wiki architectures.

1 Introduction

One of the key aspects of the Semantic Web [2, 6] is that software agents or applications are able to "understand"the meaning of contents specifically designed for them. The Semantic Web is made possible using a set of standards like RDF(S) [7, 3], OWL [1], or SPARQL [8], among others.

In the Semantic Web research area, the concept of *semantic information* represents knowledge that can be automatically analysed with no (or minimal) ambiguity. To avoid any possible ambiguity, the Semantic Web standards have been designed using logic-based formalisms and ontological representations. For example, there are a set of Description Logic *reasoners* that can be used to perform inferences with OWL models. On the other hand, different knowledge standard representations, named ontologies, have been designed to formally describe the exact meaning of a particular concept. An *ontology* is a set of formal definitions about a particular domain. Although there exist other standards and formalisms to represent ontologies, the most popular in the Web is OWL which is based in the definition of `classes`, `properties`, `individuals`, and `relationships` between them. For example, the Friend Of A Friend(FOAF) ontology can be used

* This work has been supported by TIN 2005-06885, TIN 2007-64718 and TIN 2007-65989.

to define the Person and Organization classes; the name, surname and email properties; and the *knows* relationship (applicable to individuals belonging to the Person class). The FOAF ontology comprises definitions, that is, no instances are declared for any defined class. Ontologies and data are identified by a namespace.

The evolution of the Semantic Web is directly joined to ontologies and semantic technologies success. There are currently about 11,000 ontologies available on the Internet [10, 4], and the semantic data has experimented an exponential growth for the last ten years [5]. However this high-quality information remains hidden to most end-users, developers, and even software agents, because there are only some few applications able to manage with this semantic data. Two main problems can be analysed to explain this current situation. On the one hand, the increasing difficulty to design adaptable and easily reusable Web applications where a wide set of Web technologies and programming languages, such as HTML, Javascript, CSS, DHTML, Flash, or AJAX, need to be used, converting graphical-designers in skilled programmers as pointed in [9]. On the other hand, the complexity of Semantic Web technologies requires a very specialised knowledge. For instance, the process of creating ontologies using OWL needs from domain experts and OWL specialists in order to "transfer"the experts' know-how into a specific OWL ontology. Therefore, the correct design of a semantic web application needs from a wide set of different specialised experts.

This paper proposes a new approach to solve some of the previous problems. Our approach is based on a particular methodology used to simplify the creation of Semantic Web Applications using a wiki-based approach, one of the most successful collaborative environments for the last years. Unlike common wikis, oriented to contents creation, some wikis can be used to functionality creation, in a collaborative way for developers.

This paper is structured as follows. Section 2 shows our methodological approach to design semantic applications based on wiki technologies. Section 3 describes VPOET, a semantic application that implements the previous methodology. Section 4 describes a practical case study that exploits the communications channel provided by VPOET. Section 5 shows how to get the best fitted visualisation of a semantic data element for a given user profile. Finally, Section 6 summarises the conclusions and future work.

2 Distributed Methodology for Semantic Cooperative-Based Web Applications

Interaction with human users, showing semantic data, or requesting data that will have to be converted to semantic data, is a cornerstone of the Semantic Web. Our work focuses on a technological approach, providing developers with a simple and collaborative programming framework in order to simplify the process of creation of semantic web applications. As a proof-of concept, we present a real semantic web application that uses the aforementioned framework in order to validate the technological approach. Next subsections give the detail of this approach and a concrete implementation.

2.1   Designing a Platform Based in Contribution for Semantic Applications Developers

Unlike recent efforts to create wiki-based technologies that allow editing semantic data (so-called semantic wikis, like Semantic Mediawiki, IkeWiki, or ODEWiki) in our approach we go a little bit further and allow users to create easily and collaboratively pieces of code that can be included in Semantic Web applications. This technological approach does not require developers with skills in multiple languages and technologies, but just wiki essentials, and basic skills on a programming language and semantic web technologies. For this kind of developers, and for a concrete wiki-engine called JSPWiki[1], we have created a software framework called Fortunata . This software exploits plugins, software pieces that extend a given functionality. In this case, our plugins extend the functionality of an open-software wiki. Applications designed under this architectural paradigm let developers to create functionality in a decentralised way. Traditional development centralises the source code. Therefore, extending functionality typically requires accessing the source code and compile. The result is a new version of the application. However, plugins let members of a community to contribute creating new functionality with a minimal degree of dependence. When a developer has created and tested a new plugin, the source code is sent to the wiki administrator. If the code is considered valid and safe, it is compiled and added to the wiki engine. Unlike traditional development environments, this addition does not require to check for dependencies or compiling the whole application code. Even, in our system, it can be done while the application is running. Semantic web technologies provide us an additional advantage: simpler data integration. Fortunata-based applications comprise a set of plugins managing a semantic data source. These applications can integrate easily semantic data from other Fortunata-based applications.

2.2   Applying the Architectural Aspects to Real Applications

As a result of applying this aspect, different roles appear for both developers and end-users. Figure 1 shows a clear separation between end-users, developers, and semantic agents, as well as different roles that are introduced below.

The architectural aspect results in two different kinds of developers, as are shown in figure 1. Table 1 shows the activities and requirements of these users. User1 plays the role of "semantic web applications developer", providing with Fortunata-based plugins (F-plugins in figure 1). A different kind of developer is represented by user5. She does not contribute with plugins, but takes advantage of the semantic data created by user1's applications.

---

[1] See http://jspwiki.org

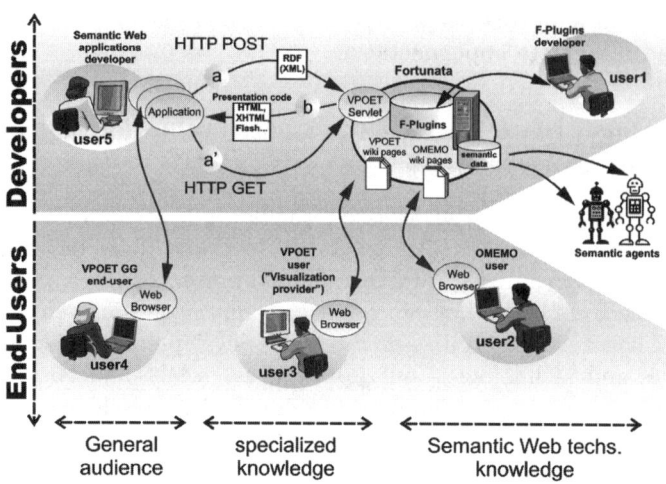

Fig. 1. Involved roles in the proposed system

As a proof-of-concept, we have created some Fortunata-based applications. In this paper we focus on VPOET. Let us see a brief description of this application and how it benefits from the methodological aspect.

VPOET enable end-users, denoted "visualisation providers" in this context, to create visualisation templates for a given ontology element, not only to show semantic data (output templates) but to request data from the user (input templates). These templates can be created by any user with basic skills in client-side technologies, such as HTML or Javascript, using simple macros provided in VPOET. Visualization providers can get information about the ontology element reading the wiki pages generated by another Fortunata-based application, or reading other manually created wiki pages referencing to these pages. In figure 1, user3 represents this kind of user.

Besides creating the visualisation template, visualisation providers indicate the features of their templates using forms, specifying details such as template type (input or output), behaviour in case of changes to the font size, sizes (preferred, minimum, maximum), code-type provided (HTML, Javascript, CSS), or dominant colours. As any other Fortunata-based application, all the generated information is published as semantic data, so that it can be used by semantic agents. Besides, a HTTP GET/POST channel has been created to get access to the semantic data. Figure 1 shows this channel in the case of VPOET, and how it is exploited by developers like user5. For testing purposes, we have exploited this channel creating a Google Gadget called GG-VPOET. End users like user4 use GG-VPOET to render a semantic data source under a concrete visualisation template. Other applications can exploit this channel. For example, we are using this channel to query for the most appropriated visualisation for a given user profile. This experimental user profile contains data about the interactive impairments of the user, its interaction device, or its aesthetic preferences.

Table 1. Description on the roles in the proposed system

Role	Activities	Requirements
user1	F-plugins developer. Uses the Fortunata framework to create semantic plugins	Basic java programming skills
user5	Semantic Web applications developer. Uses the HTTP channel provided by VPOET	Basics of HTTP in any programming language
user2	OMEMO user. Any user interested in obtaining a simple and textual description of the elements in a given ontology	None
user3	VPOET user. Client side graphical designer	Requires basics of client side technologies
user4	VPOET-GG end-user. Any user interested in providing a visualisation of a semantic data source	None

3   Using VPOET

VPOET lets users create visualisation templates for any ontology element. Although VPOET can be used by any user with basic skills in client side-side web technologies, it has been created to let professional **graphical-designers** author attractive designs capable of rendering semantic data. Users of VPOET are denoted "visualisation providers" (VPs). From an end-user point of view, this application is like any other web application, with form elements like text fields, radio buttons, or buttons. VPs just have to follow an online tutorial to start creating templates.

The process to create a template starts targeting an ontology element. For example, the next subsection reports on a use case that follows the tutorial aforementioned, in which the element **Person** from the FOAF ontology version 20050403 is targeted. The process to create the template comprises these steps:

1. Getting information about the structure of the targeted element. That is, to know which sub-elements comprise the element. The visualisation provider obtains this information reading wiki pages automatically generated by OMEMO (user2 in figure 1), other Fortunata-based application.
2. Authoring a graphical design in which semantic data will be inserted. End-users are free to use their favourite web authoring tool.
3. Choose an identifier (ID) to create a wiki page with that ID. This wiki page shows information about the VP and its templates stored.
4. The graphical design comprises a set of files (images, and client-side code such as HTML, CSS, or javascript). The client-side code is copied-pasted in the appropriated form fields. Image files or "included" files must be uploaded

Table 2. Main macros available for visualisation providers in VPOET

Macro	Arguments	Explanation
OmemoGetP	propName	It is substituted by the property value propName
OmemoBaseURL	No arguments	It is substituted by the URL of the server in which VPOET is running
OmemoConditionalVizFor	propName, designerID, designID	Renders the property propName only if it has a value, using the template indicated
OmemoGetLink	relationName	It is substituted by a link capable of displaying elements of the type pointed by the relation relationName

to the provider wiki page, or uploaded to any web server. In any case, the client code must point correctly to these files.

5. A test loop starts, using semantic-data sources (typically external to VPOET) containing instances of the targeted element.
   a) Paths (relatives or absolutes) must be substituted by means of a specific macro.
   b) Semantic data are inserted using specific macros.
   c) The design is tested against the test data sources
   d) This loop finish when the design produces a successful visualisation for all the semantic test data sources.
6. The design is characterized by its creator, providing info about the template features, such as type, colors, size policy, or font changes behavior.

Most of the effort required to create a template is located in the test loop, especially in the insertion of macros. The table 2 shows the most relevant macros available in VPOET, the arguments each macro requires, and a brief explanation of each macro.

VPOET has been designed to let its users reuse their templates. This is achieved using: (1) the conditional rendering of a property (using the macro OmemoConditionalVizFor) and (2) links capable of displaying the destination element of a relation (macro OmemoGetLink). A detailed explanation, and usage examples, can be found at http://ishtar.ii.uam.es/fortunata.

4   Using the HTTP Channel in VPOET

Although the information stored in VPOET is published as semantic data reachable through an URL that can be used by semantic agents, an additional channel to let non-semantic users access this information has been created. It has been implemented as a servlet that let users make HTTP GET/POST requests with

Table 3. Parameters accepted in the HTTP GET/POST request

Parameter	Value	Explanation/Example
action	renderOutput	Request a visualisation for the elements object in the data source given in parameter object
	renderInput	Request a visualisation to request data for the element object from the user
object	prefix.class[.ver]	Example: foaf.Person
	prefix.relation[.ver]	Example: foaf.firstName
source (GET only)	URL	URL of the semantic data source
[provider]	ID	Identifier of the visualization provider. For example: user3.test
outputFormat	HTML	Default value
	XHTML	XHTML is used by WAP 2.0 mobile phones
[userProfile] (GET only)	URL	URL of the RDF data source with the user profile

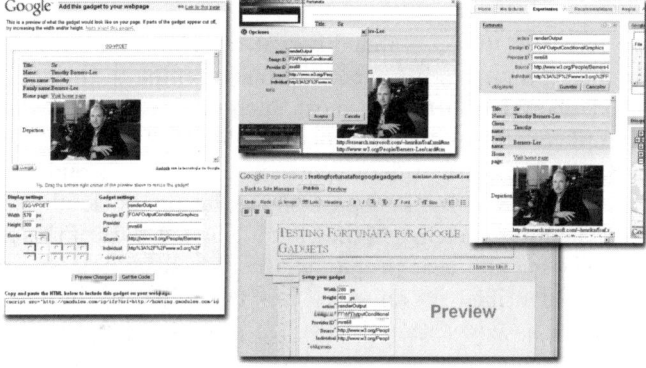

Fig. 2. Using GG-VPOET in different application oriented to end-users. In clockwise: a personal page, Google Desktop, iGoogle, and Google Pages.

variable parameters in order to facilitate queries like "get an output visualisation created by provider X for the element foaf.Person.20050603 for the semantic data at URL Y". The complete syntax is shown in Table 3.

When the GET method is used, the parameter source must be provided to indicate where semantic data source can be found. In the other hand, when POST method is used, the parameter source is not necessary because the semantic data must be contained by the HTTP message. If the parameter provider is not provided, VPOET will return the "best visualisation" given the user profile pointed by parameter userProfile. When there is no template for a requested element, a default visualisation is provided.

An Fortunata-based application, called MIG, provide users with a form (in a wiki page) to specify the user profile. As any Fortunata-based application, this information is public and accessible.

The HTTP messages with the specified syntax can be sent to VPOET by other programs (agents) written in any programming language, or by javascript applications executed in a web browser. However, browsers are more limited than other applications because they suffer security restrictions due to communication is restricted to the server which holds the web application. However, our approach do not have this problem because communications are centralised by Fortunata.

To let final users exploit this channel, a Google Gadget has been implemented, as was show in figure 1. In this figure, user4 use this gadget in its web pages, or in some Google products, such as iGoogle, Google Pages, or Google Desktop. This gadget is configured providing the same information that was provided for the test phase. Figure 2 shows this gadget in action using an output template for foaf:Person.

5   Matching the User Profile and the VPOET Semantic Templates

Let us suppose that VPOET contains different templates for `foaf.Person`, and an external application requesting a `foaf.Person` template through the HTTP channel. VPOET should return "the most adequate"template for a given user profile. An example of this matching process is depicted in figure 3.

Each ontology, identified by a namespace, is shown as a cloud. The elements of the ontology, and their individuals, are shown inside its cloud; with ontology elements and some individuals inside the cloud. The left part of this figure shows the ontology describing the user profile, characterised by namespace a. In this example, the user identified as a:$user34$ has the following profile: (1) uses a WAP2 mobile phone as interaction device, (2) prefers simple aesthetics and (3) he/she is daltonic (colour-blindness associated to red-green colours).

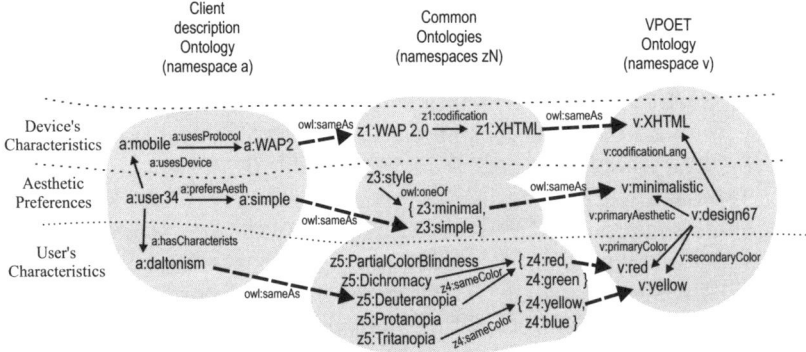

Fig. 3. Matching process to find a visualisation template from a given user profile

In centre part of figure 3, public well-known ontologies are shown. Ontology z_1 indicates that the protocol WAP2.0 is codified as XHTML. For ontology z_3, "minimal" and "simple" are different kinds of styles but semantically close. Ontology z_5 has a visual-impairments hierarchy.

The right part of figure 3 shows the VPOET ontology, with namespace v. In this ontology, the template identified as $v{:}design67$ is codified using the XHTML language, its primary aesthetic is minimalistic, and it has red and yellow as primary and secondary colours.

With just this semantic information, it is impossible to find that $v{:}design67$ is even a valid template for $a{:}user34$. An additional semantic data source is required in order to link elements belonging to different ontologies. These links use to be "sameAs" relations, shown as discontinuous bold arrows in figure 3. Joining all this semantic information, a semantic agent can make a semantic query (e.g., using SPARQL language) based in the user profile, like this one: "select a template with these characteristics: (1) codified in XTHML, (2) with minimalism as chief aesthetic, and (3) with primary colours avoiding red and green tones for text and background". For this example, the result of this query would be the design $v{:}design67$. Additional restrictions can refine the query to get the "most adequate" template for a given user profile.

6 Conclusions and Future Work

The work presented in this paper aims at providing developers with a simple and collaborative programming framework i order to simplify the process of creation of semantic web applications. Developers require (1) development environments simple and collaborative, (2) facilities for reuse of the contributed functionality, and (3) minimal dependencies between contributors. To achieve these requirements, we have taken advantage of an open source wiki-engine. We have developed a java library called Fortunata in order to facilitate developers the creation of plugins with semantic capabilities. As a proof-of-concept, some applications have been built using Fortunata. VPOET is an example of one of these applications.

From a developer's perspective, we consider that the targeted requirements concerning developers are successfully accomplished by the selected wiki-engine. However, it must be noticed that this is the result of our experience for some concrete applications. Concerning end-users, these applications are intended for a wide audience with no previous training in programming or semantic web technologies. This objective has been achieved be means of forms and simple macros, and experiments with end-users (not described in this paper) confirm it.

These are the initial steps towards a semantic agent capable of providing an automatic generation of the user interface. This agent can use the data provided by VPOET in order to adapt the user interface to the user's profile (device used, user's impairments, and aesthetic preferences). Many open aspects remains open: composition of templates, or interaction between templates, among others.

The architecture shown in this paper can provide developers with a simple but powerful infrastructure to achieve these long-term objectives.

References

1. Bechhofer, S., van Harmelen, F., Hendler, J., Horrocks, I., McGuinness, D., Patel-Schneider, P., Stein, L.: OWL Web Ontology Language Reference. Technical report (2004), http://www.w3.org/TR/2004/REC-owl-ref-20040210/

2. Berners-Lee, T., Hendler, J., Lassila, O.: The semantic web. Scientific American 284(5), 28–37 (2001)

3. Brickley, D., Guha, R.: RDF Vocabulary Description Language 1.0: RDF Schema. W3C Recommendation. Technical report (2004), http://www.w3.org/TR/rdf-schema/

4. Ding, L., Finin, T., Joshi, A., Pan, R., Cost, R., Peng, Y., Reddivari, P., Doshi, V., Sachs, J.: Swoogle: a search and metadata engine for the semantic web. In: Proceedings of the Thirteenth ACM conference on Information and knowledge management, pp. 652–659 (2004)

5. Finin, T., Ding, L.: Search engines for semantic web knowledge. In: Proceedings of XTech 2006: Building Web 2.0, Amsterdam, May 16-19 (2006)

6. Herman, I.: The Semantic Web home page. Technical report, W3C (2006), http://www.w3.org/2001/sw/

7. Klyne, G., Carroll, J.: Resource Description Framework (RDF): Concepts and Abstract Syntax. Technical report, W3C (2004), http://www.w3.org/TR/rdf-concepts/

8. Prud'hommeaux, E., Seaborne, A.: Sparql query language for RDF. Technical report, W3C Recommendation (2008)

9. Rochen, R., Rosson, M., Pérez, M.: End user Development of Web Applications, ch. 8, pp. 161–182. Springer, Heidelberg (2006)

10. Warren, R.H., Baker, C.O.J.: Ontologies: Where are we at? In: Knowledge-Based Bioinformatics Workshop, Canada poster (2005)

Are Many Heads Better Than One—On Combining Information from Multiple Internet Sources

Jakub Stadnik[1], Maria Ganzha[2], and Marcin Paprzycki[2]

[1] Warsaw University of Technology, Warsaw, Poland
[2] Systems Research Institute Polish Academy of Sciences, Warsaw, Poland
{Maria.Ganzha,Marcin.Paprzycki}@ibspan.waw.pl

Summary. In this paper we look into three approaches, based on: *Game Theory, Auction* and *Consensus* methods, to combine information from multiple sources. As originally introduced, they are conceptualized using an agent metaphor and implemented using a JADE agent platform. Preliminary performance comparison completes the presentation.

1 Introduction

Since different Internet search engines produce different results for the same query, we can say that they "see" the world differently. The question then arises: how to combine answers from different sources in such a way that the obtained answer would be "better" than when using only a single source? What suggest combining "advice" from multiple sources is a standard situation, when a panel of experts is used to address a problem. Combining multiple suggestions can be achieved, among others, utilizing a *Consensus method* [5, 2, 4], *Game Theory* and *Auctions* [8, 7]. These approaches have been originally proposed as based on software agents. While this is somewhat spurious (proposed functionalities can be achieved without agents), we follow predecessors and use JADE agent platform to implement combining information from multiple Internet sources.

This note is organized as follows. In the next section we introduce the three approaches to information joining. We follow with preliminary experimental results and their analysis.

2 System Setup

Proposed system can be split into two main parts: *Client Module* (the interface) and the *Main Agent* (system manager). *Client Module* is responsible for interacting with the end-user. The *Main Agent* receives requests from the *Client Module* and manages agents for information retrieval and combining results.

At the beginning there is only the *Main Agent* (*MA*) waits for a query and the processing algorithm of choice. When the input is received the MA creates as

C. Badica et al. (Eds.): Intel. Distributed Comput., Systems & Appl., SCI 162, pp. 177–186, 2008.
springerlink.com

many *Search Agents* (*SA*) as there are selected search engines (user can specify how many and/or which search engines to include, or a default number will be used). Each *SA* is assigned a different search engine. SA's query the database using the content of the query and information about selected algorithm, to retrieve the weights set, which are used during data processing. Weights are the ranks of search engines; computed for a given algorithm based on previous results for a given query. Their belong to interval $(0,1)$ depending on how the algorithm evaluated result set of a particular search engine. If engine performed badly— results were not satisfactory in the sense of the algorithm; it is assigned a smaller weight than the engine results of which were considered as better ones (in the sense of the algorithm). If this is the first time for a given query, all ranks are set to 1. Those weights are used during the ranking processes to "boost" URLs originating from engines, which contributed better results in previous runs for that query and algorithm.

Next, *SA*s execute the query and return their results to the *MA*, which processes them according to the selected algorithm. When the processing is finished, the MA sends the final results to the web application to be displayed. Note that if the algorithm was able to find the best result, the result list is displayed and knowledge base is updated instantaneously. The search engine which yielded the top result is ranked as the best and other engines are ranked according to how close they were to this engine. If, however, algorithm was not able to yield a "satisfactory" answer; application displays "an answer" and an option to provide feedback (subjective evaluation performed by the user). Feedback (if received) ranks search engines. In the knowledge base we store—for each query, search engine and method of answer processing, and the engine rank.

2.1 Common Algorithms

There is a number of algorithms that are used by multiple approaches. First, the algorithm for the initial URL ranking. This initial ranking is performed before the *Game Theory* and *Auction* methods (not the *Consensus* method) start their computations. Its purpose is to calculate confidence values of *Search Agents* about retrieved URLs. In general, the confidence value is calculated as follows: $|result\ set\ of\ agent| - position\ of\ the\ URL\ in\ resulting\ set$. However, the *Game Theory* and the *Auction* methods require that each result set contains the same URLs (in any order). If this is not the case they break, since comparison of ranks certain URLs cannot be performed. Therefore, this algorithm updates the result sets with missing URLs. It also determines if the main computational parts of the two approaches can even be performed. The rule is as follows: if for all pairs of result sets A and B the $A \cap B = \varnothing$ then the main part of the *Game Theory* and *Auction* cannot start. Thus, if a result set has no common URL with any other result set it is removed from the process as being not suitable for the algorithms which require every URL to be in every result set. The pseudo-code of the algorithm is as follows.

URL ranking algorithm for *Game Theory* and *Auction* methods

Input Map of results $\langle a^i, r^i \rangle$ provided by

$$m$$

Search Agents—each in the form $r^i = \langle U_1^i, U_2^i, \ldots, U_n^i \rangle$, where U_j^i, $j = 1, \ldots, n$, are URLs.
Output Map containing URL ranking.
BEGIN
1. for each agent in the map:
   - check if other agents result sets contain any of the URLs of the agent
   - construct matrix representing how many URLs of the agent are contained in the each result set of other agents
2. check if each agent has at least one common URL with another if not—remove it from further processing
3. if result set of every agent is disjoint with each result set of every other agent—stop algorithm
4. for each agent in map:
   - for each URL in agent result set:
     - rank the URL as follows:

$$rank(U^i) = (|r| - i) \times weight(r)$$

     , where i is a position of URL in r
     - find agents which result set does not contain the URL, update their rankings:

$$rank(U^i) = 1.0 \times weight(r)$$

     (weights calculation—listing 2.1)
5. return ranking
END

Weights calculation for *Game Theory* and *Auction* methods

Weights calculation is performed after *Game Theory* and *Auction* methods finish their main negotiation parts. This algorithm is to rank the search engines according to how the URLs from a given engine were evaluated (placed) in the final answer. The topmost URL is chosen to be the feedback result and other result sets are weighted accordingly to the number of URLs overlapping with the result set which provided the URL.

Input: Result from feedback; initial result sets
Output: Map of weights with corresponding agents
BEGIN
1. find the agent whose result set contains the "best" result, set his weight to 1
2. for all other agents:

$$find\, d(r^{(i)}, r^w)$$
$$W[i] = \frac{|r^{(i)}| - d(r^{(i)}, r^w)}{|r^{(i)}|}$$

where $d(r^{(i)}, r^w)$ is the number of different URLs between the result set of agent i and the "winner" agent
3. return weights
END

After this part is finished, ranks are stored in the knowledge base for further use. These weights are used as follows: when issuing the query for the second time for a particular method (*Game theory* or *Auction* in this case) the weight of the result set is used to decrease the rank of the URL which originates from this result set. The rank of such URL is multiplied by this weight. Thus, if it is less than 1 it is being decreased. This process gives handicap to URLs which are returned by the search engines with low weights—those contributed "not so good" results for a particular query. If the weight is equal to zero the rank is multiplied by 0.01 (to still keep it in "in the game").

3   Three Main Algorithms

3.1   Game Theory

This approach was used in the *NeurAge* system [8, 7, 1]. In its modified version, instead of voting for certain "classes of data," agents vote for URLs retrieved by search engines. Confidence values from the original algorithm have been replaced by URL ranks (obtained after above-described the pre-processing). Furthermore, in the original algorithm agents delivered a single "data class" as the answer. However, in Internet searching multiple, ranked responses are expected. Therefore, in the adapted approach, 10 "best" URLs are returned. Here, we utilize an iterative approach, where each iteration starts the selection process from the beginning, without previously selected URLs. This modification does not violate main assumptions of the algorithm [9].In general, a "game" consists of set of players, set of moves (strategies) and specification of payoffs for each combination of moves. In a normal form the game that is defined as follows:

> There is a finite set P of players, which we label $\{1, 2, ..., m\}$. Each player k has finite number of pure strategies (moves) $S_k = \{1, 2, ..., n_k\}$ A pure strategy profile is an association of strategies to players, that is m-tuple $\sigma = (\sigma_1, \sigma_2, ..., \sigma_m)$ such that $\sigma_1 \in S_1$, $\sigma_2 \in S_2, ..., \sigma_m \in S_m$ Let strategy profiles be denoted by Σ A payoff function is a function $F : \Sigma \rightarrow \Re$ which intended representation is the award given to a single player as the outcome of the game. Accordingly to specify a game the payoff function has to be specified for each player in the player set $P = \{1, 2, ..., m\}$.

> **Definition 1.** A game in normal form is a structure (P, S, F) Where $P = \{1, 2, ..., m\}$ is a set of players, $S = (S_1, S_2, ..., S_m)$ is a mtuple of pure strategy sets, one for each player and $F = (F_1, F_2, ..., F_m)$ is a m-tuple of payoff.

In the game considered here, components are as follows: (a) players are agents, (b) possible moves are to change or to keep the URL, (c) payoffs for those moves are defined as a 2×2 matrix. Each agent is assigned two values: for the keeping the URL and for changing it. Those values may or may not change each round of the game, depending on the outcome of the previous round.

At the beginning of the process, results obtained by the *Manager Agent* from *Search Agents* are filtered, ranked and updated (see above). The URL ranking represents confidence in a specific URL. In each round of the game two agents with highest ranked URLs have two possibilities: to keep their answer or to change it. If the keep action has higher value than the change action, the agent will be assigned the action to keep its URL for the next round. If, however, the agent is assigned the action to change its URL and the second agent is assigned the action to keep its URL, the latter is considered a *winner* of the round and the former is considered a *loser*—it and its result set are discarded from further considerations. Then the next round starts (without the agent and its result set; removed in previous round) and so on, until there is only one agent and its top URL is the winner. Process is then repeated, without the URL that was selected in the previous "big" round (this URL is removed from all result sets; recall that all sets have all URL's included; see above). Game continues until 10 (best) URLs are selected.

3.2 Auction-Based Approach

Auction-based approach was originally used in the *NeurAge* system [7], and was adapted to return 10 distinct URLs (rather than a single result). In each round of the auction each agent has its "product" (URL) assigned. Afterward, the "cost" for each assigned URL is calculated. Costs are compared and agent with the highest cost is considered to be a loser. Afterward, the confidence values for selected URLs are updated by subtracting the cost from their value. Henceforth, the next round takes place. If the agent that was marked before as a loser, loses again, it and its result set are discarded from further auctions. After removal of a twice-looser, process enters the next round, and continues until a single agent remains with its selected answer. This process is repeated 10 times and after each round the URL that was just selected, is removed from result-sets of all agents.

Here, we present an example of flow of one round of the *Auction* method:

Input: Map containing URL rankings.
Output: 10 URLs.
BEGIN
1. repeat until there are 10 URLs in answer list
2. repeat until one agent remains
3. find highest ranked URLs for all agents and pair them $(A^{(i)}, U^{(i)})$
4. calculate costs for each agent:

$$cost(A^{(i)}) = \frac{\sum_{i=1, i \neq j}^{m} (rank(A^{(i)}, U^{(i)}) - rank(A^{(i)}, U^{(j)}))}{10}$$

where $U^{(i)}$ is URL from pair $(A^{(i)}, U^{(i)})$ (highest ranked URL for agent $A^{(i)}$ and $U^{(i)}$ is a highest ranked URL for agent $A^{(i)}$
5. find agent with highest cost—a loser; it may happen that all agents have the same costs—if it occurs twice the agent which is assigned the URL initially ranked as the lowest is considered a loser and thus removed from further negotiation, if it is so, go to 7.
6. if the agent is a loser twice in a row remove it from further auctions
7. update URL rankings for all agents as follows:

$$rank(A^{(i)}, U^{(i)}) = rank(A^{(i)}, U^{(i)}) - cost(A^{(i)})$$

where the pair $(A^{(i)}, U^{(i)})$ is found at the beginning; at this point the winning URL can be changed
8. add URL to the answer list
9. remove the URL from all answer sets

3.3   Consensus Method

The *Consensus* method was used previously in the AGWI system [5, 3, 6, 2, 4]. Its aim is to combine a set of answers into a final joint answer. The difference in the modified approach are as follows. The algorithm for measuring distances between result sets was adapted (to use the Levenshtein method). Furthermore, in the AGWI system there were more search engines than there were Search Agents (and thus only some of them were selected to be used). Here, there are as many Search Agents as there search engines.

In the *Consensus method*, result sets are evaluated and a combined result set (without repeating URLs) is created. Next, for each URL its average position in all result sets is calculated. In what follows, the combined result sets are sorted according to the average position of each URL. Then the consensus answer is found and its consistency checked. Before performing the calculation, however, the result sets and consensus are normalized; only a specific number of top URLs are incorporated into the answer. This number is of size of the smallest non-zero result set. To check consistency of the consensus answer, average of distances between result sets and average of distances between each result set and the consensus answer are evaluated. If the average of distances is bigger than average of distances of result sets and the consensus, then consensus answer is consistent; if not, the consensus answer is not consistent. The following listing presents the pseudo code of algorithm for finding the consensus answer.

> **Input:** Map of results provided by m Search Agents. Map containing weights for result sets.
> **Output:** *Consensus* answer.
> **BEGIN**
> 1. create set **URLS** from all URLs from all result sets (without repetitions)
> 2. for each $U \in$ **URLS**
>   - create array $<t_1, t_2, \ldots, t_n>$, where t_i is position on which U appears in $r^{(i)}$;
>   - if U does not appear in $r^{(i)}$ then set t_i as the length of the longest ranking increased by 1
>   - divide each t_i by $weight(r^{(i)})$; if $weight(r^{(i)}) = 0$ divide by 0.01
>   - calculate average $t(U)$ of values $(t_1, t_2, \ldots, t_n)$
> 3. consensus answer is obtained by ordering elements of according to values
> **END**

Having checked the consistency of the result set, the algorithm decides on the next step. When the consistency is low, the answer containing all results is returned and feedback is requested. If the consistency is high, 10 first URLs from the consensus answer are presented.

Depending on the outcome of the consistency check the different entry point is used for the weight calculation algorithm. If the consistency of the consensus is high, agent whose result set has the smallest distance to the consensus is selected as the agent whose weight will be equal to 1 and the algorithm in following listing does not require the feedback URL as an input—thus, step 1 is omitted. If the consistency is low, the first step of the algorithm must be performed to find the "winner" agent.

> **Input:** Result from feedback; initial result sets
> **Output:** Map of weights with corresponding agents
> **BEGIN**
> 1. find the agent whose result set contains the best result from feedback, set his weight to 1

2. for all other agents:

$$find\, d(r^{(i)}, C)$$

$$W[i] = \frac{|r^{(i)}| - d(r^{(i)}, C)}{|r^{(i)}}$$

where $d(r^{(i)}, C)$ is the the Levenshtein distance

3. return weights

END

4 Initial Experimental Results

In our initial set of experiments three queries were issued for the testing purposes: "consensus decision making", "consensus decision making for conflict solving", and "is consensus decision making for conflict solving good enough or maybe Game theory or auction is better".

The idea was to take three queries which are related to the same topic; however first was to be simple, second intermediately complex, and third was to be very complex, while retaining coherence. There were 5 search engines queried. Four of them were English-language-based: *Google, Ask.com, Live, Yahoo!* and one of Polish origin—*Interia*, which in fact is a Google based engine; however very often it produces results which differ from its parent engine. System was set-up to return 20 results for each query. In this note, due to the lack of space, in Table 1 we present only two "performance measures;" the *Set Coverage* and the and *URL to URL coverage* for each of the three approaches, for each of the queries. The Set Coverage measures how many URLs from the final result are contained in the result set returned by a search engine regardless of the position of the URL. In other words, this measure tells us if there is a relationship between the combined answer and answers returned separately by each search method. The and URL to URL coverage measures how many URLs were at the same position in both results (of the algorithm and that of the search engine).

Let us observe that as the query becomes more complex, the coverage drastically decreases. This can be explained by the fact that the responses generated by various search engines have less and less in common. Therefore, regardless of the method used, the final answer set becomes a collection of "separate links" chosen from each individual answer-set. This trend is even more drastic in the URL to URL comparison. Here, already for the intermediate query practically no URL is in the same location in the answer set as it is in any of the search engines. This indicates also that this performance measure is not particularly useful for the application in question.

As expected, results returned by *Interia* and *Google* are very similar, with both performance measures varying, randomly favoring either one of them. Interestingly, these two search engines seem to have best performance for the complex query. However, this may be a result of collusion, where two similar search engines "dominate" views of the others. This observation provides also a warning, that the selection of the "groups of experts" has to provide as "orthogonal" view of the subject as possible. Otherwise, regardless of the method used,

Table 1. Summary of experimental results

Auction method, simple query					
Auction	Ask.com	Live	Interia	Yahoo	Google
Set Coverage	60%	40%	110%	60%	70%
URL to URL	0%	10%	20%	30%	20%
Game theory method, simple query					
Game theory	Ask.com	Live	Interia	Yahoo	Google
Set Coverage	60%	60%	70%	80%	60%
URL to URL	30%%	10%	0%	50%	0%
Consensus method, simple query					
Consensus	Ask.com	Live	Interia	Yahoo	Google
Set Coverage	70%	50%	80%	70%	80%
URL to URL	20%	20%	10%	20%	10%
Auction method, intermediate query					
Auction	Ask.com	Live	Interia	Yahoo	Google
Set Coverage	10%	10%	50%	10%	40%
URL to URL	0%	10%	0%	0%	0%
Game theory method, intermediate query					
Game theory	Ask.com	Live	Interia	Yahoo	Google
Set Coverage	60%	40%	30%	40%	30%
URL to URL	40%	0%	10%	0%	10%
Consensus method, intermediate query					
Consensus	Ask.com	Live	Interia	Yahoo	Google
Set Coverage	50%	30%	70%	40%	60%
URL to URL	0%	20%	0%	0%	0%
Auction method, very complex query					
Auction	Ask.com	Live	Interia	Yahoo	Google
Set Coverage	0%	0%	30%	0%	40%
URL to URL	0%	0%	10%	0%	20%
Game theory method, very complex query					
Game theory	Ask.com	Live	Interia	Yahoo	Google
Set Coverage	0%	40%	30%	10%	50%
URL to URL	0%	0%	0%	10%	0%
Consensus method, very complex query					
Consensus	Ask.com	Live	Interia	Yahoo	Google
Set Coverage	10%	20%	90%	20%	40%
URL to URL	0%	0%	30%	0%	0%

the returned combined answer may be dominated by a few experts that see the problem similarly.

Observed results suggest also that the consensus method does what its name suggests—delivers response that is closest to consensus. This can be seen particularly in the case of the complex query, where for the consensus method the Set Coverage is non-zero also for search engines other than *Interia* and *Google*.

Overall, on the basis of all of our experiments (also these not reported here), we can state that: (1) Results delivered by the *Auction method* are highly dependent on each individual result set and do not represent well the "combined view" of all search engines. No matter if the URL is in many result sets, it may not make it to the final (combined) result. Instead, returned are "winning" URLs, which appear in a single result sets. (2) *Game Theory* method also does not seem to create a combined view of initial answers. However, if a URL is at of of top places of more than one result set, it is very likely to be incorporated into the final result set (even though it may be locate much lower than its average position). (3) The *Consensus method* returned the results which represent the most common view of participating search engines. However in three tested cases all returned result sets were inconsistent(!) according to consensus theory itself. This happens due to the high "position dispersion" of URL's throughout the result sets. There are situations where a URL is, for instance, on the 1st place in one result set, on the 9th place in another result set, and on the 5th in the next. For this result, the Levenshtein distance between response sets is relatively large and thus the final result set is inconsistent. Nevertheless, if one was not to take the consistency into account (as in its current form it may not be a useful measure after all), the *Consensus method* provided results which could be claimed to be "the best overall."

5   Concluding Remarks

In this note we discussed three methods for combining results from multiple Internet sources and presented initial evaluation of their performance. Our results indicate that each method leads to a different combined answer set. Out of these methods, the *Consensus method* seems to generate the most "common" view of the initial answers, while the remaining two methods tend to favor certain answers over others. This is particularly the case for the *Auction theory*. We are currently performing additional experiments with the three methods and starting to look more qualitatively into obtained answer sets (to establish their value for actual users).

Acknowledgment

Work was supported from the "Funds for Science" of the Polish Ministry for Science and Higher Education for years 2008-2011, as a research project.

References

1. Canuto, A.M.P., Abreu, M.: Analyzing the benefits of using a fuzzy neuro model in the accuracy of the neurage system: an agent-based system for classification tasks. In: Proceedings of the International Joint Conference on Neural Networks, pp. 2951–2958 (2006)
2. Nguyen, N.: Consensus system for solving conflicts in distributed systems. Journal of Information Sciences 147, 91–122 (2002)
3. Nguyen, N.: Processing inconsistency of knowledge at semantic level. Journal of Universal Computer Science 11(2), 285–302 (2005)
4. Nguyen, N.: Methods for achieving susceptibility to consensus for conflict profiles. Journal of Intelligent and Fuzzy Systems: Applications in Engineering and Technology 17(3), 219–229 (2006)
5. Paprzycki, M., Nguyen, N.T., Ganzha, M.: A Consensus-Based Multi-agent Approach for Information Retrieval in Internet. In: Alexandrov, V.N., van Albada, G.D., Sloot, P.M.A., Dongarra, J. (eds.) ICCS 2006. LNCS, vol. 3993, pp. 208–215. Springer, Heidelberg (2006)
6. Nguyen, N., Małowiecki, M.: Consistency measures for conflict profiles. In: GI 1973. LNCS, vol. 1, pp. 169–186. Springer, Heidelberg (2004)
7. Santana, L., Canuto, A., Abreu, M.: Analyzing the performance of an agent-based neural system for classification tasks using data distribution among the agents. In: Proceedings of the International Joint Conference on Neural Networks, pp. 2951–2958 (2006)
8. Santana, L., Canuto, A., Junior Xavier, J., Campos, A.: A comparative analysis of data distribution methods in an agent based neural system for classification tasks. In: Proceedings of the Sixth International Conference on Hybrid Intelligent Systems, vol. 9 (2006)
9. Szymanska, E.: Personal communication

Formal Modeling and Verification of Real-Time Multi-Agent Systems: The REMM Framework*

Francesco Moscato[1], Salvatore Venticinque[2], Rocco Aversa[2], and Beniamino Di Martino[2]

[1] Dep. of Computer Science and Systems, Univ. of Naples Federico II
    francesco.moscato@unina.it
[2] Dep. of Computer Science Engineering, Second Univ. of Naples
    {rocco.aversa,salvatore.venticinque,beniamino.dimartino}@unina2.it

Summary. Multi Agent Systems represent a new approach for modeling complex and distributed systems. Many efforts of software engineering aim at providing methodologies and tools for designing and developing MAS. However formal verification of MAS dependability is still an open issue. Here we focus on modeling, design and verification of real-time properties in MASs. We propose a methodology that supports developers in different phases of MAS developing cycle. We also present an integrated environment that allows for UML design, code generation, time constraints verification and testing of soft-real time MASs. A case of study is described to demonstrate an application of such methodology and the utilization of developed tools.

1 Motivation

During last years new challenges in software engineering arose because of the spread of distributed and reliable systems. Due to the complexity and the distribution of novel architectures, solutions have been built by autonomous and proactive components. *Multi-agent system*(MAS) represent a model for designing and developing these systems [1, 2]. Several methodologies have been proposed for MAS design and development [3, 4]. However software engineering has not provided yet any approaches to model and verify dependability of such kinds of systems in design and testing phases. A set of agents,which execute under real-time constraints [5], can be modeled as real-time MAS. For such systems, it is critical to verify, since design phase, if the provided solution will be able to satisfy certain time constraints. In our model agents have to achieve some goals within certain deadlines. If at run-time unpredictable events happen, and timing constraints cannot be satisfied by current strategy, a reconfiguration is needed. New plans must be evaluated to meet again deadlines, when it is possible. Unachievable goals should be withdrawn to reduce the system workload.

In such a context, verification and validation of MAS properties and on-line reconfiguration are the open challenges we are addressing.

* This work has been supported by LC3 - LABORATORIO PUBBLICO-PRIVATO DI RICERCA SUL TEMA DELLA COMUNICAZIONE DELLE CONOSCENZE CULTURALI - National Project of Ministry of Research MIUR DM17917.

C. Badica et al. (Eds.): Intel. Distributed Comput., Systems & Appl., SCI 162, pp. 187–196, 2008.
springerlink.com

Here we present a methodology and an integrated framework, which support design and development of MAS, where the correct achievement of system goals depends by soft real-time requirements. The methodology will be applied to a case study where it is necessary to develop intelligent bots for the Unreal Tournament game.

Second section introduces the architecture of a real-time MAS and present an overview of our framework. Section 3 describes a language for modeling real-time BDI agents and their interactions. In section 4 the proposed methodology is presented in detail. In section 5 we provide an example of application of our methodology to a relevant case study. Finally conclusions are due.

2   An Overview of the REMM Framework

In the REMM framework, agents are modeled according to the BDI logic: they are characterized by beliefs, desires and intentions. Agents can collaborate or compete to achieve their goals. Real-time requirements to be satisfied in the Execution Environment must be transported in the MAS by granting that agents goals are achieved within a certain deadline. It means that to pursue a goal:

1. For each necessary goal must exist at least a plan that can complete within the deadline.
2. Agents must schedule and execute only those plans which can complete within a maximum amount of time.

A Verifier checks that a goal can be achieved, by a plan, within a certain deadline. During the design phase, it grants the correctness of the project. At run time it performs an on-line checking.

A planner composes and schedules plans of an agent in order to grant at each moment the reachability of pursued goals.

To develop such a system we must define: (1) *A real-time agent formalism.* We called Real-Time AML (RT-AML) our extension of AML[13],that is a set of UML Stereotypes to model BDI agents; (2) *A tool for real-time agent modeling.*; (3) *A tool for verification of real-time properties of a MAS.*

3   Real-Time Agents Modeling Language: RT-AML

RT-AML extends AML [13], a language for describing MAS by using an UML-based approach. AML is a semi-formal language used for MAS description. The original support for describing time is poor: it only allows for definition of duration of messages passing during the interactions of several agents. It is defined by using ad-hoc stereotypes in UML 2.0. In particular, it can be, on its own, easily extended by using proper UML stereotypes and constructs. Two views of AML systems will be used in this work: the *Architecture* and the *Mental* view. The main elements of these views are *Entities* and *Agents*. The former element defines a meta-classes hierarchy for defining the main MAS components, while the latter one contains all the meta-classes used to define a single agent. In Mental view, several meta-classes are used to define the BDI logic which describes agents behavior. It is possible to define *beliefs* of an agent, the *goals* it wants to reach and the available *plans* to reach the goals.

RT-AML mainly adds the following UML stereotypes definitions to the original AML profile:

Table 1. Main RT-AML stereotypes

Name	Stereopype	AML Base Class	Description
Real-Time Agent Type	$<< agentRT >>$	UMLClass, UMLClassifierRole, UMLObject	Agent Type RT constraints
Real-Time Belief	$<< beliefRT >>$	UMLClass, UMLObject	Belief with RT constraints
Real-Time Decidable Goal	$<< dgoalRT >>$	UMLClass, UMLObject UMLObject	Decidable Goal with RT constraints

4   The Design and Verification Methodology

The methodology we present for MAS design and development is divided into four phases: *Modeling; Validation* ; *Translation; Run-time.* In the first phase a MAS system is defined by using the RT-AML language; in the second phase a RT-AML model is translated into a timed automata in order to perform verification tasks; in the third phase the RT-AML model is translated into stubs for an agents platform (JADEX [6]), which will be specialized by programmers. In the last phase the constraints defined in the MAS modeling phase are monitored at run-time in order to evaluate if design specifications are verified during real system execution. In the following the four phases will be described in detail.

4.1   Modeling Phase

A number of RT-AML static diagrams used to define agents beliefs, states and stimula are used to model agents. *(RT-)AML Activity Diagram* and *(RT-)AML Communicative Sequence Diagram* are used here to model interactions among agents. These two diagrams represent a dynamic view of the system and are critical for definition of a real-time behavior. We will focus in the following on the definition of this view since the statical one is similar to class diagram of any object oriented applications.

For each agent behavior, in RT-AML Activity Diagrams, a sequences of actions (called *Action States*), which leads from an Initial state to a Final State, must be defined. All states are characterized by real-time properties. Stimula (internal or external) can be defined by edges connecting states. Stimula are also characterized by real-time properties and can be generated by the agent itself (e.g. acting on other agents), or they are sensed by the agent in a given state to enable a (timed) state transition. All agents start from an Initial State, characterized by specific beliefs, and aim at achieving a Final State that is associated to a RT-Goal. In such a way plans are defined for agents, which pursue a

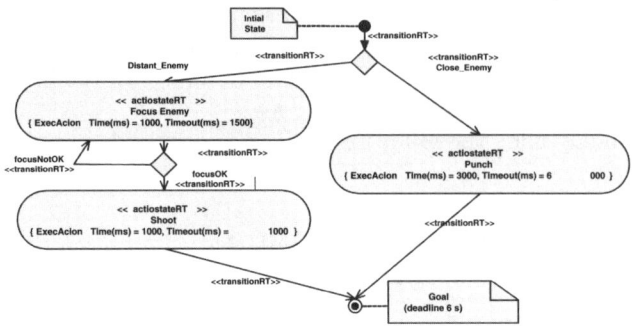

Fig. 1. Agent Plan with decision model

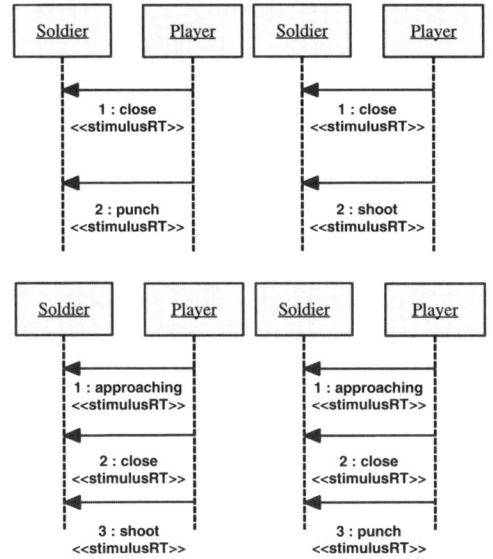

Fig. 2. Communicative Sequence Diagram

defined goal when a given set of beliefs are true. Depending on external stimula and beliefs, an agent can usually take different choices to achieve a goal. These choices are modeled by choice elements (rhombuses) in the RT-AML Activity Diagram as shown in Fig. 1.

Once the agent behaviors are defined by the RT-AML Activity Diagrams, interactions among agents have to be defined. This is done by defining RT-AML Communicative Sequence Diagrams, which describe sequences of stimula exchanged among agents. Stimula exchanged with other agents, or Self Stimula (sent from an agent to itself), are characterized by real-time properties and can be represented by $<< StimulusRT >>$ meta-class. An example of such diagrams is reported in Fig.2

4.2   Validation Phase

In this phase the RT-AML model is translated into a formalism on which model checking techniques [7] can be applied. The language chosen for the validation and verification phase is the timed automaton formalism [8, 9]. Basically, timed automata are state automata where time is defined explicitly. It allows to evaluate how much time it is possible to remain in a given state. Furthermore it is possible to define if a state transition generates an external event or if an event allows for transitions enactment. Due to lack of space, we will describe briefly how translation works and how timed automata are built from RT-AML diagrams introduced in the previous section.

All Action States in the RT-AML Activity Diagrams are then translated in States of the Timed Automata model. Each diagram has its own timed automaton, while their product automaton is representative of the whole system. State invariants and transitions guards are defined by taking into account timing properties from RT-AML model. RT-AML Sequence Diagram is used to define timed automata synchronization events.

For example part of timed automata generated by these translation rules looks like in Fig.3 The verification problem is then translated into a reachability problem: a MAS can reach a goal if the Final State representing a goal is translated into a reachable state into the timed automata model. This property can be expressed into TCTL logics [10]. Obviously all properties, which can be expressed in TCTL logics can be checked on timed automata model.

4.3   Translation Phase

In this phase the (validated) RT-AML model is translated into a series of stubs, which implement agents interfaces and their behaviors. In our methodology stubs for JADEX [6] are produced. A deep description of this phase is omitted for brevity's sake but the JADEX stubs are generated similarly to Java code from UML diagrams.

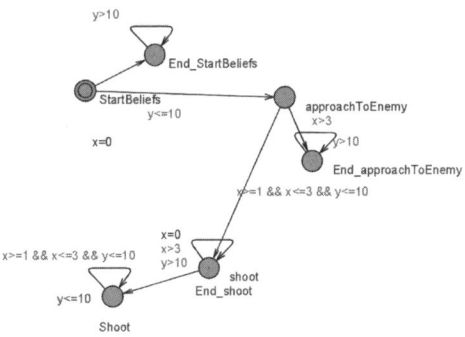

Fig. 3. Specification of a Timed Automata

4.4   Run-Time Phase

At run-time, due to unpredictable events, real-time specifications of a MAS could not be satisfied as it was expected. This may lead to situations where real-time goals cannot be reached within specified deadlines or cannot be reached at all. It is then necessary to monitor run-time execution to manage such situations. JADEX Monitor and Scheduler classes implemented in the previous phase are used at run-time to identify such events. Usually, it is possible to have three type of timing behaviors for activities in Action States (see Fig.4):

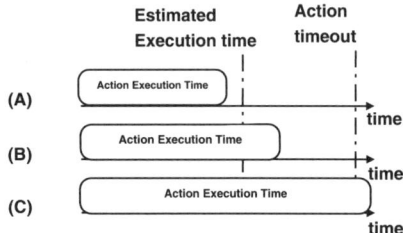

Fig. 4. Timelines

Each Action State is associated to an expected execution time and to a time-out. If a real execution of an Action State exceeds the estimated time and even its timeout, the plan state reachability set could change. In order to check continuosly the reachability of the current configuration, the original RT-AML model is modified with the new timing behavior.

Check results could inform that: goal is still reachable; goal is still reachable but not through all execution path (i.e. some execution path cause a deadline expiration); goal is never reachable. In the first case agents can continue their execution without any reconfiguration; in the second case agents may choose the current plan to reach their goal; in the last case agents have to change their plans because their goal are definitely not reachable.

4.5   The REMM Framework

In order to enact the described phases, the REMM framework has been developed. Its architecture is depicted in Fig.5. Modeling and Design Tool enacts the Modeling phase and it is implemented by using StarUML [12], in which the RT-AML metaclasses profile has been defined. The Translation tool enacts both second and third phase of the proposed methodology. It translates both RT-AML model into timed automata and into JADEX stubs. The verification tool is used for validation and verification phase. Agents Code produced by the translation tool is interfaced with a run-time environment for validation and verification also during execution as it has been explained before.

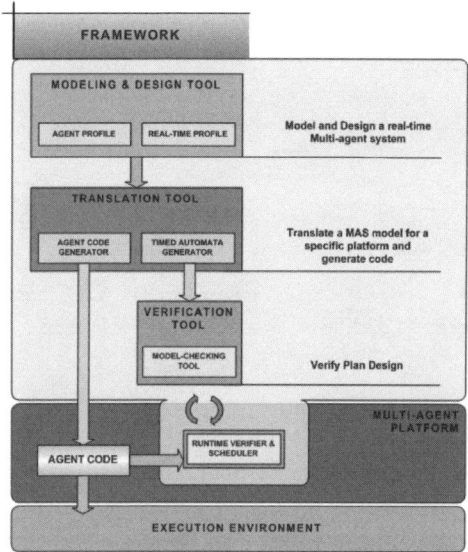

Fig. 5. REMM Architecture

5 The GameBot Example

As proof of concept we desiged and developed intelligent agents, called bots, for the GameBots [14] framework. Agents developed in GameBots are then executed into the Unreal Tournament game to play against human players. They must be provided with reactivity, proactivity and autonomy and must satisfy real-time requirements while interact with humans players and with the execution environment. Two bots are designed in this example. *AttackBot* aims at searching and shooting human players, *SupportBot* is searching for other bots for supporting purposes. A snapshoot from StarUML, of a Static view of the RT-AML model of the two bots is showed in Fig.5. Blue shapes represent agents; double circles correspond to goals, and pentagons represent plans. The two bots specialize a common agent class called GameBot that includes all GameBot interfaces. The main goal of each bot is *FindEnemy*. AttackBot pursue also other two goals: *Shoot*ing enemies (human players) and to *getCured* when its energy is low. SupportBot has aims at achieving *FindFriend* and *Heal* goals. Each Bot pursue these goals while they are *exploring* the arena to find out enemies or the other bot. When the enemy is found, the AttackBot tries to shoot him by using the *shooting* plan. It has to move properly in order to follow the enemy and kill him without wasting ammos. Each movement is an action to execute under real-time constraints and the whole shooting plan is also a real-time plan. If the bot is wounded by enemies, it calls for help (*Call-for-help*). And the other bot tries to reach it in time. Some activity diagrams, which have been showed in previous sections, are part of the complete example that is not described in detail here for lack of space.

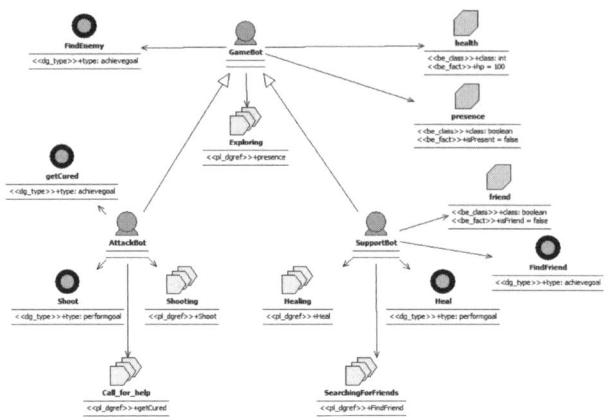

Fig. 6. Bots model

Fig. 7. Game Screenshot

Bots use run-time validation when their deadline expires, choosing other plan if goals become unreachable (for example if a bot was shooting a player that runs away, its goal of shooting becomes unreachable and exchanges its current plan and goal with SearchEnemy ones). In Fig.5 a screenshot where AttackBot tries to achieve the Shoot goal is showed.

6   Summary and Conclusion

We described in this paper a methodology we defined to support modeling and design of real time Multi Agent Systems. We developed a framework for development, verification and execution of multi agents applications. The pillars of our methodology are: a Real Time formalism for MAS modeling; a translation tool for generation of timed automata and of agent code; a run-time environment for code execution and on-line verification. Real Time formalism allow to introduce time constraints into BDI architecture and interactions of MASs. Timed Automata are used to verify if agents goals could be achieved at design phase. Agent code is automatically generated from Real Time Agent Language to be executed into a JADEX platform. An agent monitor is embedded in the code for on-line checking of goals reachability, when, at run-time, estimated parameters of model are updated with timing measured during execution. A case of study is presented. Characters implemented by agents, which play against a human user, in an interactive environment, under soft real-time constraints.

References

1. d'Inverno, M., Luck, M., Georgeff, M., Kinny, D., Wooldridge, M.: The dMARS Architechure: A Specification of the Distributed Multi-Agent Reasoning System. Journal of Autonomous Agents and Multi-Agent Systems 9(1-2), 5–53 (2004)
2. Himoff, J., Skobelev, P., Wooldridge, M.: MAGENTA Technology: Multi-Agent Systems for Industrial Logistics. In: Proceedings of the AAMAS 2005, Industry Track, Utrecht, The Netherlands (2005)
3. Kavi, K., Aborizka, M., Kung, D.: A framework for designing, modeling and analyzing agent based software systems. In: Proc. of 5th International Conference on Algorithms & Architectures for Parallel Processing, Beijing, China, October 23-25 (2002)
4. Silva, V., Lucena, C.: From a Conceptual Framework for Agents and Objects to a Multi-Agent System Modeling Language. Journal of Autonomous Agents and Multi-Agent Systems 9(1-2), 145–189 (2004)
5. Zhang, L.: Development Method for Multi-Agent Real-Time Systems. International Journal of Information Technology 12(5) (2006)
6. Pokahr, A., Braubach, L., Lamersdorf, W.: Jadex: A BDI Reasoning Engine. In: Bordini, R., Dastani, M., Dix, J., Seghrouchni, A. (eds.) Multi-Agent Programming. Kluwer, Dordrecht (2005)
7. Wooldridge, M., Huget, M.-P., Fisher, M., Parsons, S.: Model Checking Multi-Agent Systems: The MABLE Language and Its Applications. International Journal on Artificial Intelligence Tools 15(2), 195–225 (2006)
8. Wooldridge, M.: An Automata-theoretic approach to multiagent planning. In: Proceedings of the First European Workshop on Multiagent Systems (EUMAS 2003). Oxford University, Oxford (2003)
9. Bordini, R.H., Fisher, M., Visser, W., Wooldridge, M.: Verifiable multi-agent programs. In: Dastani, M., Dix, J., El Fallah-Seghrouchni, A. (eds.) PROMAS 2003. LNCS (LNAI), vol. 3067, pp. 72–89. Springer, Heidelberg (2004)
10. Behrmann, G., David, A., Larsen, K.G.: A Tutorial on Uppaal. In: Department of Computer Science. Aalborg University, Denmark (2004)
11. Bellifemine, F., Caire, G., Poggi, A., Rimassa, G.: JADE - A white paper. EXP in search of innovation - Special Issue on JADE TILAB Journal (2003)
12. Wong, S.: StarUML Tutorial [Connexions Web site] (September 10, 2007), http://cnx.org/content/m15092/1.1/

13. Trencansky, I., Cervenka, R., Greenwood, D.: Applying a UML-based agent modeling language to the autonomic computing domain. In: Companion to the 21st ACM SIGPLAN Conference on Object-Oriented Programming Systems, Languages, and Applications OOPSLA 2006, Portland, Oregon, USA, October 22 - 26, pp. 521–529. ACM, New York (2006)
14. Kaminka, G.A., Veloso, M., Schaffer, S., Sollitto, C., Adobbati, R., Marshal, A.N., Scholer, A.S., Tejada, S.: GameBots: the ever-challenging multi-agent research test-bed. Communications of the ACM (January 2002)

Email Archiving and Discovery as a Service

Frank Wagner[1], Kathleen Krebs[2], Cataldo Mega[3], Bernhard Mitschang[1],
and Norbert Ritter[2]

[1] University of Stuttgart, Universitätsstraße 38, D-70569 Stuttgart
   {fkwagner,mitsch}@ipvs.uni-stuttgart.de
[2] University of Hamburg, Vogt-Koelln-Straße 30, D-22527 Hamburg
   {kkrebs,ritter}@informatik.uni-hamburg.de
[3] IBM Deutschland Entwicklung GmbH, Schönaicher Straße 220, D-71032 Böblingen
   cataldo_mega@de.ibm.com

Summary. Corporate governance and legislative regulations are forcing companies to extend their IT infrastructure by Email Archive and Discovery (EAD) systems for compliance reasons. Praxis shows that every installation is different from another; not only in terms of the execution infrastructure, but also in terms of e.g. document and archiving procedures that map a company's own business rules. As a consequence, EAD systems have to be highly customizable to their intended usages.

For this purpose, we propose a service-oriented approach at various levels of detail that, on one hand, allows for describing EAD properties at the abstract (service) level and, on the other hand, supports the appropriate mapping of these services to the existing execution infrastructure. In this paper, we focus on the development and (architectural) design of an EAD system, which is well suited to fulfill these requirements. On the long run, we consider this solution as an important step on the way to an effective distributed and scalable approach, which, as we think, can be achieved by appropriate mechanisms of automatic workload management and dynamic provisioning of EAD services based on e.g. grid technology.

1 Introduction

Large enterprises use software to capture, manage, store, preserve and deliver content since more than 20 years. Content in this case does not only mean structured or unstructured information. The term stands for any kind of electronic artifacts: records, data, metadata, documents, websites, etc. that are related to organizational processes. This variety of heterogeneous content types involves various application areas of Enterprise Content Management (ECM). Nowadays the role of ECM is changing considerably, since more and more new requirements have to be fulfilled by content management systems (CMS). On one hand, all digital assets (content) of an enterprise need to be managed appropriately and in a comprehensive way (functional requirements). On the other hand, these huge amounts of content data need to be processed very efficiently. Besides high performance and scalability, cost effectiveness (of CMS software and hardware) is a further non-functional key requirement. By cost effectiveness we mean both reduction in storage and processing cost of steadily growing content collections as well as the administration of the underlying IT infrastructure. The latter is especially relevant, since there is a growing market for ECM solutions in the area of small and medium enterprises.

C. Badica et al. (Eds.): Intel. Distributed Comput., Systems & Appl., SCI 162, pp. 197–206, 2008.
springerlink.com   © Springer-Verlag Berlin Heidelberg 2008

Email archiving and discovery (EAD) for compliance is a more and more important facet of ECM, obviously showing the above mentioned non-functional requirements. In our research we consider EAD as a representative area in ECM, which can be used to analyze key characteristics of ECM solutions.

Several factors are driving the need for EAD:

Mailbox limitations: The size of individuals' mailboxes are almost always restricted for two reasons: 1) calculable storage costs and 2) performance of the email server. Therefore emails have to be deleted from the email servers and archived elsewhere.

Knowledge stored in the emails: Email is playing a more and more important role for the communication within and between enterprises. Obviously, there is a huge amount of value (knowledge) contained in email systems worth of being discovered.

Legislative requirements: Triggered by large financial scandals, like those about the US enterprises Enron and WorldCom (now MCI), laws are enforcing regulatory compliance when archiving emails [6, 7]. Additionally, existing laws and regulations demanding safekeeping of documents are nowadays also applied to digital documents and email communications.

Court orders: More and more companies are forced by litigation to (re)produce old emails. Some large companies are even receiving several of these requests per day. Without proper procedures in place, this is a very time-consuming, expensive and error-prone task.

Obviously EAD is a very important issue for today's enterprises. One major challenge regarding the development of an appropriate EAD system is to support the required email processing, archiving and search functions in an efficient and scalable way. These non-functional requirements are hard to fulfill facing the huge amount of data to be managed and the high numbers of emails the EAD system has to continuously cope with. In this paper, we focus on the development and (architectural) design of an EAD system, which is well suited to fulfill these requirements. We further provide a distributed architecture and corresponding empirical evaluations proving that our approach meets current performance and scalability requirements. On the long run, we consider this solution as an important step on the way to an even more effective approach, which, as we think, can be achieved by appropriate mechanisms of automatic workload management and dynamic provisioning of EAD services based on e.g. grid technology.

The remainder of the paper is organized as follows. Section 2 gives an overview of related work. Section 3 describes major use cases and requirements for an EAD system. In section 4 we present our approach to a service-based EAD and evaluate it in section 5. Finally, section 6 concludes the paper and gives an outlook.

2 Related Work

As email archiving is a very hot topic for many enterprises, there are many solutions available on the market [1]. As the publicly available information is often limited to marketing brochures, we only mention two solutions. The solution IBM and its former business partner iLumin [9, 10] once offered used a central relational database to hold the catalog. In large settings this database is a bottleneck, especially as it is also responsible for the analysis of the emails. In contrast, Symantec's solution [5] uses multiple search engines for the catalog, but they are set up in a static way.

Some research projects also deal with email archiving. The "Texas Email Repository Model" [2] mostly deals with how to operate a statewide email archive and with long-term preservation. A project within IBM research [4] considered a very stringent fraud model. They make sure that nobody in the enterprise, no matter how many privileges he has, can manipulate the data or metadata of a record after it was archived. In contrast, we are focusing more on performance and assume the environment being sufficiently secured.

The most common approach adopted in ECM systems w.r.t. scalability was scale-up on the basis of large multi-processor systems. Measurements [3, 8] performed on cluster and grid systems indicate that scale-out might be more cost effective and affordable if administration and maintenance overhead can be kept to a minimum.

3   Email Archiving and Discovery

Figure 1 shows the major components of an email system. An important component is the email server, which is managing the mailboxes of its users. It reliably stores the mails in the inbox and in folders. For emails coming from the Internet into the system, it is nowadays common to scan them for spam and viruses. Therefore email servers provide interfaces to plug-in filters. Although it is not always allowed to just drop such emails, the further processing of the email can be different.

The components we are most interested in are the archive, where mails are stored for a longer period of time according to corporate-specific policies, and the discovery service, that allows searching for information in the archive.

The archiving system has four main tasks: ingest emails, full-text index, archive, and manage them for search and retrieval. To fulfill these tasks the archive has crawlers which identify qualifying emails by interpreting predefined archive rules, then retrieve

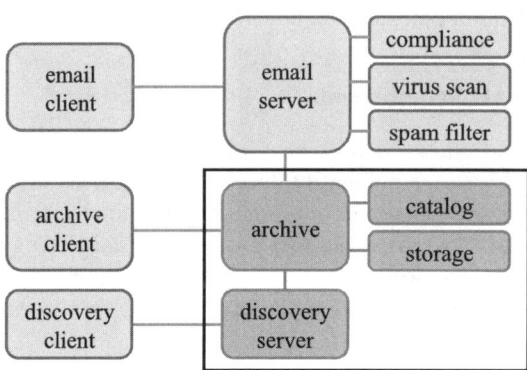

Fig. 1. Major email system components

them from the email server and pass them to parsers that extract the plain text from all kinds of document formats. Further, the email messages and attachments are full text indexed and stored in the catalog for later search and retrieval.

3.1 Use Cases

In this section we will describe the primary EAD use cases. Of course there are many more use cases. However, we want to focus on those relevant for the message of this paper, especially ingestion of the inbound message flow from the various email sources into the archive.

Use Case 1 - Ingestion of emails: There are two ways for an email to enter the system. a) A crawler regularly scans all mailboxes and archives all emails that meet specific rules. Archived emails are either completely removed from the mailbox, or are replaced by small stubs with references to the archived versions. b) The archive system captures all emails processed by the email server. For this purpose, IBM® Lotus® Domino® for example provides journaling mailboxes.

Use Case 2 - Search and retrieve: A regular user tries to retrieve one of his archived emails. He possibly remembers the sender/receiver of the email or some words/parts of the subject or body of the email. After matching emails have been found in the archive, the user may want to either directly look at some of them or re-import them into his mailbox.

Use Case 3 - Compliance and Discovery: A user equipped with higher privileges searches inside all the emails or a large subset of them, e.g. to check that internal compliance rules are met, or to find emails that might be relevant to a court case. Therefore he might search for all emails send and/or received by a given set of users within a certain period of time and containing a specific company name. After examining matching emails he may want to put a legal hold onto these emails in order to assure that they will not be deleted, and to export these emails to be further examined by e.g. lawyers.

Use Case 4 - Remove emails from the archive: The last use case covers the removal of emails. The archive therefore regularly deletes all the emails that have reached the end of their retention time, and are not on a legal hold.

3.2 Functional EAD Requirements

In contrast to 'usual' Content Management it is a special property of EAD that emails are never updated. Given the amount of data to be considered, most emails will very likely never be retrieved. They are only ingested and finally disposed at the end of their retention period. Therefore, the major concern of an EAD system design must be the optimization of the ingest process. Classification and filtering of emails as well as persistence mechanisms with different properties, e.g. for records management, are further required functions.

3.3   Non-functional EAD Requirements

Important non-functional requirements for an EAD system are reliability, authenticity and security. These properties are mandatory for an archive in a real-world environment. Without, the results of the archive are not usable in legal cases.

A further requirement is high performance, and especially high throughput. The EAD system has to support the email load expected for large enterprises.

Finally the system has to be affordable and cost-effective, especially in the small and medium businesses segment. Being able dynamically adapt to the current situation by acquiring additional resources and releasing unused resources is worthwhile.

4   Design and Engineering Issues

Concrete requirements largely vary from industry to industry and from company to company. We have chosen a service-oriented approach for various reasons: (1) customizability; (2) high performance and scalability through dynamic provisioning and corresponding resource management and (3) developing an initial concept for providing EAD as a Service.

As indicated in Figure 2, the system is separated into several hardware and software components. We use service-orientation to encapsulate all components and make them being manageable resources. This is necessary to allow dynamic provisioning and management of all EAD application components.

The application logic layer contains the EAD services meeting the functional requirements: ingest of new emails (ingest service), and making the data accessible for search and retrieval (content service). These services can be instantiated on the infrastructure as needed and the number of their incarnations can be adapted to the current load. The services at the virtualization layer coordinate the services in the application logic layer and integrate their results. The ingest and content services as well as the catalog component of our service-oriented EAD approach are described in more detail in the next sub-sections.

Our approach allows the configuration of the system at three levels: the process, the services and the application. Customization at the process level comprehends

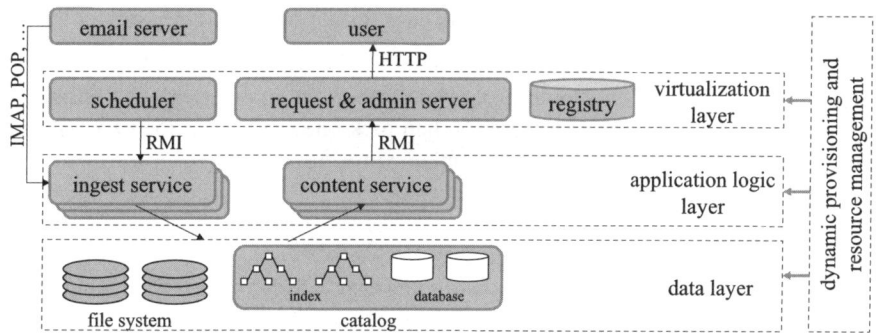

Fig. 2. Architecture of the service-oriented EAD approach

skipping individual handlers or replacing one handler by another. For example classification might not be necessary in all situations. Customization at the service level means setting service parameters. Examples are the hashing-algorithm that should be used to generate the IDs, or the merge factor used by Lucene. EAD applications can be provided following the software as a service idea. The customers only pay per usage according to agreed SLAs.

4.1   Ingestion Service

An EAD system must monitor emails entering, leaving and exchanged within an enterprise. The emails are captured, parsed, interpreted, and, if found relevant, archived. Typically such a system is composed of many components as indicated in Figure 1. One observation from existing EAD solutions is that data is often copied between different components of the system, and that some steps are performed similarly by multiple components. An example for such a step is text extraction. This has to be done for a content-based filtering to evaluate compliance rules, and for full-text indexing of emails. Therefore one central point of our approach is to reduce unnecessary copying and duplicate work in order to increase the throughput of the system.

To start a new ingest job an ingest service gets the location of the mailbox to crawl, the necessary credentials and additional configuration parameters from the scheduler. It then connects to the email server and performs several steps for each email:

1. The crawler retrieves the emails eligible for archiving from the email server and puts them into a working queue.
2. Unique IDs are generated based on the email using a hash algorithm like MD5.
3. The IDs are used to check, whether the email itself or parts of it have already been stored in the system. Duplicates are not archived again.
4. Plain text is extracted from attachments with rich media like PDF-documents or office documents.
5. The extracted plain text can be further processed with linguistic methods to augment the text information.
6. The emails are classified to determine if and how long they have to be archived.
7. The extracted plain text from headers, body and attachments is added to a Lucene full text index.
8. Finally the email is stored in the archive in a format well suited to provide 100% format fidelity with its original.

To be able to scale-up on a multiprocessor system, the crawler can be configured to instantiate multiple threads. The threads then pick one email after the other from a shared working queue and drive them through the process.

4.2   Content Service

The catalog and the archived data are distributed over several nodes. Managing them and making them accessible is the duty of the content services. Search requests from a user are first processed by the request server which then forwards the request to the relevant content services. The content service then processes the request on its local indexes and returns the result to the request server. Requests to retrieve emails from the archive are processed in a similar way. Additionally the content services are used to manage the local data, e.g. to merge full-text indexes or to delete emails when their retention period is over.

4.3   Catalog

An important component of any archiving system is the catalog. It is an – at least logically – centralized component that manages the state of the system. It essentially contains the following information:

- Specification of the items stored in the archive. It lists all the fields of the item, their names, data types and some other properties.
- Information about the actual ingestion process. The steps that have to be performed for each item and the rules that have to be applied.
- Metadata about each item in the archive as defined by the item type. This is the data that is generated by the ingestion process and can then be searched on by the user in order to find an item.
- User and access control. Although the users may be managed externally for example in an LDAP server, it is still necessary to specify who is allowed to do what.
- The state of the system. The catalog keeps track of the existing indexes, their state, and some summary data about their content.

There are three big differences between the metadata extracted from the emails and the other listed information: a) the rate at which the metadata is modified is much higher than for the others. For each email that is archived a new entry is added. And this typically happens more often than everything else in the system. b) This metadata is by orders of magnitude the largest part of the total information. c) In an EAD system, as in any archive, the item itself does not change. Therefore the extracted metadata is never modified.

Emails can be considered as semi-structured data with a very simple structure (only a few header fields). Large parts of the email, the body and the attachments, are free text, and even the content of the header fields, like the subject and maybe even the email addresses, are more or less free text. This characteristic makes search engines well suited for managing the metadata of an email archiving system.

The main problem of search engines is their deficiencies with respect to the update of individual fields of a document. To overcome this deficiency, we are following a hybrid approach. The aim is to store the volatile information in a DBMS where it can be efficiently updated, and to store the non-volatile information in the search engine.

To avoid expensive joins between DBMS and search engine, it is crucial that most operations are performed on only one of the two systems. Especially the two most important and critical operations, ingest and search, should only by executed by the search engine. The amount of information that needs to be retrieved from the DBMS, for example for access control, is small. It is sufficient to retrieve it once at the beginning of an ingest process or a search session.

Search engines typically do not provide ACID transactions. The indexes are updated incrementally in batches. This leads to a lag in the searchable information. But for an

EAD system this is not a problem, as long as all information will be searchable within a customer-specific time frame. This also opens the possibility to ingest the emails independently on several machines, and to integrate the results from time to time.

5   Design Evaluation

To evaluate our service-oriented approach we implemented a subset of the functionality required for an EAD system using a set of EAD services. These services and initial performance measurements are discussed in this section.

5.1   Architecture

In a service-oriented architecture the atomic element is the service. This indicates that all components of the EAD system have to be transformed into or encapsulated by a service. Our service-oriented design is shown in figure 3. All included services are part of a scalable service pool. Besides the mainly static components scheduler and registry, a set of EAD services (factory, content service and ingest service) is present in the service pool.

The second main component of the system is the Content Repository. In this design it is a logical component that is combined from a distributed catalog, implemented as full-text search indices, and an archive which is responsible for the actual archiving of the original email documents. The repository abstraction layer offers an integrated access to the content repository, without requiring knowledge about the distribution.

5.2   Measurements

In this section we are presenting some results of initial performance measurements. In these tests the emails (enron email dataset as test dataset) are read from a file system.

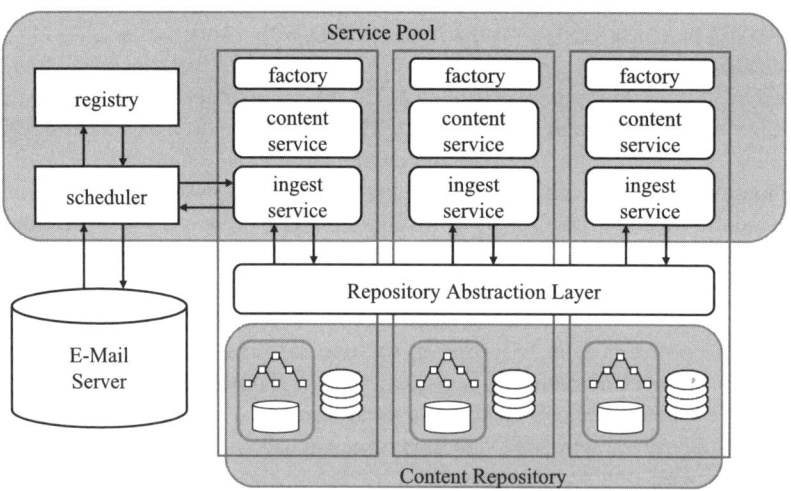

Fig. 3. Architecture of our current prototype

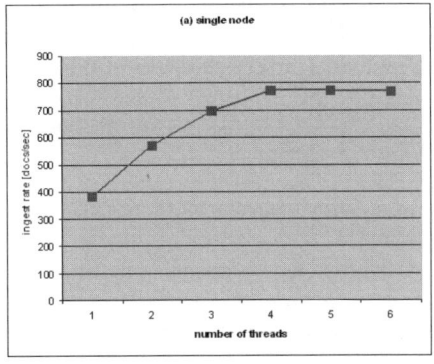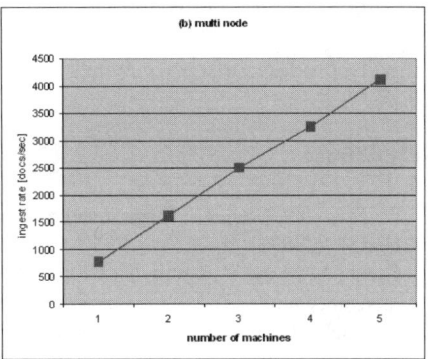

Fig. 4. Ingest performance measurements

The currently implemented reduced ingestion process consists of indexing the emails using Lucene and storing them in the file system. As this test setup is very simplistic, the results are only meant to show the tendency.

The multi-threaded implementation of the crawler allows for utilizing the CPU power of a single computer very well. Figure 4(a) shows that on the test machines with two dual-core processors the throughput increases up to four threads. Resource monitoring shows that by further increasing the thread number the CPUs get exhausted, and the throughput stagnates.

The way the catalog is separated into a DBMS part and a search engine part facilitates a straightforward system evolution towards a distributed approach with EAD system being distributed over multiple nodes. This distributed system still has one coordinator-DBMS, but may comprise a varying number of machines with largely independent workers and search engines. Initial experiments (see Figure 4(b)) indicate a nearly linear scalability in the number of nodes. Therefore the architecture presented in this paper is a promising approach towards an affordable and adaptive system that can scale-out as necessary. The details of the distribution and the partitioning and as well as the orchestration of the services for scalability are beyond the scope of this paper.

6 Conclusion and Outlook

In this paper we presented an architectural approach for the implementation of an service-oriented EAD system. The functionalities are implemented as services that can be instantiated as needed on the infrastructure. This facilitates the adaption of the EAD system to different customer requirements and workloads. The system provides a promising performance while yet fulfilling the requirements towards an EAD system.

Current and future work encompasses the management and provisioning of the system. This includes various issues over all layers, e.g. monitoring and adjusting the system so the assured service level agreements are fulfilled.

Trademarks: IBM, Lotus and Domino are registered trademarks of the International Business Machines Corporation in the United States, other countries or both. Other company, product, or service names may be trademarks or service marks of others.

References

1. DiCenzo, C., Chin, K.: Magic quadrant for e-mail active archiving, 2007. Technical report, Gartner (2007)
2. Green, M., Soy, S., Gunn, S., Galloway, P.: Coming to term – designing the texas email repository model. D-Lib Magazine 8(9) (2002)
3. Michael, M., Moreira, J.E., Shiloach, D., Wisniewski, R.W.: Scale-up x scale-out: A case study using nutch/lucene. In: Parallel and Distributed Processing Symposium, pp. 1–8 (2007)
4. Mitra, S., Hsu, W.W., Winslett, M.: Trustworthy keyword search for regulatory-compliant records retention. In: Proceedings of the 32nd International Conference on Very Large Data Bases, Seoul, Korea, pp. 1001–1012. ACM Press, New York (2006)
5. Symantec. Symantec enterprise vault introduction and planning. Technical report, Symantec (2006)
6. Thickins, G.: Compliance: Do no evil – critical implications and opportunities for storage. Byte and Switch Insider 2(5) (2004)
7. U.S. Department of the Interior. It's in the mail: Common questions about electronic mail and official records (2006)
8. Yu, H., Moreira, J.E., Dube, P.: Performance studies of a websphere application, trade, in scale-out and scale-up environments. In: Parallel and Distributed Processing Symposium, pp. 1–8 (2007)
9. Yung, W.W.: Explore the IBM mail management and compliance solution. developerWorks (2005)
10. Zhu, W.-D., Friedrich, T., Hogg, R., Maletz, J., McBride, P., New, D.: E-mail Archiving and Records Management Integration Solution Guide. IBM Redbooks (January 2006)

Similarity of DTDs Based on Edit Distance and Semantics*

Aleš Wojnar, Irena Mlýnková, and Jiří Dokulil

Charles University in Prague, Czech Republic
ales.wojnar@gmail.com, irena.mlynkova@mff.cuni.cz,
jiri.dokulil@mff.cuni.cz

Summary. In this paper we propose a technique for evaluating similarity of XML schema fragments. Contrary to existing works we focus on structural level in combination with semantic similarity of the data. For this purpose we exploit the idea of edit distance utilized to constructs of DTDs which enables to express the structural differences of the given data more precisely. In addition, in combination with the semantic similarity it provides more realistic results. Using various experiments we show the behavior and advantages of the proposed approach.

1 Introduction

The eXtensible Markup Language (XML) [3] has already become a standard for data representation and manipulation and, thus, it appears in most areas of information technologies. A possible optimization of XML-based methods can be found in exploitation of similarity of XML data. The most common areas of exploitation of data similarity are clustering, dissemination-based applications (e.g. [1]), schema integration systems (e.g. [7]), data warehousing, e-commerce, semantic query processing etc. But despite the amount of existing similarity-based approaches is significant, there is still a space for both improvements and new ways of similarity exploitation.

In this paper we focus on similarity of XML schema fragments expressed in DTD language [3] and, in particular, on persisting disadvantages of the existing approaches. The key emphasis is currently put on the semantic similarity of schema fragments reflecting the requirements of corresponding applications. And if the approaches consider DTD structure, they usually analyze only simple aspects such as, e.g., leaf nodes or child nodes of roots of the fragments. Therefore, we focus on more precise analysis of the structure, but, on the other hand, we still preserve the exploitation of semantic similarity. For this purpose we combine and adapt to DTD constructs two verified approaches – edit distance and semantics of element/attribute names.

The paper is structured as follows: Section 2 overviews the related works. Section 3 describes the proposed approach and Section 4 results of related experiments. Section 5 provides conclusions and outlines future work.

---

* This work was supported in part by the National Programme of Research (Information Society Project 1ET100300419).

C. Badica et al. (Eds.): Intel. Distributed Comput., Systems & Appl., SCI 162, pp. 207–216, 2008.
springerlink.com

2   Related Work

The number of existing works in the area of XML data similarity evaluation is nontrivial. We can search for similarity among XML documents, XML schemes, or between the two groups. We can distinguish several levels of similarity, such as, e.g., structural level, semantic level or constraint level. Or we can require different precision of the similarity.

In case of document similarity we distinguish techniques expressing similarity of two documents D_A and D_B using edit distance, i.e. by measuring how difficult is to transform D_A into D_B (e.g. [10]) and techniques which specify a simple and reasonable representation of D_A and D_B, such as, e.g., using a set of paths, that enables their efficient comparison and similarity evaluation (e.g. [12]). In case of similarity of a document D and a schema S there are also two types of strategies – techniques which measure the number of elements which appear in D but not in S and vice versa (e.g. [2]) and techniques which measure the closest distance between D and "all" documents valid against S (e.g. [9]). And finally, methods for measuring similarity of two XML schemes S_A and S_B combine various supplemental information and similarity measures such as, e.g., predefined similarity rules, similarity of element/attribute names, equality of data types, similarity of schema instances or previous results (e.g. [4, 5]). But, in general, the approaches focus mostly on semantic aspects of the schema fragments, whereas structural ones are of marginal importance.

3   Proposed Algorithm

The proposed algorithm is based mainly on the work presented in [10] which focuses on expressing similarity of XML documents D_A and D_B using tree edit distance, i.e. the amount of operations necessary to transform D_A to D_B. The main contribution of the algorithm is in introducing two new edit operations *InsertTree* and *DeleteTree* which allow manipulating more complex structures than a single node. And repeated structures can be found in a DTD as well if it contains shared or recursive elements. But, contrary to XML documents that can be modeled as trees, DTDs can, in general, form general cyclic graphs. Hence, procedures for computing edit distance of trees need to be utilized to DTD graphs. In addition, not only the structural, but also the semantic aspect of elements is very important. Therefore, we will also concern semantic similarity of element/attribute names.

Algorithm 1. Main body of the algorithm

Input: XSD_A, XSD_B
Output: Edit distance between XSD_A and XSD_B
1: T_A = ParseXSD(XSD_A);
2: T_B = ParseXSD(XSD_B);
3: $Cost_{Graft}$ = ComputeCost(T_B);
4: $Cost_{Prune}$ = ComputeCost(T_A);
5: **return** EditDistance(T_A, T_B, $Cost_{Graft}$, $Cost_{Prune}$);

The method can be divided into three parts depicted in Algorithm 1, where the input DTDs are firstly parsed (line 1 and 2) and their tree representations are constructed. Next, costs for tree inserting (line 3) and tree deleting (line 4) are computed. And in the final step (line 5) we compute the resulting edit distance, i.e. similarity, using dynamic programming.

3.1 DTD Tree Construction

The key operation of our approach is tree representation of the given DTDs. Neverthe-less, the structure of a DTD can be quite complex – the specified content models can contain arbitrary combinations of operators (i.e. "|" or ",") and cardinality constraints (i.e. "?", "*" or "+"). Therefore, we firstly simplify the complex regular expressions using a set of transformation rules.

Simplification of DTDs. For the purpose of simplification of DTD content models we can use various transformation rules. Probably the biggest set was defined in [11], but these simplifications are for our purpose too strong. Hence, we use only a subset of them as depicted in Figures 1 and 2.

I-a) $(e_1\|e_2)^* \to e_1^*, e_2^*$
I-b) $(e_1, e_2)^* \to e_1^*, e_2^*$
I-c) $(e_1, e_2)? \to e_1?, e_2?$
I-d) $(e_1, e_2)^+ \to e_1^+, e_2^+$
I-e) $(e_1\|e_2) \to e_1?, e_2?$

II-a) $e_1^{++} \to e_1^+$	II-b) $e_1^{**} \to e_1^*$
II-c) $e_1^*? \to e_1^*$	II-d) $e_1?^* \to e_1^*$
II-e) $e_1^{+*} \to e_1^*$	II-f) $e_1^{*+} \to e_1^*$
II-g) $e_1?^+ \to e_1^*$	II-h) $e_1^+? \to e_1^*$
II-i) $e_1?? \to e_1^?$	

Fig. 1. Flattening rules   **Fig. 2.** Simplification rules

The rules ensure that each cardinality constraint operator is connected to a single element and avoid usage of "|" operator, though at the cost of a slight information loss.

DTD Tree. Having a simplified DTD, its tree representation is defined as:

Definition 1. *A DTD Tree is an ordered rooted tree $T = (V, E)$, where*

1. *V is a finite set of nodes, s.t. for $\forall v \in V$, $v = (v_{Type}, v_{Name}, v_{Cardinality})$, where v_{Type} is the type of a node (i.e. attribute, element or #PCDATA), v_{Name} is the name of an element/attribute, and $v_{Cardinality}$ is the cardinality constraint operator of an element/attribute,*
2. *$E \subseteq V \times V$ is a set of edges representing relationships between elements and their attributes or subelements.*

An example of a DTD and its tree representation (after simplification) is depicted in Figure 3.

Shared and Repeatable Elements. The structure of a DTD does not have to be purely tree-like. There can occur both shared elements which invoke undirected cycles and re-cursive elements which invoke directed cycles. In case of a shared element we easily

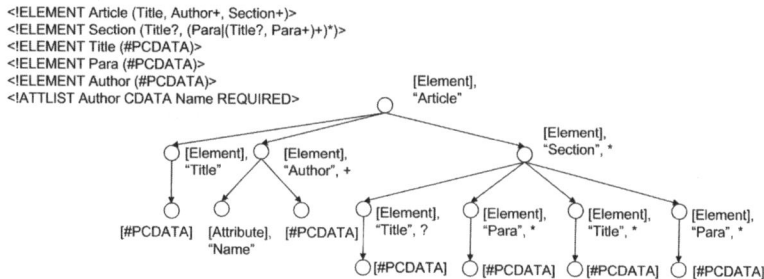

```
<!ELEMENT Article (Title, Author+, Section+)>
<!ELEMENT Section (Title?, (Para|(Title?, Para+)+)*)>
<!ELEMENT Title (#PCDATA)>
<!ELEMENT Para (#PCDATA)>
<!ELEMENT Author (#PCDATA)>
<!ATTLIST Author CDATA Name REQUIRED>
```

Fig. 3. An example of a DTD and its tree representation

create its separate copy for each sharer. But, in case of recursive elements the same idea would invoke infinitely deep trees. However, we exploit the observation of a statistical analysis of real-world XML data [8] that the amount of repetitions is in general very low – less than 10. Actually, for our method it is not important exactly how many occurrences we use because each of them can be transformed using a single edit operation.

3.2   Tree Edit Operations

Having the above described tree representation of a DTD, we can now easily utilize the tree edit algorithm proposed in [10]. For a given tree T with a root node r of degree m and its first-level subtrees $T_1, T_2, ..., T_m$, the tree edit operations are defined as follows:

Definition 2. *Substitution$_T$(r_{new}) is a node substitution operation applied to T that yields the tree T' with root node r_{new} and first-level subtrees $T_1, ..., T_m$.*

Definition 3. *Given a node x with degree 0, Insert$_T$(x, i) is a node insertion operation applied to T at i that yields the new tree T' with root node r and first-level subtrees $T_1, ..., T_i, x, T_{i+1}, ..., T_m$.*

Definition 4. *If the first-level subtree T_i is a leaf node, Delete$_T$(T_i) is a delete node operation applied to T at i that yields the tree T' with root node r and first-level subtrees $T_1, ..., T_{i-1}, T_{i+1}, ..., T_m$.*

Definition 5. *Given a tree T_j, InsertTree$_T$(T_j, i) is an insert tree operation applied to T at i that yields the tree T' with root node r and first-level subtrees $T_1, ..., T_i, T_j, T_{i+1}, ..., T_m$.*

Definition 6. *DeleteTree$_T$(T_i) is a delete tree operation applied to T at i that yields the tree T' with root node r and first-level subtrees $T_1, ..., T_{i-1}, T_{i+1}, ..., T_m$.*

Transformation of a source tree T_A to a destination tree T_B can be done using various sequences of the operations. But, we can only deal with so-called *allowable* sequences, i.e. the relevant ones. For the purpose of our approach we only need to modify the original definition [10] as follows:

Definition 7. *A sequence of edit operations transforming a source tree T_A to a destination tree T_B is allowable if it satisfies the following two conditions:*

1. *A tree T may be inserted only if tree similar to T already occurs in the source tree T_A. A tree T may be deleted only if tree similar to T occurs in the destination tree T_B.*

2. *A tree that has been inserted via the InsertTree operation may not subsequently have additional nodes inserted. A tree that has been deleted via the DeleteTree operation may not previously have had nodes deleted.*

While the original definition requires exactly the same nodes and trees, we relax the requirement only to similar ones. The exact meaning of the similarity is explained in the following text and enables to combine the tree edit distance with other approaches. Also note that each of the edit operations is associated with a non-negative cost.

3.3 Costs of Inserting and Deleting Trees

Inserting (deleting) a subtree T_i can be done with a single operation *InsertTree* (*Delete-Tree*) or with a combination of *InsertTree* (*DeleteTree*) and *Insert* (*Delete*) operations. To find the optimal variant the algorithm uses pre-computed cost for inserting T_i, $Cost_{Graft}(T_i)$, and deleting T_i, $Cost_{Prune}(T_i)$. The procedure can be divided into two parts: In the first part *ContainedIn* list is created for each subtree of T_i; in the second part $Cost_{Graft}$ and $Cost_{Prune}$ are computed for T_i. For our purpose we modify procedure defined in [10] to involve similarity.

Similarity of Elements/Attributes. Similarity of elements/attributes can be evaluated using various criteria. Since the structural similarity is solved via the edit distance, we focus on semantic, syntactic and cardinality-constraint similarity.

Semantic similarity is a score that reflects the semantic relation between the meanings of two words. We exploit procedure *SemanticSim* described in [5] which determines ontology similarity between two words w_1 and w_2 by iterative searching a thesaurus and comparing w_1 with synonyms of w_2.

Syntactic similarity of element/attribute names is determined by computing the edit distance between their labels. For our purpose the classical Levenshtein algorithm [6] is used that determines the edit distance of two strings using inserting, deleting or replacing single characters.

And finally, we consider *similarity of cardinality constraints* of elements specified by the cardinality compatibility table depicted in Table 1.

The overall similarity of elements/attributes e_1 and e_2 is computed as $Sim(e_1, e_2) = Max(SemanticSim(e_1, e_2), SyntacticSim(e_1, e_2)) \times \alpha + CardinalitySim(e_1, e_2) \times \beta$, where $\alpha + \beta = 1$ and $\alpha, \beta \geqslant 0$.

Table 1. Cardinality compatibility table

	*	+	?	none
*	1	0.9	0.7	0.7
+	0.9	1	0.7	0.7
?	0.7	0.7	1	0.8
none	0.7	0.7	0.8	1

ContainedIn **Lists.** The procedure for determining element/attribute similarity is used for creating *ContainedIn* lists which are then used for computing $Cost_{Graft}$ and $Cost_{Prune}$. The list is created for each node of the destination tree and contains pointers to similar nodes in the source tree.

The procedure for creating *ContainedIn* lists is shown in Algorithm 2. Since creating of lists starts from leaves and continues to root, there is recursive calling of procedure at line 2. At line 4 we find all similar nodes of n in tree T_A and add them to a temporary list. If n is a leaf node, the *ContainedIn* list is created. For a non-leaf node we have to filter the list with lists of its descendants (line 6). At this step each descendant of n has to be found at corresponding position in descendants of nodes in the created *ContainedIn* list. More precisely, let $v_A \in n_{ContainedIn}$, $children_{v_A}$ is the set of v_A descendants, and c is a child of n. Then $c_{ContainedIn} \cap children_{v_A} \neq \emptyset$, otherwise v_A is removed from $n_{ContainedIn}$.

Algorithm 2. CreateContainedInLists(T_A, n)

Input: tree T_A, root n of T_B
Output: *CointainedIn* lists for all nodes in T_B
1: **for all** *child* of n **do**
2:    CreateContainedInLists(T_A, *child*);
3: **end for**
4: $n_{ContainedIn}$ = FindSimilarNodes(T_A, n);
5: **for all** *child* of n **do**
6:    $n_{ContainedIn}$ = FilterLists($n_{ContainedIn}$, $child_{ContainedIn}$);
7: **end for**
8: Sort($n_{ContainedIn}$);

Costs of Inserting Trees. When the *ContainedIn* list with corresponding nodes is created for node r, the cost for inserting the tree rooted at r can be assigned. The procedure is shown in Algorithm 3. The *foreach* loop computes sum, sum_{d_0}, for inserting node r and all its subtrees. If *InsertTree* operation can be applied (*ContainedIn* list of r is not empty), sum_{d_1}, is computed for this operation at line 8. The minimum of these costs are finally denoted as $Cost_{Graft}$ for node r.

Algorithm 3. ComputeCost(r)

Input: root r of T_B
Output: $Cost_{Graft}$ for T_B
1: $sum_{d_0} = 1$;
2: **for all** *child* of r **do**
3:    ComputeCost(*child*);
4:    sum_{d_0} += $Cost_{Graft}$(child);
5: **end for**
6: $sum_{d_1} = \infty$;
7: **if** $r_{ContainedIn}$ is not empty **then**
8:    $sum_{d_1} = ComputeInsertTreeCost(r)$;
9: **end if**
10: $Cost_{Graft}(\text{r}) = Min(sum_{d_0}, sum_{d_1})$;

Costs of Deleting Trees. Since the rules for deleting a subtree from source are same as rules for inserting a subtree into destination tree, costs for deleting trees are obtained by the same procedures. We only switch tree T_A with T_B in procedures *CreateContainedInLists* and *ComputeCost*.

3.4 Computing Edit Distance

The last part of the algorithm, i.e. computing the edit distance, is based on dynamic programming. At this step the procedure decides which of the operations defined in Section 3.2 will be applied for each node to transforming source tree T_A to destination tree T_B. This part of algorithm does not have to be modified for DTDs so the original procedure presented in [10] is used. (We omit the formal algorithm for the paper length.)

3.5 Complexity

In [10] it was proven that the overall complexity of transforming tree T_A into tree T_B is $O(|T_A||T_B|)$. In our method we have to consider procedures for constructing DTD trees and for evaluating similarity. Constructing a DTD tree can be done in $O(|T_A|)$ for tree T_A. Complexity of finding similarity depends on procedures *SemanticSim*, *SyntacticSim* and *CardinalitySim*. *SyntacticSim* is computed for each pair of elements in trees T_A and T_B, so its complexity is $O(|T_A||T_B||\omega|)$, where ω is maximum length of an element/attribute label. *CardinalitySim* is also computed for each pair of elements, however, with constant complexity, i.e. in $O(|T_A||T_B|)$. Complexity of *SemanticSim* depends on the size of the thesaurus, so the overall complexity is $O(|T_A||T_B||\Sigma|)$, where Σ is the set of words in the thesaurus. And it also determines the complexity of the whole algorithm.

4   Experiments

To analyze the behavior of the proposal we have performed various experiments with both real-world and synthetic XML data.

Real-World XML Data. In the first test we have used seven different real-world DTDs. First five DTDs (**c1, c2, ..., c5**) represent an object CUSTOMER, but in more or less different ways. Next two DTDs represent different objects – TVSCHEDULE (**tv**) and NEWSPAPER (**np**). The parameters have been set to default values and both structural and semantic similarities have been exploited. The resulting similarities are depicted in Table 2.

Expectably, DTDs representing the same object, i.e. CUSTOMER, have higher mutual similarities (the average similarity is 0.44) than similarities among DTDs representing different objects (the average for NEWSPAPER DTD is 0.13 and average for TVSCHEDULE DTD is only 0.03). The only one exception is between CUSTOMER1 and NEWSPAPER due to their structural similarity.

In the second test we have used the same DTDs, but we have evaluated their similarities regardless semantic similarity. As we can see in Table 3, the resulting values

Table 2. Structural and semantic similarity of real-world DTDs

	c1	c2	c3	c4	c5	tv	np
c1	1	0.57	0.43	0.19	0.71	0.08	0.42
c2	0.57	1	0.57	0.45	0.48	0.10	0.11
c3	0.43	0.57	1	0.39	0.36	0.01	0.13
c4	0.19	0.45	0.39	1	0.21	0.00	0.00
c5	0.71	0.48	0.36	0.21	1	0.00	0.11
tv	0.08	0.10	0.01	0.00	0.00	1	0.00
np	0.42	0.11	0.13	0.00	0.11	0.00	1

Table 3. Similarity of real-world DTDs without semantic similarity

	c1	c2	c3	c4	c5	tv	np
c1	1	0.45	0.23	0.09	0.57	0.00	0.13
c2	0.45	1	0.50	0.42	0.32	0.00	0.00
c3	0.23	0.50	1	0.30	0.15	0.00	0.00
c4	0.09	0.42	0.30	1	0.20	0.00	0.00
c5	0.57	0.32	0.15	0.20	1	0.00	0.00
tv	0.00	0.00	0.00	0.00	0.00	1	0.00
np	0.13	0.00	0.00	0.00	0.00	0.00	1

are lower, however, the trend between same and different objects is same as in the first test.

Semantic Similarity. In the next set of tests we have focused on various parameters of the similarity measure using synthetic data. Firstly, we have dealt with semantic similarity. For this purpose, we defined three DTDs (see Figure 4) with exactly the same structure, but different element names. In addition, element names of the first and second DTD have similar meaning while the element names of the third DTD have no lexical meaning. The results are depicted in Table 4.

```
<!ELEMENT PERSON (DOMICILE, WORK)>       <!ELEMENT USER (RESIDENCE, JOB)>        <!ELEMENT AAA (BBB, DDD)>
<!ELEMENT DOMICILE (STATE, TOWN)>        <!ELEMENT RESIDENCE (COUNTRY, CITY)>    <!ELEMENT BBB (EEE, FFF)>
<!ELEMENT STATE (#PCDATA)>               <!ELEMENT COUNTRY (#PCDATA)>            <!ELEMENT DDD (#PCDATA)>
<!ELEMENT TOWN (#PCDATA)>                <!ELEMENT CITY (#PCDATA)>               <!ELEMENT EEE (#PCDATA)>
<!ELEMENT WORK (#PCDATA)>                <!ELEMENT JOB (#PCDATA)>                <!ELEMENT FFF (#PCDATA)>
<!ATTLIST PERSON SURNAME CDATA           <!ATTLIST USER LASTNAME CDATA           <!ATTLIST AAA CCC CDATA
           #REQUIRED>                               #REQUIRED>                               #REQUIRED>
```

Fig. 4. Synthetic DTDs for analysis of semantic similarity

Table 4. Influence of semantic similarity

Semantic similarity	✓	✗
PERSON x USER	0.92	0.40
PERSON x AAA	0.33	0.33

Table 5. Comparing different costs of edit operations

Cost	1	5	10	100
USER1 x USER2	0.92	0.74	0.52	0.52

As we can see, there is a significant difference in comparing the first two DTDs – they were correctly identified as almost similar when we used semantic similarity. Consequently, despite the semantic similarity is a time-consuming task due to necessary searching through thesaurus, it is enables to acquire more precise similarity values.

Edit Distance Operations. In the last test we have focused on two key edit operations used for transforming DTD trees, *InsertTree* and *DeleteTree*, proposed for transforming repeating structures of a tree. For the purpose of the test we have defined two similar DTDs depicted in Figure 5, whereas one of them involves shared elements. We have performed their similarity evaluation with different costs of edit operations *InsertTree* and *DeleteTree*.

```
<!ELEMENT USER (CUSTOMER, EMPLOYEE)>          <!ELEMENT USER (CUSTOMER)>
<!ELEMENT CUSTOMER (USERDATA, ORDERS)>         <!ELEMENT CUSTOMER (USERDATA, ORDERS)>
<!ELEMENT EMPLOYEE (USERDATA, POSITION)>       <!ELEMENT USERDATA (ID, NAME, BIRTHDAY)>
<!ELEMENT USERDATA (ID, NAME, BIRTHDAY)>       <!ELEMENT ORDERS (#PCDATA)>
<!ELEMENT ORDERS (#PCDATA)>                    <!ELEMENT ID (#PCDATA)>
<!ELEMENT POSITION (#PCDATA)>                  <!ELEMENT NAME (#PCDATA)>
<!ELEMENT ID (#PCDATA)>                        <!ELEMENT BIRTHDAY (#PCDATA)>
<!ELEMENT NAME (#PCDATA)>
<!ELEMENT BIRTHDAY (#PCDATA)>
```

Fig. 5. Synthetic DTDs for analysis of edit operations

As we can see in Table 5, in first two cases the operations were really used, but in the last two comparisons the costs for the operations were too high and the repeating tree structures were transformed using a sequence of single-node edit operations. Hence, the DTDs were correctly identified as similar only when the costs of the operations were set sufficiently low. This observation is similar to the observation made for edit distance algorithms used for XML documents.

5   Conclusion

The aim of this paper was a proposal of an algorithm for evaluating XML schema similarity on structural level. In particular, we have focused on DTDs which are still more popular than other languages for schema specification. We have combined two approaches and adapted them to DTD-specific structure – edit distance and semantic similarity. The exploitation of edit distance enables to analyze the structure of DTDs more precisely, whereas the semantic similarity enables to get more precise results, though at the cost of searching a thesaurus.

In our future work we will focus mainly on further improvements of our approach, such as other edit operations (e.g. moving a node or adding/deleting a non-leaf node) or XML Schema definitions that involve new constructs (e.g. unordered sequences of elements) as well as plenty of syntactic sugar.

References

1. Altinel, M., Franklin, M.J.: Efficient Filtering of XML Documents for Selective Dissemination of Information. In: VLDB 2000, pp. 53–64. Morgan Kaufmann, San Francisco (2000)
2. Bertino, E., Guerrini, G., Mesiti, M.: A Matching Algorithm for Measuring the Structural Similarity between an XML Document and a DTD and its Applications. Inf. Syst. 29(1), 23–46 (2004)
3. Bray, T., Paoli, J., Sperberg-McQueen, C.M., Maler, E., Yergeau, F.: Extensible Markup Language (XML) 1.0 (Fourth Edition). W3C (2006)
4. Do, H.H., Rahm, E.: COMA – A System for Flexible Combination of Schema Matching Approaches. In: VLDB 2002, pp. 610–621. Morgan Kaufmann, Hong Kong (2002)
5. Lee, M.L., Yang, L.H., Hsu, W., Yang, X.: XClust: Clustering XML Schemas for Effective Integration. In: CIKM 2002, pp. 292–299. ACM Press, New York (2002)
6. Levenshtein, V.I.: Binary Codes Capable of Correcting Deletions, Insertions and Reversals. Soviet Physics Doklady 10, 707 (1966)
7. Milo, T., Zohar, S.: Using Schema Matching to Simplify Heterogeneous Data Translation. In: VLDB 1998, pp. 122–133. Morgan Kaufmann, San Francisco (1998)

8. Mlynkova, I., Toman, K., Pokorny, J.: Statistical Analysis of Real XML Data Collections. In: COMAD 2006, New Delhi, India, pp. 20–31. Tata McGraw-Hill Publishing, New York (2006)
9. Ng, P.K.L., Ng, V.T.Y.: Structural Similarity between XML Documents and DTDs. In: ICCS 2003, pp. 412–421. Springer, Heidelberg (2003)
10. Nierman, A., Jagadish, H.V.: Evaluating Structural Similarity in XML Documents. In: WebDB 2002, Madison, Wisconsin, USA, pp. 61–66 (2002)
11. Shanmugasundaram, J., Tufte, K., Zhang, C., He, G., DeWitt, D.J., Naughton, J.F.: Relational Databases for Querying XML Documents: Limitations and Opportunities. In: VLDB 1999, pp. 302–314. Morgan Kaufmann, San Francisco (1999)
12. Zhang, Z., Li, R., Cao, S., Zhu, Y.: Similarity Metric for XML Documents. In: FGWM 2003, Karlsruhe, Germany (2003)

Timer-Based Composition of Fault-Containing Self-stabilizing Protocols

Yukiko Yamauchi[1], Sayaka Kamei[2], Fukuhito Ooshita[1],
Yoshiaki Katayama[3], Hirotsugu Kakugawa[1], and Toshimitsu Masuzawa[1]

[1] Graduate School of Information Science and Technology, Osaka University
{y-yamaut,f-oosita,kakugawa,masuzawa}@ist.osaka-u.ac.jp
[2] Department of Information Engineering, Graduate School of Engineering,
Hiroshima University
s-kamei@se.hiroshima-u.ac.jp
[3] Graduate School of Computer Science and Engineering,
Nagoya Institute of Technology
katayama@nitech.ac.jp

Summary. Self-stabilizing protocols provide autonomous recovery from finite number of transient faults. Fault-containing self-stabilizing protocols promise not only self-stabilization but also quick recovery and small effect from small scale of faults. In this paper, we introduce a timer-based composition of fault-containing self-stabilizing protocols that preserves the fault-containment property of source protocols. Our framework can be applied to a larger subclass of fault-containing self-stabilizing protocols than existing compositions [1].

Keywords: fault-containment, self-stabilization, composition, timer.

1 Introduction

Large scale networks that consist of a large number of processes communicating with each other have been developed in these years. It is necessary to take measures against faults (e.g. memory crash at processes, topology change, etc.) when we design distributed protocols for large scale networks. A *self-stabilizing protocol* converges to a legitimate configuration from any arbitrary initial configuration. Self-stabilization was first introduced by Dijkstra [2]. Since then, many self-stabilizing protocols have been designed for many problems [3, 4, 5]. The stabilization property provides autonomous adaptability against any number of transient faults that corrupt memory contents at processes. In practice, the adaptability to small scale faults is important because catastrophic faults rarely occur. However, self-stabilization does not promise efficient recovery from small scale of faults and sometimes the effect of a fault spreads over the entire network.

When a fault corrupts f processes in a legitimate configuration, we call the obtained configuration an f-*faulty configuration*. An f-*fault-containing self-stabilizing protocol* promises self-stabilization and fault-containment [6, 7, 8]: starting from an f'-faulty configuration ($f' \leq f$), it reaches a legitimate

C. Badica et al. (Eds.): Intel. Distributed Comput., Systems & Appl., SCI 162, pp. 217–226, 2008.

configuration in the time and with the number of processes affected proportional to f or less.

Hierarchical composition of multiple protocols is expected to extend the application of existing protocols and to ease the design of new protocols. In a hierarchical composition of two (or more) protocols, the output of one protocol (called the lower protocol) is used as the input to the other (called the upper protocol), and the obtained protocol provides the output of the upper protocol for the input to the lower protocol. Executing two (or more) different self-stabilizing protocols in parallel is well known as *fair composition* that provides hierarchical composition of the source protocols and promises self-stabilization of the obtained protocol [5]. However, a fair composition cannot preserve the fault-containment property when composing fault-containing self-stabilizing protocols.

Related work. *Global neighborhood synchronizers* are often used to implement timers at processes in designing fault-containing self-stabilizing protocols. Ghosh et al. proposed a transformer for non-reactive self-stabilizing protocols to obtain corresponding 1-fault-containing protocols [7]. An obtained 1-fault-containing protocol guarantees that the *output* of the protocol recovers quickly. However, their transformer utilizes a global neighborhood synchronizer and the effect of a fault spreads over the entire network via global synchronization. However, the expected property for fault-containment is temporal and spatial containment of the effect of faults.

Contributions. Yamauchi et al. defined composition of fault-containing self-stabilizing protocols, called *fault-containing composition* and proposed the first framework for fault-containing composition [1]. *RWFC* (Recovery Waiting Fault-containing Composition) strategy is to stop the upper protocol until the lower protocol recovers. In [1], *RWFC* strategy is implemented as follows: each process evaluates a *local predicate* to check local consistency of the current configuration of the lower protocol whenever the process wants to execute the upper protocol. If the process finds the lower protocol locally consistent, then the process executes the upper protocol. Otherwise, the process cannot execute the upper protocol. Thus, each process has to communicate with distant processes to evaluate the local predicate. Moreover, they put many restrictions on source protocols and it regulates the application of the composition framework.

In this paper, we present a novel timer-based technique for fault-containing composition. We adopt *RWFC* strategy, however, the proposed composition utilizes *recovery time* of fault-containing protocols. Recovery time is the maximum time for the system to recover from a target faulty configuration. We force the upper protocol to stop during the recovery time of the lower protocol. After that, the upper protocol can execute on the correct input from the lower protocol. Thus, the upper protocol can recover with its fault-containment property and the composite protocol promises fault-containment as a whole.

Our framework does not need communication among distant processes and relaxes the restrictions on source protocols. In [1], it is necessary that each process has to keep detecting the inconsistency of the lower protocol during the recovery of the lower protocol by communicating with distant processes while in this paper, each process has to detect the inconsistency of the lower protocol in the initial configuration by communicating direct neighbors.

We use *local timers* at processes to measure the recovery times of the source protocols. Global neighborhood synchronizers are often used to implement local timers. However, a fault-containing protocol bounds the effect of faults with *contamination radius*: the maximum distance from any faulty process to any process affected by the faulty process. We introduce a *local neighborhood synchronizer* that emulates M synchronized rounds among the k-neighbors of the initiator that initiates the synchronization.

2 Preliminary

A system is a network which is represented by an undirected graph $G = (V, E)$ where the vertex set V is a set of processes and the edge set E is a set of bidirectional communication links. Each process has a unique identity. Process p is a neighbor of process q iff there exists a communication link $(p, q) \in E$. A set of neighbors of p is denoted by N_p. Let $N_p^0 = \{p\}$, $N_p^1 = N_p$ and for each $i \geq 2$, $N_p^i = \bigcup_{q \in N_p^{i-1}} N_q \setminus \{p\}$. The set of processes denoted by N_p^i is called *i-neighbor* of p. The distance between p and q $(q \neq p)$ is denoted by $dist(p, q)$ and $dist(p, q) = j$ iff $q \notin N_p^{j-1} \wedge q \in N_p^j$.

Each process p maintains local variables and the values of all local variables at p define the local state of p. Local variables are classified into three classes: input, output, and inner. The input variables indicate the input to the system and they are not changed by the system. The output variables are the output of the system for external observers. The inner variables are internal working variables used to compute output variables.

We adopt *locally shared memory model* as a communication model: each process p can read the values of the local variables at $q \in N_p \cup \{p\}$. A protocol at each process p consists of a finite number of *guarded actions* in the form of $\langle guard \rangle \rightarrow \langle action \rangle$. A $\langle guard \rangle$ is a boolean expression involving the local variables of p and N_p, and an $\langle action \rangle$ is a statement that changes the values of p's local variables (except input variables). A process with a guard evaluated *true* is called *enabled*. We adopt *distributed daemon* as a scheduler: in a computation step, distributed daemon selects a nonempty set of enabled processes and these processes execute the corresponding actions. The evaluation of guards and the execution of the corresponding action is *atomic*: these computations are done without any interruption. A configuration of a system is represented by a tuple of local states of all processes. An *execution* is an infinite sequence of configurations $E = \sigma_0, \sigma_1, \sigma_2, \cdots$ such that σ_{i+1} is obtained by applying one computation step to σ_i or σ_{i+1} is the final configuration.

Distributed daemon allows *asynchronous* executions. In an asynchronous execution, the time is measured by computation steps or *rounds*. Let $E = \sigma_0, \sigma_1, \sigma_2, \cdots$ be an asynchronous execution. The first round $\sigma_0, \sigma_1, \sigma_2, \cdots, \sigma_i$

is the minimum prefix of E such that for each process $p \in V$ if p is enabled in σ_0, either p's guard is disabled or p executes at least one step in $\sigma_0, \sigma_1, \sigma_2, \cdots \sigma_i$. The second and latter rounds are defined recursively by applying the definition of the first round to the remaining suffix of the execution $E' = \sigma_{i+1}, \sigma_{i+2}, \cdots$.

A *problem* (task) T is defined by a legitimate predicate on configurations. A configuration σ is *legitimate* iff σ satisfies the legitimate predicate. In this paper we treat *non-reactive* problems: no process changes its state after the system reaches a legitimate configuration, e.g. spanning tree construction, leader election, etc. We say a distributed protocol $P(T)$ solves T in a configuration iff the configuration satisfies the legitimate predicate $L(P(T))$. The input (output) of $P(T)$ is represented by the conjunction of input (output, respectively) variables at each process. We omit T if T is clear.

Definition 1. *Self-stabilization*
Protocol P is self-stabilizing iff it satisfies the following two properties:

Stabilization: *starting from any arbitrary initial configuration, it eventually reaches a legitimate configuration.*
Closure: *once it reaches a legitimate configuration, it remains in legitimate configurations thereafter.*

A *transient fault* corrupts some processes by changing the values of their local variables arbitrarily. A configuration is f-*faulty* iff the minimum number of processes such that we have to change their local states (except inner variables) to make the configuration legitimate is f. We say process p is *faulty* iff we have to change p's local state to make the configuration legitimate and otherwise *correct*.

Definition 2. f-*fault-containment*
A self-stabilizing protocol is f-fault-containing iff it reaches a legitimate configuration from any f'-faulty configuration ($f' \leq f$) with the number of processes that change their states according to the fault and the time to reach a legitimate configuration depending on f (not $|V|$).

We simply denote an f-fault-containing self-stabilizing protocol as an f-fault-containing protocol. The performance of an f-fault-containing protocol is measured by stabilization time, recovery time, and contamination radius:

Stabilization time: the maximum (worst) number of rounds to reach a legitimate configuration from an arbitrary initial configuration.
Recovery time: the maximum (worst) number of rounds to reach a legitimate configuration from an f'-faulty configuration ($f' \leq f$).
Contamination radius: the maximum distance from any faulty process to the process that changes its local state according to the faulty process during the recovery from an f'-faulty configuration ($f' \leq f$).

A hierarchical composition of two protocols P_1 and P_2 is denoted by $(P_1 * P_2)$ where the variables of P_1 and those of P_2 are disjoint except that the input to P_2 is the output of P_1. We define the output variables of $(P_1 * P_2)$ is the output variables of P_2. A legitimate configuration of $(P_1 * P_2)$ is defined by $L((P_1 * P_2))$ where $L(P_1 * P_2) = L(P_1) \wedge L(P_2)$.

Definition 3. *Fault-containing composition*

Let P_1 be an f_1-fault-containing protocol and P_2 be an f_2-fault-containing proto-col. A hierarchical composition $(P_1 * P_2)$ is a fault-containing composition of P_1 and P_2 iff $(P_1 * P_2)$ is an $f_{1,2}$-fault-containing protocol for some $f_{1,2}$ such that $0 < f_{1,2} \leq \min\{f_1, f_2\}$.

In a hierarchical composition, the input to P_2 can be corrupted by a fault when the fault corrupts the output variables of P_1. However, the input to P_1 can be seen as the system parameters, e.g. topology, ID of each process, etc.

Assumption 1. *For a hierarchical composition $(P_1 * P_2)$, the input to P_1 is not corrupted by any fault.*

We consider a subclass of fault-containing protocols Π such that each f-fault-containing protocol $P \in \Pi$ satisfies Assumption 2, 3, and 4. Many exist-ing fault-containing protocols [6, 8] satisfy Assumption 2, 3, and 4.

Assumption 2. *The legitimate configuration of P is uniquely defined by the input variables.*

Consider a composition $(P_1 * P_2)$ of an f_1-fault-containing protocol P_1 and an f_2-fault-containing protocol P_2. Starting from an f'-faulty configuration $(f' \leq \min\{f_1, f_2\})$, if the output of P_1 after P_1 reaches a legitimate config-uration is different from what it was before the fault, then the input to P_2 changes and the output of P_2 may change drastically to adopt it. Then, P_2 can-not guarantee fault-containment. Because the input to P_1 is not changed by any fault (Assumption 1), Assumption 2 guarantees that P_1 recovers to the unique legitimate configuration and ensures the possibility of fault-containment of P_2 in the composite protocol.

Assumption 3. *The legitimate predicate $L(P)$ for P is represented in the form $L(P) \equiv \forall p \in V : cons_p(P)$. The predicate $cons_p(P)$ involves the local variables at p and its neighbors, and it is defined over the values of output, inner, and input variables.*

We say process p is *inconsistent* iff $cons_p(P)$ is *false*, otherwise *consistent*. Because we work on non-reactive problems, the predicate $cons_p(P)$ is evaluated *false* when process p is enabled.

Assumption 4. *In an f'-faulty configuration $(f' \leq f)$, if a faulty process p is a neighbor of correct process(es), at least one correct process q neighboring to p or p itself evaluates $cons_q(P)$ (or $cons_p(P)$) false.*

For a faulty process p and a neighboring correct process q, $cons_p(P)$ ($cons_q(P)$, respectively) involves the local variables at q and p. Because p is faulty, there can be some inconsistency between the local state of p and that of q.

3   The Composition Framework

Let P_1 be an f_1-fault-containing protocol and P_2 be an f_2-fault-containing protocol. Our goal is to produce $f_{1,2}$-fault-containing protocol $(P_1 * P_2)$ for $f_{1,2} = \min\{f_1, f_2\}$.

Fair composition of fault-containing protocols cannot preserve the fault-containment property. When a fault corrupts the output variables of P_1 at f processes ($f \leq f_{1,2}$), during the recovery of P_1, P_2 can be executed in parallel to adopt the changes in the output variables of P_1. The number of contaminated processes in P_1 may become larger than f_2 and this causes the number of processes that change their local states in P_2 becomes larger than f_2. These processes can change its state repeatedly until P_1 recovers. If more than f_2 processes change their states repeatedly in P_2, then P_2 cannot guarantee fault-containment even though f (the number of the processes that the original fault corrupts) is smaller than f_2.

We implement the *RWFC* strategy with *local timers* at processes. Let r_1 (r_2, respectively) be the recovery time of P_1 (P_2, respectively) and c_1 (c_2, respectively) be the contamination radius of P_1 (P_2, respectively). We implement timers at processes with a *local neighborhood synchronizer* that synchronizes the processes in $\max\{c_1, c_2\}$-neighbors for each faulty process for $(r_1 + r_2)$ rounds. We first define the specification of the local neighborhood synchronizer in Section 3.1 and show our composition framework in Section 3.2. Finally, we present an implementation of the local neighborhood synchronizer in Section 3.3.

3.1   Specification of the Local Neighborhood Synchronizer

In this section we define a specification of our local neighborhood synchronizer for fault-containing composition $(P_1 * P_2)$. For P_i, r_i represents the recovery time and c_i represents the contamination radius where $i = 0, 1$.

Specification 1. *Each process p maintains a counter variable t_p that takes an integer in $[0..(r_1 + r_2)]$. The local neighborhood synchronizer is self-stabilizing and in a legitimate configuration, $t_p = 0$ holds at $\forall p \in V$.*

The local neighborhood synchronizer is implemented with a typical technique of synchronizers [7]. We say a process is *s-consistent* iff its counter variable differs at most one with those at all its neighbors involved in the synchronization. Synchronization is realized by making each counter variable s-consistent and then decrementing it with preserving the s-consistency.

Synchronization radius is the maximum distance from any faulty process and a process involved in the synchronization caused by the faulty process. From Assumption 4, the distance between a process that finds inconsistency in the source protocols and any contaminated process is at most $k_{1,2} = \max\{c_1, c_2\} + \max\{f_1, f_2\} + 1$. To involve all $\max\{c_1, c_2\}$-neighbors of each faulty process, it is necessary to involve all $k_{1,2}$-neighbors for each faulty process into the synchronization.

A *counter sequence* of process p is the sequence of the value of t_p from an initial configuration.

Specification 2. *Starting from an f-faulty configuration ($f \leq f_{1,2}$), the local neighborhood synchronizer should provide the following five properties:*

Containment: *synchronization radius is $O(k_{1,2})$.*
Synchronization: *each processes involved in the synchronization decrements its counter variable with keeping s-consistency.*
Correct sequence: *a counter sequence $v_p^0, v_p^1, \cdots$ of any correct process p involved in the synchronization has a prefix $v_p^0, v_p^1, \cdots, v_p^{i-1}, v_p^i$ for some i such that $v_p^0 = v_p^1 = \cdots = v_p^{i-1} = 0$ and $v_p^i = r_1 + r_2$.*
Faulty sequence: *a counter sequence $v_q^0, v_q^1, \cdots$ of any faulty process q has a suffix $v_q^i, v_q^{i+1}, \cdots, v_q^j, \cdots$ for some i and j such that $v_q^k - v_q^{k+1} \leq 1$ for $i \leq k \leq j$ and $v_q^j = v_q^{j+1} = \cdots = 0$.*
Termination: *the local neighborhood synchronizer reaches a legitimate configuration in $(r_1 + r_2 + O(1))$ rounds.*

We do not assume that faulty processes decrement their counter variables from $(r_1 + r_2)$. From Assumption 4, when a faulty process is surrounded by other faulty processes, it cannot determine whether it is correct or not.

Specification 3. *The following APIs should be available at each process $p \in V$ for the application of the local neighborhood synchronizer:*

call_start_synch_NS: *when this function call is executed at process p, it starts the synchronization involving $k_{1,2}$-neighbors of p. These processes decrements their counter variables from $(r_1 + r_2)$ to 0 with keeping s-consistency and the system reaches the legitimate configuration in $O(r_1 + r_2)$ rounds.*
call_exec_NS: *when this function call is executed at process p, if p is enabled in the local neighborhood synchronizer, then it executes one of the corresponding actions and if p decrements t_p, this function call returns true, otherwise false. If p is not enabled, then p does nothing and this function call returns $\perp$.*

3.2   The Framework *FC-LNS*

Our composition framework *FC-LNS* (*Fault-containing Composition with the Local Neighborhood Synchronizer*) is shown in Protocol 3.1. Process p executes the guarded actions of the local neighborhood synchronizer by executing call_exec_NS, and whenever it decrements t_p, p executes the source protocols by executing the procedure $A(t_p)$ that selects which source protocol is executed at p. If p finds inconsistency in P_1 when $0 < t_p \leq r_2$ or in P_1 or P_2 when $t_p = 0$, then it initiates the synchronization of the local neighborhood synchronizer by executing call_start_synch_NS. Thus, p and its $k_{1,2}$-neighbors execute P_1 untill P_1 reaches the legitimate configuration. After that, they executes P_2 on the correct output from P_1 and P_2 reaches the legitimate configuration with its fault-containment property.

Protocol 3.1. *FC-LNS*

Procedure $A(t_p)$ **for process** p
    if $(r_2 \leq t_p < r_1 + r_2)$ **then execute** P_1
    else execute P_2;
Actions for any process p
    $true \quad \longrightarrow$
    if call_exec_NS $= true$ **then** $A(t_p)$;
    if $\{(0 < t_p \leq r_2) \wedge \neg cons_p(P_1)\} \vee \{(t_p = 0) \wedge (\neg cons_p(P_1) \vee \neg cons_p(P_2))\}$
        then call_start_synch_NS

Theorem 1. *FC-LNS provides a* $\min\{f_1, f_2\}$*-fault-containing protocol* $(P_1 * P_2)$ *for an* f_1*-fault-containing protocol* P_1 *and* f_2*-fault-containing protocol* P_2*. For each* $i = 1, 2$*, let* c_i *be the contamination radius of* P_i *and* r_i *be the recovery time of* P_i*. The contamination radius of the obtained protocol is* $O(\max\{c_1, c_2\} + \max\{f_1, f_2\})$*. The recovery time of the obtained protocol is* $O(r_1 + r_2)$*.*

Due to page restriction, we omit the detailed proof.

3.3 Local Neighborhood Synchronizer

In this section we present an implementation of the local neighborhood synchronizer *LNS* that meets the specifications in Section 3.1.

For any given M and k, *LNS* provides the synchronization of M rounds among k-neighbors of the *initiator*. The synchronization consists of three phases. In the first phase, an initiator arises and the shortest path tree rooted at the initiator is constructed to involve all the k-neighbors of the initiator into the synchronization. Then, in the second phase, the synchronized count-down of counter variables takes place among k-neighbors of the initiator. In the third phase, the shortest path tree is released from the root to the leaves.

Each process p has two variables, t_p and d_p: t_p is the counter variable and d_p is the depth variable which is used to construct the shortest path tree. In a legitimate configuration, $t_p = 0 \wedge d_p = 0$ holds at $\forall p \in V$.

Let p be an initiator. Each process $q \in N_p^k$ constructs the shortest path tree by setting $d_q = k - dist(p, q)$ where $dist(p, q)$ denotes the distance between p and q. The parent(s) of q is any neighbor $r \in N_q$ where $d_r = d_q + 1$. A process $s \in N_q$ is a child of q iff $d_s = d_q - 1$.

The protocol *LNS* is shown in Protocol 3.2. Parameter $Predicate_p^{init}$ is a predicate that involves local variables at p and all its neighbors and parameter $Action_p^{dec}$ is a set of actions that changes the value of local variables at p except t_p and d_p. The first phase starts when some process p finds inconsistency in the timer variables or depth variables of itself and its neighbors, and executes S_1. After p executes S_1, each process $q \in N_p^k$ executes S_2 (and S_3 if necessary) and q is involved in the shortest path tree by setting $t_q = M$ and $d_q = k - dist(p, q)$. After $t_q = M \wedge d_q = k - dist(p, q)$ holds at q and all its neighbors get involved in the shortest path tree, q goes into the second phase. In the second phase, q decrements t_q by executing S_4. The synchronization is realized by decrementing t_q with keeping the s-consistency ($D_q(2) = true$). The execution of S_4 starts

Protocol 3.2. *LNS* $(Predicate_p^{init}, Action_p^{dec})$

Predicates

$safe\_d_p \equiv \{d_p = k\} \vee \{d_p = 0\} \vee \{(0 < d_p < k) \wedge (\exists q \in N_p : d_q - d_p = 1)\}$

$OK\_d_p \equiv safe\_d_p \wedge (\forall q \in N_p : |d_p - d_q| \leq 1)$

$safe\_t_p \equiv \{t_p = 0\} \vee \{t_p = M\} \vee \{(\exists q \in N_p : |t_p - t_q| \leq 1) \wedge$
$\qquad (\forall q \in N_p : (t_q = 0 \wedge (t_p = M \vee d_p = 0)) \vee (|t_p - t_q| \leq 1))\}$

$OK\_t_p \equiv safe\_t_p \wedge [\{t_p = 0\} \vee \{(d_p > 0) \wedge (\forall q \in N_p : |t_p - t_q| \leq 1)\} \vee$
$\qquad \{(d_p = 0) \wedge (\exists q \in N_p : d_q = 1 \wedge |t_p - t_q| \leq 1)\}]$

$init_p \equiv I_p(1) \vee I_p(2) \vee I_p(3)$
$\quad I_p(1) \equiv \{(t_p > 0) \vee (d_p > 1)\} \wedge \neg(t_p = M \wedge d_p = k) \wedge$
$\qquad \{\forall q \in N_p : t_q = 0 \wedge d_q = 0\}$
$\quad I_p(2) \equiv \{(0 < d_p < k) \wedge (\forall q \in N_p : d_p \geq d_q) \wedge (\exists q \in N_p : d_q > 0)\}$
$\quad I_p(3) \equiv \{\neg safe\_t_p \wedge (t_p \neq M \vee d_p \neq k) \wedge (\forall q \in N_p : t_p \geq t_q)\}$

$raise_p \equiv R_p(1) \wedge R_p(2)$
$\quad R_p(1) \equiv (t_p \neq M)$
$\quad R_p(2) \equiv \{\exists q \in N_p : (t_q = M) \wedge (d_q > 0) \wedge \neg((t_p = M - 1) \wedge (d_p = d_q - 1)) \wedge$
$\qquad (\forall r \in N_p : d_r < d_p \rightarrow t_r = M)\}$

$maxd_p \equiv M_p(1) \wedge M_p(2)$
$\quad M_p(1) \equiv (t_p > 0) \wedge (d_p \neq k)$
$\quad M_p(2) \equiv (\max_{q \in N_p}\{d_q\} \neq 0) \wedge (\max_{q \in N_p}\{d_q\} - 1 > d_p)$

$dec_p \equiv OK\_d_p \wedge OK\_t_p \wedge D_p(1) \wedge D_p(2)$
$\quad D_p(1) \equiv (t_p > 0) \wedge (\forall q \in N_p : t_p \geq t_q)$
$\quad D_p(2) \equiv (\forall q \in N_p : t_p = t_q \rightarrow d_p \geq d_q)$

$clrd_p \equiv C_p(1) \wedge C_p(2)$
$\quad C_p(1) \equiv (t_p = 0) \wedge \{\forall q \in N_p : (t_q = 0)\}$
$\quad C_p(2) \equiv (d_p > 0) \wedge \{\forall q \in N_p : (d_p \geq d_q) \vee (d_q = 0)\}$

Actions for any process p

$S_1 \quad init_p \vee Predicate_p^{init} \quad \longrightarrow \quad t_p = M; d_p = k$

$S_2 \quad raise_p \quad \longrightarrow \quad t_p = M;$
$\qquad$ **if** $(\neg((\max_{q \in N_p}\{d_q\} = k - 1) \wedge (d_p \neq k)))$ **then** $d_p = \max_{q \in N_p}\{d_q\} - 1$

$S_3 \quad maxd_p \quad \longrightarrow \quad d_p = \max_{q \in N_p}\{d_q\} - 1$

$S_4 \quad dec_p \quad \longrightarrow \quad t_p = t_p - 1; Action_p^{dec}$

$S_5 \quad clrd_p \quad \longrightarrow \quad d_p = 0$

---

from the initiator to leaves. In the third phase, after all the neighbors finish the count-down $(C_q(1) = true)$, q executes S_5 and sets $d_q = 0$. The execution of S_5 also starts from the initiator to leaves and the shortest path tree is released. Eventually, the third phase ends and $t_q = 0 \wedge d_q = 0$ holds at $\forall q \in V$.

LNS satisfies Specification 1 and 2 with $M = r_1 + r_2$ and $k = k_{1,2}$. (Due to page restriction, we omit the detailed proof.) APIs of *LNS* defined by Specification 3 is given as the parameter $Predicate_p^{init}$ and $Action_p^{dec}$ where $Predicate_p^{init} = \{(0 < t_p \leq r_2) \wedge \neg cons_p(P_1)\} \vee \{(t_p = 0) \wedge \neg(cons_p(P_1) \wedge cons_p(P_2))\}$ and $Action_p^{dec} = A(t_p)$.

4 Conclusion

We proposed a novel timer-based fault-containing composition. To implement timers, we designed a local neighborhood synchronizer protocol that is very useful in fault-containment. Though we discussed only the theoretical aspects of fault-containing composition, simulation experiments are one of the most interesting issues. Our next goal is to establish a composition framework for various types of source protocols preserving their fault-tolerance.

Acknowledgement. This work is supported in part by Global COE (Centers of Excellence) Program of MEXT, Grant-in-Aid for Scientific Research ((B)19300017, (B)17300020, (B)20300012, and (C)19500027)) of JSPS, Grand-in-Aid for Young Scientists ((B)18700059 and (B)19700075) of JSPS, Grant-in-Aid for JSPS Fellows (20-1621), and Kayamori Foundation of Informational Science Advancement.

References

1. Yamauchi, Y., Kamei, S., Ooshita, F., Katayama, Y., Kakugawa, H., Masuzawa, T.: Composition of fault-containing protocols based on recovery waiting fault-containing composition framework. In: Datta, A.K., Gradinariu, M. (eds.) SSS 2006. LNCS, vol. 4280, pp. 516–532. Springer, Heidelberg (2006)
2. Dijkstra, E.W.: Self-stabilizing systems in spite of distributed control. Communications of ACM 17(11), 643–644 (1974)
3. Chen, N.S., Yu, H.P., Huang, S.T.: A self-stabilizing algorithm for constructing spanning trees. Information Processing Letters 39, 147–151 (1991)
4. Huang, S.T., Chen, N.S.: Self-stabilizing depth-first token circulation on networks. Distributed Computing 7(1), 61–66 (1993)
5. Dolev, S., Israeli, A., Moran, S.: Self-stabilization of dynamic systems. In: Proceedings of WSS 1989 (1989)
6. Ghosh, S., Gupta, A.: An exercise in fault-containment: self-stabilizing leader election. Information Processing Letters 59(5), 281–288 (1996)
7. Ghosh, S., Gupta, A., Herman, T., Pemmaraju, S.V.: Fault-containing self-stabilizing algorithms. In: Proceedings of PODC 1996, pp. 45–54 (1996)
8. Ghosh, S., He, X.: Fault-containing self-stabilization using priority scheduling. Information Processing Letters 73, 145–151 (2000)

Calibrating an Embedded Protocol on an Asynchronous System

Yukiko Yamauchi[1], Doina Bein[2], Toshimitsu Masuzawa[3], Linda Morales[4], and I. Hal Sudborough[5]

[1] Graduate School of Information Science and Technology, Osaka University
  y-yamaut@ist.osaka-u.ac.jp
[2] Department of Computer Science, University of Texas at Dallas
  siona@utdallas.edu
[3] Graduate School of Information Science and Technology, Osaka University
  masuzawa@ist.osaka-u.ac.jp
[4] Department of Computer Science, University of Texas at Dallas
  lmorales@utdallas.edu
[5] Department of Computer Science, University of Texas at Dallas
  hal@utdallas.edu

Summary. Embedding is a method to obtain new distributed protocols for other topologies from existing protocols designed for specific topologies. But the fault tolerance of the original protocol is rarely preserved in the protocol embedded in the target topology, called *embedded protocol*. Specifically, transient faults can affect intermediate processes along the path in the target topology that corresponds to a link in the original topology.

In this paper, we propose to analyze and model the communication of the embedded protocol as unreliable communication along the links of the original protocol. We propose a particular type of unreliable channel called *almost reliable* channel and we show an implementation of these channels for embedding a protocol into another topology.

Keywords: asynchronous system, channel system, embedding, fault tolerance.

1 Introduction

Processes in distributed systems can be modeled as finite state machines, called *Communicating Finite State Machines (CFSM)*, that communicate using message-exchange over unbounded, unidirectional, error-free FIFO channels. Neighboring processes are linked by two such channels, one for sending and one for receiving. A channel is said to be *FIFO (First-In First-Out)* if it preserves the order in which the messages were sent through it.

CFSM systems [2, 3, 4, 8, 13, 14] have the power of Turing machines when the cardinality of the message alphabet is at least two. Decidability for specific problems (termination detection, reachability, etc.) is studied for either general case of CFSM or CFSM with particular type of channels (reliable, lossy, fair lossy) [1, 5, 6, 7, 10, 11, 12, 15].

C. Badica et al. (Eds.): Intel. Distributed Comput., Systems & Appl., SCI 162, pp. 227–236, 2008.

There are two particular types of FIFO channels mostly used. A *fair lossy* channel has a limited amount of lossiness [9]: If infinite many messages are sent, then infinite many messages are received. A *reliable* channel does not create or duplicate messages, and every message sent is eventually received. Independently, Finkel [5] and Abdulla and Jonsson [1] have modeled as finite automata the communication protocols using FIFO channels that allow loss of messages.

A *transient fault* corrupts the memory of some process but it does not affect its program (code). We study how we can model the effect of transient faults over the communication paths using message-passing on an embedded topology.

Embedding one (original) topology onto a target topology is often used to ease the design of distributed protocols. Yamauchi et al. proposed a ring embedding on trees that preserves the fault-tolerance of original protocols designed for rings in trees [16]. A one-to-one node embedding preserves some fault-tolerance capability since a faulty process in one topology corresponds to a faulty process in the other topology. When the *dilation* (the maximum distance in the target topology between two neighboring nodes in the original topology) is greater than one, i.e. a link between processes in the original topology becomes a simple path in the target topology, the communication in the target topology can be modeled as an unreliable channel in the original topology. We cope with faults that affect the intermediate processes along a path. For coping with transient faults at the endpoints there are already established, fault-tolerant methods as self-stabilizing, snap-stabilization, or fault-containing.

Let an original topology G_v be embedded on a target topology G_r with reliable channels. If a transient fault corrupts some processes in the target topology, we cannot ensure a reliable communication on G_v. For any pair of adjacent processes in G_v, the path between them in the target topology G_r will be modeled as a single channel. Thus a transient fault that affects the communication of the embedded protocol (that corrupts some or all the intermediate processes along that path) can be modeled as the reliability of the direct channels in the original protocol.

We propose a particular type of unreliable channels, channels with failures [9], where a failure is considered either a message loss or a duplication, called *almost reliable* channels. We show the usefullness of this type of channels by proposing a transformer that takes as input a protocol P_v running on a topology G_v with almost reliable channels and a one-to-one node embedding λ of G_v onto another topology G_r, and outputs a protocol P_r running on topology G_r with reliable channels that solves the same task as the original protocol. Also our proposed transformer preserves the fault-tolerant property of the original protocol, namely if the original protocol P_v is fault-tolerant, then the output protocol P_r will have the fault-tolerant characteristic preserved. At the price of message duplication (some messages can be received twice), we ensure no message loss and no message creation. Furthermore, by attaching sequence numbers to messages we can exclude message duplication.

The paper is organized as follows. The communication model and the embedding are given in Section 2. In Section 3, we define the almost reliable channels, the embedding transformation problem, and Transformer $\mathcal{T}_\lambda$. A sketch of the proof of correctness is given in Section 4. We finish with concluding remarks in Section 5.

2 Models

A distributed system is represented by a graph $G = (V, E)$ where the vertex set $V = \{p_1, p_2, \cdots, p_n\}$ is a set of n processes and the edge set E is a set of bidirectional channels. Process p_i is a neighbor of process p_j iff there exists a communication link $(p_i, p_j) \in E$.

We consider the message-passing model of communication: A process p_i communicates with some adjacent process p_j by invoking $send(m)$ to p_j. Process p_j receives some message m when it terminates the invocation of $receive(m)$. Action $receive(m)$ at p_i triggers some internal computation at p_i that changes the state of p_i. Each action of $send(m)$, $receive(m)$, and any internal computation is executed atomically. For any two messages m_1 and m_2 sent on a FIFO channel e, if $send(m_1)$ occurs earlier than $send(m_2)$, then $receive(m_1)$ occurs earlier than $receive(m_2)$.

Each process p_i holds a set of local variables and the values of all local variables of p_i define the local state of p_i. A *transient fault* can affect many processes by changing the values of the processes' variables to arbitrary ones, but it does not affect the code of the processes (the set of actions to be executed when $send()$ or $received()$ is invoked). In other words, for any process the data area may be corrupted by a transient fault, but not the program area.

Each process executes a protocol P that consists of a finite number of statements called *guarded actions*, of the form $\langle label \rangle :: \langle guard \rangle \rightarrow \langle action \rangle$. The guard of an action is a Boolean expression involving the variables of the process and of its neighboring processes. The action can be executed only if its guard evaluates to *true*. A process with at least one enabled guard is called *enabled*. A distributed daemon will select non-deterministically a non-empty subset of enabled processes to execute one of the enabled actions. We assume that the actions are atomically executed: the evaluation of a guard and the execution of the corresponding action, if it is selected for execution, are done in one atomic step.

Let $\lambda : E_v \rightarrow 2^{E_r}$ be an one-to-one node-embedding of a topology $G_v = (V, E_v)$ onto a target topology $G_r = (V, E_r)$. (The set of nodes for both graphs G_v and G_r is isomorphically the same.) Let k be the dilation of the embedding λ which is the maximum distance in G_r between two adjacent nodes in G_v.

When p_i sends some message m to its neighbor p_j in G_v, message m should be relayed from p_i to p_j in G_r. The embedding λ provides a routing function at each process p, $f_\lambda^p : V \times V \rightarrow V$ in G_r, that enables message exchange between neighboring processes in G_v. Let f_λ be the collection of the routing functions of all processes. For each process p_k, the routing function $f_\lambda^{p_k}(p_i, p_j)$ specifies the process r to which p_k should relay a message either generated locally at p_k (when $p_i = p_k$) or received from p_i on p_k's channels, and addressed to p_j. If p_k receives a message addressed to p_k, it *delivers* the message in G_v. Otherwise, p_k forwards the message further. In both cases the message is stored locally at p_k, and the local copy will be sent further.

Every message received at process p_i and not addressed to p_i will be forwarded to the correct neighbor by sending it to the neighbor defined by $f_\lambda^{p_i}$. We assume that it takes at most δ_{p_i} rounds for a message to be forwarded, where δ_{p_i} is the degree of process p_i (the number of communication links incident to p_i). This delay is caused by the embedding and is not related to the transformation problem.

The *k-majority* (for short, majority) is a value that appears at least k times in a set of $2k - 1$ values. If we have less than $2k - 1$ values, or there is no value that appears at least k times, then the majority is undefined ($\perp$).

Definition 1. Almost Reliable channels
A FIFO channel (p_i, p_j) is almost reliable if it satisfies the following conditions:
- For any message m sent by p_i to p_j, if there is no fault at p_i during the send action and at p_j during the receive action, m is eventually received but may be duplicated.
- p_j does not receive any message that was not sent by p_i, if there is no fault at p_i and p_j (no message creation).

If there is a fault at endpoints during the transmission of a message m, an almost reliable channel guarantees nothing about m.

Definition 2. Transformation problem for reliable channels
Design a transformer $\mathcal{T}_\lambda$ that takes as input a protocol $\mathcal{P}_v$ (original protocol), designed for an asynchronous system with almost reliable channels G_v (original topology), and the one-to-one node embedding λ, and gives as output the protocol $\mathcal{P}_r$ (embedded protocol) that runs on an asynchronous system with reliable channels modeled as the graph G_r (target topology).

3   Calibrating the Network

If the following three conditions are satisfied in $\mathcal{P}_v$:

(1) message m is sent by p_i in $\mathcal{P}_v$ to p_j along the link in the original topology $e = (p_i, p_j) \in E_v$ with Action $send(m)$,
(2) the message m is delivered at process p_j in $\mathcal{P}_v$ with Action $receive(m)$,
(3) there are no faults at processes p_i and p_j during the transmission,

then, in Protocol $\mathcal{P}_r$, message m sent by process p_i along the embedded path $ee = \lambda(e) = (p_i, \ldots, p_j)$ is delivered at process p_j, independent on whether there are transient faults at intermediate processes on the embedded path.

We implement the following transformer $\mathcal{T}_\lambda$ (Algorithm 3). For each process p_i, the atomic guarded actions $send(m)$ and $receive(m)$ of the original protocol $\mathcal{P}_v$ are replaced by the atomic actions $snd(m)$ (Action S) and $rcvd(m)$ (Action R) that will be executed by the same process in Protocol $\mathcal{P}_r$.

Each process p_i holds a variable counter c_{p_i} with non-negative integer values, a variable mj_{prev} that stores the previously delivered majority, and a stack $stack$ that stores the messages addressed to p_i. The size of the stack must be at least $3k$.

The guard $snd(m)$ is *true* whenever process p_i sends a message m to process p_j. In topology G_v, p_i sends m along the link $e = (p_i, p_j)$, while in the target topology G_r, process p_i sends m $2k - 1$ times along the path $ee = (p_i, \ldots, p_j) = \lambda(p_i, p_j)$. When p_i sends $2k - 1$ copies of message m in P_r, we assume that each intermediate process along the path $ee = (p_i, \ldots, p_j) = \lambda(p_i, p_j)$ holds at any time at most one copy of m.

The guard $rcvd(m)$ is *true* when p_i receives a message m which is addressed to itself. Whenever process p_j counts and stacks $2k - 1$ messages (addressed to it) from process

---

Algorithm 3.1. *Algorithm* $\mathcal{T}_\lambda$ *(Fault-Tolerance-Preserving Embedding)*

Communication Actions for any process i:
S $snd(m)$:: **for** j=1 **to** $2k - 1$ **do** $send(m)$

R $rcvd(m)$ :: /* store value m read in the stack and increment c_i
          $push(m, stack_i)$
          c_i++
          /* test whether majority of the top $2k - 1$ values from $stack_i$ needs to be computed */
          if $c_i \geq 2k - 1$ **then**
              $mj = majority(stack_i, 0, 2k - 1)$ /* compute maj. of the last $2k - 1$ values */
              if $mj = \perp$ **then**
                  /* the maj. is $\perp$, look for maj. in $stack$ by removing the top j values */
                  $j = 1$
                  while $((mj = majority(stack_i, j, 2k - 1)) == \perp \wedge j < 2k - 1)$ **do** $j++$
                  if $mj = mj_{prev}$ **then** $mj = \perp$ /* wait for more data */
              if $mj \neq \perp$
                  $mj_{prev} = mj$ /* store the latest defined maj. */
                  $receive(mj)$
                  $c_i = 0$

---

p_i, it computes the k-majority on the latest $2k - 1$ messages. If the result is defined (not $\perp$) then it is delivered at p_j. Else, if a new majority (different from the previous stored one) can be obtained from the stack by removing $1, 2, \ldots, k - 1$ messages, then that majority is delivered at p_j. If the previous majority if obtained by such operation, then process p_i waits for more messages to come (it delays the delivery of a message). Once some process p computes the majority over the received data, if the result is defined ($\neq \perp$), it will be *delivered* to p.

4   Proof of Correctness

We show that transient faults that may occur at the intermediate processes along the path *ee* do not prevent the delivery of message m to process p_j. In Transformer $\mathcal{T}_\lambda$, sending $2k - 1$ copies of some message m is not necessarily executed atomically at a process p_i. The receiving process of the $2k - 1$ copies (corrupted or not) of m executed at the destination p_j is non-atomic. Transformer $\mathcal{T}_\lambda$ guarantees that message m will be relayed at p_j if there is no fault at p_i and p_j during the transmission of the $2k - 1$ copies. Message m may be duplicated at p_j but cannot be lost (Theorem 1).

Whenever a message addressed to p_i is received at p_i, c_{p_i} is incremented and the message is pushed into $stack_i$. Whenever c_{p_i} passes the threshold $2k - 1$, the majority function $majority(stack, int, 2k - 1)$ is applied which returns the majority of the data starting from intth data to $(int + 2k - 1 - 1)$th data (the 0th data means the top data in the stack $stack_i$.): e.g. $majority(stack_i, 0, 2k - 1)$ returns the majority of top $2k - 1$ data in $stack_i$. If $majority(stack_i, 0, 2k - 1)$ returns a definite result ($\neq \perp$), then it is delivered at p_i and c_{p_i} is reset to 0. An undefined result ($\perp$) caused by either arbitrary initialization or faults at intermediate nodes delays the relaying of data to p_i. Then p_i looks into $stack_i$ for some defined majority (by eliminating the top element, the top two elements, ..., the top $k - 1$ elements) and if the value is not delivered at p_i, p_i delivers the value. Otherwise it delays delivering the message (decision).

We show that a decision will be taken in finite time (Lemma 2). Starting from an arbitrary initialization or after a transient fault, during any $2k - 1$ consecutive rounds, a decision will be taken at least once (Theorem 1).

When majority is applied at every process when the counter is $2k - 1$ and the result is defined ($\neq \perp$), we say that the network is *calibrated*.

We make the following assumption for transient faults.

Assumption 1. *For any pair of processes p_i and p_j that are adjacent in G_v, between the time p_i sends the first copy of a message m and the time p_j receives the last copy of m (corrupted or not), any process is affected by a fault at most once.*

This assumption is translated for the original protocol $\mathcal{P}_v$ as follows: Between the sending of a message m by p_i and receiving of m by p_j, any process can have its data corrupted at most once (which is always assumed to be the case for any protocol). In the following, we consider the case where p_i and p_j are not corrupted by faults since the almost reliable channel guarantees nothing if p_i or p_j is corrupted.

In case a transient fault affects some processes in the topology G_r, the number of corrupted copies of the same message sent by p_i and received by p_j in G_r is no greater than the number of corrupted intermediate processes on the path in G_r between p_i and p_j. Since the dilation of the embedding is k, the maximum number of intermediate processes that can be affected by transient faults is $k - 1$. If $send(m)_{ij}$ was executed by Protocol $\mathcal{P}_v$ (message m was sent by node p_i along the link $e = (p_i, p_j)$) in the original topology, that corresponds to an execution of $snd(m)_{ij}$ at p_i in Protocol $\mathcal{P}_r$ (message m was sent (at least) $2k - 1$ times along the path $ee = (p_i, \ldots, p_j) = \lambda(e)$) in the target topology. Thus, in a sequence of $2k - 1$ consecutive data that are copies of the same message m, at most $k - 1$ data can be corrupted.

Definition 3. *Let S be some (in)finite string obtained by concatenating the $2k-1$ copies (corrupted or not) of message m, followed by the copies of the next message, sent by some process p_i and addressed to and received by some process p_j. Then, there exists two unique positions k_l^S and k_r^S in S, (position 1 is when the first copy of m is received by p_j), such that the majority computed on S returns m iff it is computed between the positions $k_l^S .. k_r^S$ of S.*

Observation 1 follows directly from Definition 3.

Observation 1. *For any $k > 1$, $k \leq k_l^S \leq k_r^S \leq 2k - 1 + k - 1$.*

The values for k_l^S and k_r^S depend on the position(s) of the corrupted data in the $(2k - 1)$-string.

Observation 2. *If no data is corrupted, then $k_l^S = k$, $k_r^S = 2k - 1 + k - 1 = 3k - 2$, and the majority$(stack_i, 0, 2k - 1)$ is applied between in the position of k ($k_l^S \leq k \leq k_r^S$) at least once and returns m.*

Proof. Majority returns m iff the majority is computed any time after k copies have been received, but no later than when at most $k - 1$ copies of the next message are received.

The majority $majority(stack_i, 0, 2k-1)$ is computed at least once in the $2k-1$ *receive* actions.

String	k_l	k_r
m m m m m	3	7
m' m m m m	4	7
m m' m m m	4	7
m m m' m m	4	6
m m m m' m	3	6
m m m m m'	3	6

String	k_l	k_r
m' m'' m m m	5	7
m' m m'' m m	5	6
m' m m m'' m	5	6
m' m m m m''	4	6
m m' m'' m m	5	5

String	k_l	k_r
m m' m m'' m	5	5
m m' m m m''	4	5
m m m' m'' m	5	5
m m m' m m''	4	5
m m m m' m''	3	5

Fig. 1. At most 2 messages out of 5 are corrupted

Let S be the string where the first $k - 1$ data are correct copies of the message m, the next $k - 1$ data are corrupted copies of the m, and the last data is a correct copy of m. Then $k_l^S = k_r^S = 2k - 1$. Another "bad" string is an alternating sequence of non-corrupted copies and corrupted copies, starting with a non-corrupted copy. In this case, $k_l^S = k_r^S = 2k - 1$ also.

If l is the number of faults that have occurred, then there are $\binom{2k - 1}{l}$ possible strings with l copies corrupted, thus the total number of possible combinations is

$$\sum_{t=1}^{k-1} \binom{2k - 1}{t} = 2^{2k-2}.$$

In Figure 1 we present a particular example when $k = 3$ and $l \le 2$.

Property 1. *If the majority computed when t $(0 < t \le 2k - 2)$ copies of message m have been received returns $\perp$, and by analyzing the stack no new majority can be computed, then $\exists\, t_0 > 0$ such that the majority computed when $t + t_0$ copies of m have been received returns m $(t + t_0 \le 2k - 1)$.*

Proof. Out of $2k - 1$ copies of m, at most $k - 1$ copies can be corrupted by a single fault at intermediate processes.

Assume that the majority computed when t messages have been received returns $\perp$ and assume that m should be the new majority. We have two cases, depending on the value of t:

1. If $t < k$ then, by examining the stack, only the previous majority can be computed. But then there are $2k - 1 - t > k - 1$ copies of m still to be received. The latest once all $2k - 1$ copies of m (corrupted or not) have been received, the majority computed at that moment returns m, thus $t + t_0 \le 2k - 1$.
2. If $t \ge k$ then, by examining the stack either the old majority (if less than k correct copies of m have been received) or the new majority can be obtained (if at least k correct copies of m exists already in the stack). If the new majority can be obtained then $t_0 = 0$. If only the old majority can be obtained, it implies that the process has to delay delivering a data, and it continues to increment its counter. The latest once all $2k - 1$ copies of m (corrupted or not) have been received, the majority computed at that moment returns m, thus $t + t_0 \le 2k - 1$.

Lemma 1. *For any node p_i, except for a finite prefix of values, the values of the counter c_{p_i} when the majority is applied and returns a defined value ($\neq \perp$) represent an infinite series IS with bounded values, that is non-decreasing if no fault occurs. If a fault occurs, then the values of this series may "re-start" with a value higher than $2k - 1$ after the fault.*

Proof. In finite time, starting from an arbitrary initialization, the majority will return a value $\neq \perp$. Let l_i be the first value of the counter c_{p_i} when such majority is computed. From Observation 2 and Property 1 we obtain that $2k - 1 \leq l_i \leq 3k - 2$. At that moment the counter c_{p_i} is reset to 0, and the next majority will be computed when c_{p_i} equals to $2k - 1$ and it remains so. (The process becomes *calibrated*). A fault at p_i may cause the majority to be computed before all $2k - 1$ values are received, thus the series IS may have a sudden increase in value after the fault.

The following corollary derives directly from Lemma 1:

Lemma 2. *A process p_i will decide in finite time.*

Proof. The series IS is non-increasing and the values are lower bounded by $2k - 1$ and upper bounded by $3k - 2$. Thus for p_i, the number of times a decision is delayed by a fault is finite.

If $c_{p_i} > 2k - 2$ and majority cannot be decided, then after at most $k - 1$ messages there will be a decided majority. Assume that process p_i had a decided majority when c_{p_i} was previously 0. Majority will return a defined value at least when all the $2k - 1$ copies are received, if not earlier.

Theorem 1. *Between two consecutive transient faults, for any message m sent by p_i in Protocol P_v for p_j, in Protocol $P_r = \mathcal{T}_\lambda(P_v)$, message m sent by p_i will be relayed at p_j if there is no fault at p_i or p_j during the communication of the $2k - 1$ copies of m. In any sequence of $2k - 1$ messages received by some process p_i, a decision will be taken at p_i at least once.*

Proof. Transformer $\mathcal{T}_\lambda$ solves the transformation problem.

Let e be some link in the original topology where some message m was sent by protocol $\mathcal{P}_v$ (original protocol) that runs on an asynchronous system with almost reliable channels modeled as the graph G_v (target topology).

If message m is sent along the link $e = (p_i, p_j)$, then $2k-1$ copies of m are received by process p_j running protocol $\mathcal{P}_r$. If majority at p_j, applied after $2k - 1$ time units, returns $\perp$, then it takes longer than $2k - 1$ time units for p_j to have the message m delivered to itself.

By Property 1, message m is eventually delivered at p_j. Let p_j sends some message m' to its neighbor p_k in P_v (with the link $e_{next} = (p_j, p_k)$). Until m is not delivered at p_j, p_j does not do any internal computation to change its state. Thus p_k may receive more than $2k - 1$ copies of the message m' sent by p_j before message m is delivered to p_j.

The maximum delay of a message m sent by p_i to p_j in P_v is $(3k - 1)$. A delay of a message m means the time between the time p_i sends the first copy of m and the time p_j receives the last (corrupted or not) copy of m. This is because p_j receives at most $(k - 1)$ arbitrary data till it receives the first (corrupted or not) copy of m and after that

p_j receives $2k - 1$ (corrupted or not) copy of m. Before p_j receives the last copy of m, it delivers non-$\perp$ value by applying the majority function. If process p_k receives l copies of some message m', $2k - 1 \leq l < 3k - 1$, then message m' will be delivered at p_k at most twice.

Thus we have constant-duplication of messages, but no message creation, and neither message loss.

The following corollary holds almost directly:

Corollary 1. *If a message exchange between any two neighboring processes in G_r takes at most δ_v rounds, the delay in executing an action at process p_i in protocol P_r is of at most $(3k - 1) * \delta_p$ rounds.*

Proof. We do not have a message duplicated more than twice in a row for the same message.

As described above, the transformed protocol P_r allows message duplication (i.e., a message may be delivered twice at the destination process). This requires that the original protocol P_v should work on a topology G_v with *almost reliable* channels. However, by attaching a sequence number to each message, we can easily avoid the message duplication of P_r, and thus, we can accept as an input a (weaker) protocol that works on a topology G_v with (completely) reliable channels.

5   Conclusion

We propose the notion of almost reliable channels. We show how the almost reliable channels are used in embedding a protocol designed for one topology into another topology. Our implementation guarantees the lossless of message communication (even in the presence of transient faults) that preserves the reliability of the original protocol at the price of constant-message duplication. Our next goal is to investigate other methods to embed protocols for some specific topologies that are lossless, non-creating, non-duplicating of messages, and fault-contained.

Acknowledgement

This work is supported in part by Global COE (Centers of Excellence) Program of MEXT, Grant-in-Aid for Scientific Research ((B)19300017) of JSPS, and Grant-in-Aid for JSPS Fellows (20-1621).

References

1. Abdulla, P.A., Jonsson, B.: Verifying programs with unreliable channels. In: Proceedings of the 8th IEEE International Symposium on Logic in Computer Science, pp. 160–170 (June 1993)
2. Aggarwal, S., Gopinath, B.: Special issue on tools for computer communication systems. IEEE Transactions on Software Engineering 14(3), 277–279 (1988)

3. Bochmann, G.V.: Finite state description of communication protocols. Computer Networks 2, 361–371 (1978)
4. Brand, D., Zafiropulo, P.: On communicating finite-state machines. Journal of ACM 30(2), 323–342 (1983)
5. Finkel, A.: Decidability of the termination problem for completely specified protocols. Distributed Computing 7, 129–135 (1994)
6. Finkel, A., Rosier, L.: A survey on decidability results for classes of fifo nets. In: Rozenberg, G. (ed.) APN 1988. LNCS, vol. 340, pp. 106–132. Springer, Heidelberg (1988)
7. Finkel, A., Sutre, G.: Decidability of reachability problems for classes of two counters automata. In: Reichel, H., Tison, S. (eds.) STACS 2000. LNCS, vol. 1770, pp. 346–357. Springer, Heidelberg (2000)
8. Gouda, M.: To verify progress for communicating finite state machines. IEEE Transactions on Software Engineering 10(6), 846–855 (1984)
9. Lynch, N.A.: Distributed algorithms. Morgan Kaufmann Publishers, Inc., San Francisco (1996)
10. Pachl, J.: Reachability problems for communicating finite state machines. Technical Report CS-82-12, University of Waterloo (1982)
11. Rosier, L., Yen, H.: Boundedness, empty channel detection, and synchronization for communicating finite automata. Theoretical Computer Science 44, 69–105 (1986)
12. Schnoebelen, P.: The verification of probabilistic lossy channel systems. In: Baier, C., Haverkort, B.R., Hermanns, H., Katoen, J.-P., Siegle, M. (eds.) Validation of Stochastic Systems. LNCS, vol. 2925, pp. 445–466. Springer, Heidelberg (2004)
13. Sunshine, C.: Formal techniques for protocol specification and verification. Computer Journal 12(9), 20–27 (1979)
14. Sunshine, C.: Formal Modeling of Communication Protocols. In: Schoemaker, S. (ed.) Computer Networks and Simulation, vol. II, pp. 141–165. Elsevier, North-Holland Publishing, Amsterdam (1982)
15. Vuong, S.T., Cowan, D.D.: Reachability analysis of protocols with FIFO channels. Computer Communication Review 13(2), 49–57 (1983)
16. Yamauchi, Y., Masuzawa, T., Bein, D.: Ring embedding preserving the fault-containment. In: Proceedings of the 7th International Conference on Applications and Principles of Information Science, pp. 43–46 (2008)

Part III

Short Papers

On Establishing and Fixing a Parallel Session Attack in a Security Protocol

Reiner Dojen[1], Anca Jurcut[1], Tom Coffey[1], and Cornelia Gyorodi[2]

[1] Department of Electronic & Computer Engineering, University of Limerick, Ireland
reiner.dojen@ul.ie, anca.jurcut@ul.ie, tom.coffey@ul.ie
[2] Department of Computer Science, University of Oradea, Romania
cgyorodi@uoradea.ro

Summary. Nowadays mobile and fixed networks are trusted with highly sensitive information, which must be protected by security protocols. However, security protocols are vulnerable to a host of subtle attacks, such as replay, parallel session and type-flaw attacks. Designing protocols to be impervious to these attacks has been proven to be extremely challenging and error prone.

This paper discusses various attacks against security protocols. As an example, the security of the Wide-Mouthed Frog key distribution protocol when subjected to known attacks is discussed. Significantly, a hitherto unknown attack on Lowe's modified version of the Wide-Mouthed Frog protocol is presented. Finally, a correction for the protocol to prevent this attack is proposed and discussed.

Keywords: Security protocols, protocol flaws, parallel session attack.

1 Introduction

With the ever increasing use of distributed applications for meeting customer demands, there is a commensurate increase in the reliance of electronic communication over networks. Since many of these applications require the exchange of highly sensitive information over computer networks, the need for security protocols to ensure the protection and integrity of data is critical. Basic cryptographic protocols allow protocol principals to authenticate each other, to establish fresh session keys for confidential communication and to ensure the authenticity of data and services. Building on such basic cryptographic protocols, more advanced services like non-repudiation, fairness, electronic payment and electronic contract signing are achieved.

In this paper we introduce attacks against security protocols. As an example, the Wide-Mouthed Frog key distribution protocol is discussed and known attacks on the protocol are detailed. Additionally, a hitherto unknown attack on Lowe's modified version of the Wide-Mouthed Frog protocol is presented. A correction for the protocol to prevent the attack is proposed and discussed.

C. Badica et al. (Eds.): Intel. Distributed Comput., Systems & Appl., SCI 162, pp. 239–244, 2008.
springerlink.com

2   Attacks Against Security Protocols

A security protocol should enforce the data exchange between honest principals, while the dishonest ones should be denied any benefit of it. However, security protocols can contain weaknesses that make them vulnerable to a range of attacks such as replay, parallel session and type-flaw attacks. Many security protocols have been found to contain weaknesses, which have subsequently been exploited, for example [1][2][3][4][5][6][7][8][9]. This highlights the difficulty of designing effective security protocols.

A replay attack is one of the most common attacks on authentication and key-establishment protocols. If the messages exchanged in an authentication protocol do not carry appropriate freshness identifiers, then an adversary can get himself authenticated by replaying messages copied from a legitimate authentication session. Examples of replay attacks include [1][2][5][6].

A parallel session attack requires the parallel execution of multiple protocol runs, where the intruder uses messages from one session to synthesise messages in the other session. Examples of parallel session attacks on security protocols include [1][3][4][7]. There are several forms of parallel session attacks. In an oracle attack the intruder starts a new run of the protocol and uses one of the principals as an oracle for appropriate answers to challenges in first protocol run. A man-in-the-middle attack occurs when two principals believe they are mutually authenticated, when in fact the intruder masquerades as one principal in one session and as the other principal in another. A multiplicity attack is a parallel session attack where the principals disagree on the number of runs they have successfully established with each other.

A type flaw attack involves the replacement of a message component with another message of a different type by the intruder. Examples of type flaw attack on security protocols include [4][8][9].

3   The Wide-Mouthed Frog Protocol

To illustrate the difficulty of designing effective security protocols, the Wide-Mouthed Frog protocol, published by Burrows, Abadi and Needham [2], is discussed. It was proposed for the distribution of a fresh key between peer entities. It assumes the use of symmetric key cryptography, a trusted server and synchronized clocks. Anderson and Needham claimed a replay attack on the protocol [10] and another attack was discovered by Lowe, who modified the protocol to prevent these attacks[11]. Further, the authors of this paper present a hitherto unknown attack on the protocol. The presence of flaws in the protocol in spite of the repeated revisions demonstrates the difficulty of designing effective security protocols.

Fig. 1 depicts the steps of the Wide-Mouthed Frog protocol: In step 1 A sends timestamp Ta, the identity of B and the new session key Kab, encrypted with Kas to server S. The server forwards the session key, along with A's identity and timestamp Ts to B using Kbs. Table 1 explains the used notation.

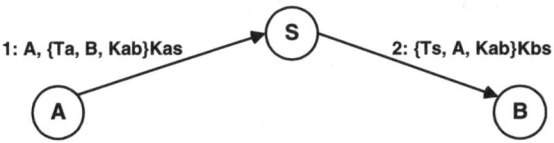

Fig. 1. The Wide-Mouthed Frog Protocol

Table 1. Notation for Protocol Descriptions

{m}k	message m encrypted with key k
A,B	legitimate principals
S	trusted server
Kas/Kbs	symmetric key shared between A & S and B & S
I(A)/I(B)	intruder masquerading as principal A/B
Ta/b/s	timestamp created by A/B/S
Na/b	once off random number (nonce) generated by principal A/B

To-date, two attacks on the Wide-Mouthed Frog protocol have been published. Anderson and Needham claim a replay attack [10] and Lowe a multiplicity attack [11].

3.1   A Replay Attack on the Wide-Mouthed Frog Protocol

The replay attack on the Wide-Mouthed Frog protocol enables an attacker to keep the session key Kab alive [10]. It assumes that the server does not keep a list of all recent keys.

The attack - depicted in Fig. 2 - works as follows: After intercepting the message exchange of the protocol run i, the intruder starts a new run of the protocol (ii) masquerading as B and sends to the server as message ii.1 a copy of message i.2. Intercepting message ii.2, the intruder can start run iii of the protocol and masquerade as A and so on. As a result the session key Kab is kept alive by the intruder.

```
i.1. A -> S : A, {Ta, B, Kab}Kas
i.2. S -> B : {Ts, A, Kab}Kbs
   ii.1. I(B) -> S : B, {Ts, A, Kab}Kbs
   ii.2. S -> A : {Ts', B, Kab}Kas
      iii.1. I(A) -> S : A, {Ts', B, Kab}Kas
      iii.2. S -> B : {Ts'', A, Kab}Kbs
      ....
```

Fig. 2. Replay attack on the Wide-Mouthed Frog Protocol

3.2   A Multiplicity Attack on the Wide-Mouthed Frog Protocol

Although Burrows, Abadi, and Needham claimed that both time stamps and nonces ensure freshness of message exchanges in the Wide-Mouthed Frog protocol [2], Lowe demonstrated that time stamps only ensure recentness of a message [11]. The protocol is thus susceptible to a multiplicity attack.

In this attack an intruder records the second message and replays it to B. Alternatively, the intruder could replay the first message to the server. As a result, B is lead into thinking that A is trying to establish a second session, whereas A has established only one session. The attack involves two interleaved runs of the protocol, as shown in Fig. 3. For this attack to succeed it is assumed that the principals do not keep lists of recent keys.

```
i.1. A -> S : A, {Ta, B, Kab}Kas
i.2. S -> B : {Ts, A, Kab}Kbs
  ii.2. I(S) -> B : {Ts, A, Kab}Kbs
```

Fig. 3. Multiplicity Attack on the Wide-Mouthed Frog Protocol

3.3   Lowe's Modified Wide-Mouthed Frog Protocol

In Lowe's modified version of the Wide-Mouthed Frog protocol [11] a nonce handshake between A and B (cf. Fig. 4) has been added to prevent the presented attacks. To-date no weaknesses of Lowe's modified versions of the Wide-Mouthed Frog protocol have been published.

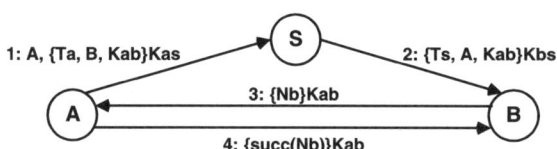

Fig. 4. Lowe's Modified Wide-Mouthed Frog Protocol

4   A New Attack on Lowe's Modified Protocol

We now present a hitherto unknown parallel session attack on Lowe's modified Wide-Mouthed Frog protocol, which uses A as an oracle. Consequently, the intruder impersonates principal B to A. Fig. 5 details the attack, which assumes that A initiates the protocol. The intruder intercepts message i.2 and starts a new protocol run as B, using message i.2 as part of ii.1. The honest principal A decrypts message ii.2, extracts key Kab and responds by sending message ii.3 which contains a freshly generated nonce Na encrypted with this session key. The intruder then replays ii.3 as its next message in run i (i.3.). In response, A returns the encrypted successor of Na (i.4.), which the intruder replays as final message in run ii (ii.4).

```
i.1. A -> S : A, {Ta, B, Kab}Kas
i.2. S -> I(B) : {Ts, A, Kab}Kbs
    ii.1. I(B) -> S : B, {Ts, A, Kab}Kbs
    ii.2. S -> A : {Ts', B, Kab}Kas
    ii.3. A -> I(B) : {Na}Kab
i.3. I(B) -> A : {Na}Kab
i.4. A -> I(B) : {succ(Na)}Kab
    ii.4. I(B) -> A : {succ(Na)}Kab
```

Fig. 5. New Attack on Lowe's Modified Wide-Mouthed Frog Protocol

Consequently, A believes that a session has been established with B in run i and also believes that B has established a session in run ii, even though B is in fact absent. Therefore, the protocol does not ensure authentication.

The protocol suffers from two weaknesses: One is the symmetry of the first two encrypted messages of the protocol, which allows the intruder to replay the second message of the first protocol run (i.2) as the initial message of the second run (ii.1). The other and major weakness is that it is not possible to establish the source of the second last message.

4.1 Fixing the Flaw

In order to fix the protocol, the source of message 3 needs to be established. This can be achieved by including the sender's identity in the third message of the protocol as presented in Fig. 6.

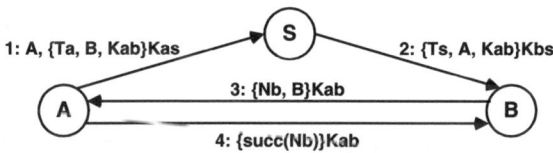

Fig. 6. Proposed Correction to Wide-Mouthed Frog Protocol

As message 3 now includes the encrypted identity of B (originator of the message), an intruder cannot successfully replay the message in a parallel session with switched roles of initiator and responder: Message ii.3 in Fig. 5 becomes {Na, A}Kab, whereas message i.3 requires {Na, B}Kab. Thus, the presented attack is prevented. Further, formal verification of the proposed protocol using the OFMC tool [12] provides confidence in its correctness.

5 Conclusion

A security protocol should enforce the data exchange between honest principals, while the dishonest ones should be denied any benefit of it. However, security protocols can contain weaknesses that make them vulnerable to a range of attacks, such as replay, parallel session and type-flaw attacks. Designing protocols to be resistant against these attacks has been proven very challenging as is highlighted by the large number of security protocols that have been found to contain exploitable weaknesses.

This paper analysed the Wide-Mouthed Frog key distribution protocol and known attacks on the protocol were presented. Additionally, a hitherto unknown attack on Lowe's modified version of the Wide-Mouthed-Frog protocol that allows an intruder to impersonate an legitimate principal was detailed. A correction for the protocol to prevent the attack was proposed and discussed.

Acknowledgement

This work was funded by the Irish Research Council for Science, Engineering and Technology (IRCSET Embark Initiative) and Science Foundation Ireland - Research Frontiers Programme (SFI RFP07 CMSF 631).

References

1. Boyd, C., Mathuria, A.: Protocols for authentication and key establishment. Springer, Berlin (2003)
2. Burrows, M., Abadi, M., Needham, R.: A logic of authentication. ACM Transactions on Computer Systems TOCS 8(1), 18–36 (1990)
3. Lowe, G.: An attack on the Needham-Schroeder public key authentication protocol. Information Processing Letters 56(3), 131–136 (1995)
4. Lowe, G.: Some new attacks upon security protocols. In: Proceedings of Computer Security Foundations Workshop VIII. IEEE Computer Society Press, Los Alamitos (1996)
5. Denning, D., Sacco, G.: Timestamps in key distributed protocols. Communication of the ACM 24(8), 533–535 (1981)
6. Aura, T.: Strategies against replay attacks. In: Proceedings of the 10th IEEE Computer Society Foundations Workshop, Rockport, MA, pp. 59–68 (June 1997)
7. Nam, J., Kim, S., Park, S., Won, D.: Security analysis of a nonce-based user authentication scheme using smart cards. IEICE Transactions Fundamentals 90(1), 299–302 (2007)
8. Hwang, T., Lee, N.Y., Li, C.M., Ko, M.Y., Chen, Y.H.: Two attacks on Neumann-Stubblebine authentication protocols. Information Processing Letters 53, 103–107 (1995)
9. Heather, J., Lowe, G., Schneider, S.: How to prevent type flaw attacks on security protocols, pp. 255–268. IEEE Computer Society, Los Alamitos (2000)
10. Anderson, R., Needham, R.: Programming Satan's Computer. In: van Leeuwen, J. (ed.) Computer Science Today. LNCS, vol. 1000, pp. 426–440. Springer, Heidelberg (1995)
11. Lowe, G.: A family of attacks upon authentication protocols. Technical Report 1997/5, Dept. Mathematics & Computer Science, University of Leicester (1997)
12. Basin, D., Mdersheim, S., Vigan, L.: OFMC: A symbolic model checker for security protocols. Int. Journal of Information Security 4(3), 181–208 (2005)

Constructing Security Protocol Specifications for Web Services

Genge Bela, Haller Piroska, and Ovidiu Ratoi

"Petru Maior" University of Targu Mures, Electrical Engineering Department
Nicolae Iorga St., No. 1, Mures, RO-200440, Romania
{bgenge,phaller,oratoi}@engineering.upm.ro

Summary. In order to integrate new security protocols, existing systems must be modified accordingly, which often means interrupting system activity. We propose a solution to this problem by developing an ontology model which provides semantic to security protocol operations. The proposed model is based on a formal specification model and is integrated in existing Web service description technologies.

1 Introduction

Security protocols are widely used today to provide secure communication over insecure environments. By examining the literature we come upon various security protocols designed to provide solutions to specific problems ([10]). With this large amount of protocols to chose from, distributed heterogenous systems must be prepared to handle multiple security protocols.

Existing technologies, such as the Security Assertions Markup Language ([12]) (i.e. SAML), WS-Trust ([15]) or WS-Federation ([14]) provide a unifying solution for the authentication and authorization issues through the use of predefined protocols. By implementing these protocols, Web services authenticate users and provide authorized access to resources. However, despite the fact that existing solutions provide a way to implement security claims, these approaches are rather static. This means that in case of new security protocols, services supporting the mentioned security technologies must be reprogrammed.

In this paper, we propose a more flexible solution to this problem by developing an ontology model aiming at the automatic discovery and execution of security protocols. An ontology is a "formal, explicit specification of a shared conceptualization" ([1]), consisting of concepts, relations and restrictions. Ontologies are part of the semantic Web technology, which associates semantic descriptions to Web services.

In order to construct the proposed ontology model, we first create an enriched formal specification model for security protocols. In addition to the information provided by existing formal models, such as the SPI calculus ([2]), the strand space model ([3]) or the operational semantics ([4]), we also include explicit processing operations. These operations are then translated to semantic concepts and properties in the proposed ontology model.

C. Badica et al. (Eds.): Intel. Distributed Comput., Systems & Appl., SCI 162, pp. 245–250, 2008.
springerlink.com © Springer-Verlag Berlin Heidelberg 2008

The paper is structured as follows. In section 2 we construct a formal security protocol specification model. Based on this, in section 3 we describe the proposed ontology model and an example implementation. In section 4 we connect our work to others. We end with a conclusion and future work in section 5.

2   Security Protocol Specifications

Existing security protocol specifications limit themselves to the representation of operations and message components that are vital to the goal of these protocols: exchanging messages in a secure manner. One of the most simplest form of specification is the *informal* specification. For example, let us consider Lowe's modified version of the BAN concrete Andrew Secure RPC ([5]):

$$A \rightarrow B\colon A, N_a$$
$$B \rightarrow A\colon \{N_a, K, B\}_{K_{AB}}$$
$$A \rightarrow B\colon \{N_a\}_K$$
$$B \rightarrow A\colon N_b$$

By running the protocol, two participants, A and B, establish a fresh session key K. The random number N_a ensures freshness of the newly generated key, while N_b is sent by participant B to be used in future sessions. Curly brackets denote symmetrical key encryption. Throughout this paper we use the term "nonce", which is a well-known term in the literature, to denote random numbers.

Participants running security protocols usually exchange message components belonging to well-defined categories. We model these categories using the following sets: R, denoting the set of participant names; N, denoting the set of nonces and K, denoting the set of cryptographic keys.

The message components exchanged by participants are called *terms*. *Terms* may contain other terms, encrypted or not. Encryption is modeled using *function names*. The definition of *function names* and *terms* is the following:

$$
\begin{aligned}
FuncName \ ::= & \text{sk} \quad (secret\,key) \\
& | \ \text{pk} \quad (public\,key) \\
& | \ \text{h} \quad (hash\,or\,keyed\,hash)
\end{aligned}
\qquad
\begin{aligned}
\mathcal{T} ::= & . \mid \mathsf{R} \mid \mathsf{N} \mid \mathsf{K} \mid (\mathcal{T}, \mathcal{T}) \\
& \mid \{\mathcal{T}\}_{FuncName(\mathcal{T})}
\end{aligned}
$$

Terms that have been encrypted with one key can only be decrypted by using either the same key (when dealing with symmetric encryption) or the inverse key (when dealing with asymmetric encryption). To determine the corresponding inverse key, we use the $\_^{-1} : \mathsf{K} \rightarrow \mathsf{K}$ function.

As opposed to regular specifications where the user decides on the meaning of each component, for our goal to be achievable, we need to include additional information in the specification so that protocols can be executed without any user intervention.

We use the term "protocol header" to denote a set of sections needed for the interpretation of the information that follows. The header we propose consists of three sections: *types, precondition* and *effect*. The predicates defined for each section are given in table 1. Using the defined terms, we now define several functions to operate on them,

Table 1. Predicate definitions used to construct the protocol header

Section	Predicate	Definition	Description
types	*part*	R^*	Participant list
	nonce	N^*	Nonce list
	key	K^*	Key list
	term	$\mathcal{T}^*$	Term list
precondition	*shared_key*	$R \times R \times K$	Shared key between two participants
	init_part	R	Initializing participant
	resp_part	R	Respondent participant
effect	*key_exchange*	$\mathcal{T}^*$	Key exchange protocol
	authentication	$\mathcal{T}^*$	Authentication protocol

Table 2. Function definitions used to construct the protocol body

Function	Definition	Example Usage : Result
gennonce	$R \rightarrow N$	*gennonce*$(A) : N_a$
genkey	$R \rightarrow K$	*genkey*$(A) : K$
encrypt	$R \times \mathcal{T} \times \mathcal{T} \rightarrow \mathcal{T}$	*encrypt*$(A, (A, N_a), K_{ab}) : \{A, N_a\}_{sk(K_{ab})}$
decrypt	$R \times \mathcal{T} \times \mathcal{T} \rightarrow \mathcal{T}$	*decrypt*$(A, \{A, N_a\}_{sk(K_{ab})}, K_{ab}^{-1}) : (A, N_a)$

resulting the "protocol body". These functions are used to provide a detailed description of atomic operations specific to term construction and extraction. Sending and receiving operations are handled by *send* : $R \times R \times \mathcal{T} \rightarrow \mathcal{T}$ and *recv* : $R \times R \times \mathcal{T} \rightarrow \mathcal{T}$ functions.

The list of proposed functions is given in table 2, which can be extended with other functions if needed.

3   Ontology Model and Semantic Annotations

Based on the formal protocol construction from the previous section we have developed an ontology model that serves as a common data model for describing semantic operations corresponding to security protocol executions. The core ontology (figure 1) defines a security protocol constructed from four domains: *Cryptographic specifications*, *Communication*, *Term types* and *Knowledge*. Interrupted lines denote ontology import, empty arrowed lines denote sub-concept association and filled arrowed lines denote functional relations (from domain to range) between concepts.

The proposed ontology has been developed in the *Protégé* ontology editor ([9]). It provides semantic to protocol operations such as generating new terms (i.e. key or nonce), verifying received terms, sending and receiving terms.

The knowledge concept plays a key role in the automatic execution process. It's purpose is to model the stored state of the protocol between exchanged messages. For example, after generating a new nonce, this is stored in the knowledge of the executing

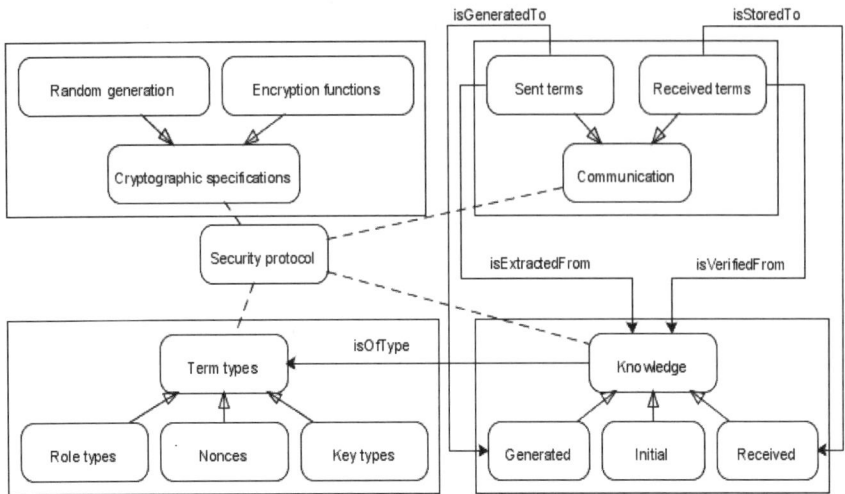

Fig. 1. Core ontology for describing security protocols

participant. When a term is received containing the same nonce, the participant must verify it's validity by comparing the stored value with the received one.

The ontology model provides semantic for the purpose of each sent and received term. However, it does not provide description of the mechanisms that would allow terms to be exchanged by parties. It also does not provide a specification of the preconditions, effects, cryptographic details or exchanged message sequences.

In our implementations, we used WSDL with Semantic annotations ([11]) (i.e. WSDL-S) to handle the aspects that are not addressed by the ontology model. The WSDL standard defines the *portType* section to denote a collection of messages. We annotate messages defined by *portType* using the *wssem:modelReference* extension attribute. Part of an example annotated XML schema representation of an encrypted message is the following:

```
<complexType name="encType">
    <sequence>
        <element name="role" type="tns:roleType"
         wssem:modelReference="http://www.owl-ontologies.com/sec.owl#RoleA">
        </element>
        <element name="nonce" type="tns:nonceType"
         wssem:modelReference="http://www.owl-ontologies.com/sec.owl#NonceA>
        </element>
        <element name="keyRef" type="tns:sharedKeyType"
         wssem:modelReference="http://www.owl-ontologies.com/sec.owl#KeyAB>
        </element>
        <element name="encAlg" type="string"
         wssem:modelReference="http://www.owl-ontologies.com/sec.owl#EncAES>
        </element>
    </sequence>
</complexType>
```

Effects and preconditions are added using the already existing *wssem:precondition* and *wssem:effect* elements. Part of the WSDL-S defining a security protocol which requires a shared key between users and as an effect produces a session key is the following:

```
<wsdl:portType name="EncComm">
  <wssem:precondition name="SharedKey"
  wssem:modelReference="http://www.owl-ontologies.com/sec.owl#SharedKeyAB"/>
  <wssem:precondition name="SessionKey"
  wssem:modelReference="http://www.owl-ontologies.com/sec.owl#SessKeyAB"/>
</wsdl:portType>
```

4   Related Work

There are several proposals and already established standards in the web service community dealing with security aspects. We will briefly describe three approaches that have the most in common with our proposal: SAML (Security Assertion Markup Language), WS-Federation (Web Service-Federation) and WS-Security (Web Service-Security).

The Security Assertion Markup Language is an "XML framework for exchanging authentication and authorization information" ([12]) between entities. This is achieved by the use of assertions, based on XML constructions, that denote authentication and authorization sequences achieved by entities. Based on these assertions, service providers can decide whether clients are authorized or not to access the requested services. SAML also provides a set of XML constructions called "profiles" to describe the required message exchange for transferring assertions. However, these are predefined messages that must be implemented by all services and protocol participants. Our proposal includes a semantic description that allows executing security protocols containing message structures that are not predefined.

A similar proposal to the SAML framework is WS-Federation ([14]). As mentioned by the authors of WS-Federation, the goals achieved are mainly the same as in the case of SAML. Major differences relate to the fact that it extends the WS-Trust model ([15]) and it provides a set of composable protocols. These components that differentiate it from the SAML framework, however, do not compete with our proposal.

The WS-Security ([13]) proposes a standard set of SOAP extensions to implement message integrity and confidentiality. It describes how to encode binary security components such as keys, random numbers or X.509 tokens. The WS-Services specification is thus a transport layer for the actual execution of protocols and not the description of the involved messages. The WS-Services could thus be used in conjunction with our proposal to encode binary data included in protocol messages.

5   Conclusions and Future Research

In this paper we presented an ontology model which provides semantic to security protocol execution operations. The ontology model is based on a formal specification model that ensures a detailed description through the use of a protocol header and a protocol body. Based on these descriptions, client applications can execute new security protocols based only on the proposed description.

The major advances in the field of protocol composition ([7, 8]) provide the means to create new protocols from existing security protocols. The analysis process of the composed protocols has been reduced to a syntactical analysis ([6]) which could be used to create protocols in real time. As future research we intend to combine the specification and ontology model proposed in this paper with composition operators in order to create protocols based on multiple specifications.

References

1. Studer, R., Benjamins, V., Fensel, D.: Knowledge Engineering: Principles and Methods. In: Data and Knowledge Engineering, pp. 161–197 (1998)
2. Abadi, M., Gordon, A.D.: A Calculus for Cryptographic Protocols: the spicalculus. In: 4th ACM Conference on Computer and Communications Security, pp. 36–47 (1997)
3. Fabrega, F.J.T., Herzog, J.C., Guttman, J.D.: Strand Spaces: Why is a security protocol correct? In: Proc. Of the 1998 Symposium on Security and Privacy, pp. 66–77 (1998)
4. Cremers, C., Maw, S.: Operational semantics of security protocols. In: Leue, S., Systa, T. (eds.) Scenarios: Models, Transformations and Tools. LNCS, vol. 3466, pp. 66–89. Springer, Heidelberg (2005)
5. Lowe, G.: Some new attacks upon security protocols. In: Proc. of the 8th Computer Security Foundations Workshop (1996)
6. Genge, B., Ignat, I.: Verifying the Independence of Security Protocols. In: Proc. of the 3rd International Conference on Intelligent Computer Communication and Processing, Romania, pp. 155–163 (2007)
7. Datta, A., Derek, A., Mitchell, J.C., Roy, A.: Protocol Composition Logic. In: Electronic Notes in Theoretical Computer Science, pp. 311–358 (2007)
8. Hyun-Jin, C.: Security protocol design by composition. Technical report Nr. 657, UCAM-CL-TR-657, Cambridge University, UK (2006)
9. Noy, N.F., Crubezy, M., et al.: Protege-2000: An Open-Source Ontology-Development and Knowledge-Acquisition Environment. In: AMIA Annual Symposium Proceedings (2003)
10. Security Protocol Open Repository (2008), http://www.lsv.ens-cachan.fr/spore/
11. World Wide Web Consortium. Web Service Semantics WSDL-S Recommendation (November 2005), http://www.w3.org/TR/wsdl/
12. Organization for the Advancement of Structured Information Standards. SAML V2.0 OASIS Standard Specification (November 2007), http://saml.xml.org/
13. Organization for the Advancement of Structured Information Standards. OASIS Web Services Security (WSS) TC (November 2006), www.oasis-open.org/committees/wss/
14. IBM. Web Services Federation Language Specification (December 2006), http://www.ibm.com/developerworks/library/specification/ws-fed/
15. Organization for the Advancement of Structured Information Standards. WS-Trust v1.3 OASIS Standard (March 2007), http://docs.oasis-open.org/ws-sx/ws-trust/200512/ws-trust-1.3-os.html/

On the Foundations of Web-Based Registries for Business Rules

Adrian Giurca[1], Ion-Mircea Diaconescu[1], Emilian Pascalau[1], and Gerd Wagner[1]

Brandenburg University of Technology, Germany
{giurca,m.diaconescu,pascalau,wagnerg}@tu-cottbus.de

Summary. In the last eight years, registries for e-business, such as ebXML or UDDI, enabling enterprise of any size and in any geographical location to conduct their businesses on the World Wide Web, were developed. Applications in domains such as insurance (for example, insurance rating), financial services (loans, claims routing and management, fraud detection), government (tax calculations), telecom customer (care and billing), e-commerce (personalizing the user's experience, recommender systems, auctions), and so on benefit greatly from using rule engines. Therefore, sharing rulesets becomes a necessity for many B2B businesses. This work presents a basic architecture of building a Web-based registry for rules. The main goal of the registry is to allow rulesets discovery. Registry entries contain both required ruleset related data (such as ruleset URI or properties describing their intended scope) and optional metadata covering additional properties such as *last modified date*.

1   Introduction and Motivation

Registries and repositories such as UDDI [11] and ebXML [10, 9] are developed to enrich the actual need for a standardization in e-business. They come with necessary infrastructure for defining business processes, business information, business entities, business relationships, e-commerce patterns, catalog of common business processes, message services, and business services. Nowadays, in many cases, businesses behavior is expressed naturally through business rules[14], [15]. There are several rule platforms such as Drools, ILog, Jess, FLORA 2, Jena Rules, SWRLTab, but there is no standard way for defining business rules. The standardization is a concern for both OMG and W3C. The first produced Production Rule Representation (PRR), Beta 1[12] and the second Basic Logic Dialect [2]. Moreover, the W3C Rule Interchange Format WG goal is to produce W3C recommendations for rules interchange on the Semantic Web. Users encode and store their rules either in PRR or RIF representation. Since RIF is an interchange format, the interoperability goal is achieved. However, the need of a rule registry remains actual and necessary. Such a registry will help users to discover business rulesets and reuse them in their applications. This paper proposes an architecture for a Web-based rule registry. Any registry implementation must comply with at least the following functionalities:

- Ruleset registration - Registering a ruleset makes it available for public/private use by other applications.

C. Badica et al. (Eds.): Intel. Distributed Comput., Systems & Appl., SCI 162, pp. 251–255, 2008.
springerlink.com   © Springer-Verlag Berlin Heidelberg 2008

- Ruleset discovery - The searching capabilities of the registry should permit to users and to applications to find out the appropriate ruleset to their needs.

2   The Registry Entry Information Model

This section describes the information model of a registry entry to be maintained in a rule registry to achieve efficient rulesets discovery.

2.1   A Business Ruleset Example

For a better understanding of the goal and the result we use an example ruleset provided by Business Rule Forum – the *Driver Eligibility* ruleset. Its purpose is to establish if a driver is or isn't eligible for an auto insurance. Below is an excerpt of the ruleset expressed in natural language:

1. If young driver and driver has training certification, then is an eligible driver.
2. If senior driver and driver has training certification, then is an eligible driver.
3. If both of the following are not true, then is an eligible driver

   - Young driver;
   - Senior Driver;

Recall that this work addresses production rulesets conforming to the OMG PRR [12], and therefore rules are vocabulary based and only the standard actions i.e. *assert*, *retract*, *updateState* and *invoke* are allowed. Readers familiar with RETE algorithm [7] are aware that rules from this ruleset can be executed only if the working memory contains a specific set of facts (they have to match rule conditions). Rule platforms use specific vocabulary language to encode concepts. For example, Drools and JRules use Java beans as vocabulary (e.g. all above rules written in Drools assume the availability of the Driver bean). Also the facts representation of the *Working Memory* must conform with this vocabulary (e.g. an instance of Driver is logically interpreted as a set of facts). Therefore, a registry entry should keep at least: (a) a reference to the vocabulary representation language and (b) a reference to the rule vocabulary.

A registry entry contains the following required information:

1. A literal acting as a primary key of a registry entry encoded as the value of *id* property;
2. An URI reference to the natural description of the ruleset, allowing *human readers* to better understand if the ruleset is or is not appropriate for his purposes (an implementation solution may be by using Dublin Core [8] *description* property);
3. A literal representing the code of the business addressed by this ruleset. The implementation may use, for example, NAICS or UNSPSC codes of the corresponding business part (encoded as a value of the Dublin Core *related* property);
4. An URI reference to the ruleset vocabulary fragment corresponding to the intended initial set of facts (encoded as a value of *prereqVocabulary* property;
5. An URI reference to the ruleset vocabulary fragment corresponding to the expected results (this is a subset of the vocabulary corresponding to the fact instances affected by rule actions), encoded as a value of the *resultVocabulary* property);

6. An URI reference to the specific ruleset implementation, encoded as the value of *ruleSetID* property;
7. An URI reference to the ruleset representation language, (e.g. http://java.sun.com/beans in the case of a Java beans vocabulary). encoded as a value of the Dublin Core *type* property;
8. An URI reference to the vocabulary representation language (encoded also with the help of Dublin Core *type* property);
9. A literal encoding the mime-type of the vocabulary representation (e.g. text/html, application/xml) encoded as a value of the Dublin Core *format* property;
10. A literal encoding the mime-type of the ruleset representation (encoded also with the help of Dublin Core *format* property);

In addition an entry may offer the following non-mandatory information:

1. the *creator* (using Dublin Core *creator* or FOAF [6] *maker*);
2. *contributors* (using Dublin Core *contributor* property);
3. the *release date* (using Dublin Core *date* property);
4. the *version* (using Dublin Core *hasVersion* property);
5. the *submission date* (using Dublin Core *dateSubmitted* property);
6. the *last modified date* (using Dublin Core *modified* property);
7. the ruleset applicable *licence* (using Dublin Core *licence* property);
8. the *publisher* (using Dublin Core *publisher* property);
9. *positive usages* and *negative usages* (values of *positiveUsage* and *negativeUsage* respectively);
10. known *translators* to other rule languages (values of *translator* property);

All those properties are optional, but they are strongly recommended this allowing a much sharp discovery of the rulesets.

The example below (using Turtle notation[4]) describes a *DriverEligibility* ruleset entry in the registry. The next section describes how to use SPARQL [13] queries to extract information from the registry.

```
@prefix ex: <http://www.example.com/>
@prefix wrr: <http://www.tu-cottbus.de/rules/registry/entry#>

wrr:regEntry1 rdf:type wrr:RuleSetEntry;
              wrr:id "DriverEligibility";
              dc:creator ex:JohnSmith;
              dc:description ex:DriverEligibilityDesc
              dc:related "52-NAICS";
              wrr:vocabularyReference _:VocRef01;
              wrr:ruleSetReference _:RSRef01;
              wrr:positiveUsage ex:pExample1
              wrr:negativeUsage ex:nExample1
              wrr:negativeUsage ex:nExample2

_:VocRef01 rdf:type wrr:Vocabulary;
           dc:type <http://www.w3.org/TR/rdf-schema/>
```

```
        dc:format "application/rdf+xml"
        wrr:vocabularyID ex:UservDE_Vocabulary;
        wrr:prereqVocabulary ex:UservDE_PrereqVocabulary;
        wrr:resultVocabulary ex:UservDE_ResultVocabulary;

_:RSRef01 rdf:type wrr:RuleSet;
        dc:type <http://jena.sourceforge.net/>
        dc:format "text/html"
        wrr:ruleSetID ex:UservDE_RuleSet;
```

The choice of RDF(S)[5] for the entry representation is due to its conformance with at least the following criteria:

- The registry refers distributed resources on the Web;
- RDF(S) offers a well established semantics[1] of distributed and shared data on the Web [1];
- Well established metadata standards such as Dublin Core and FOAF use RDF Schemas;
- RDF(S) offers a query language for Semantic Web i.e. SPARQL [13];
- Reasoning on the resources is possible by adding web rules and inference engines (such as Jena Rules).

3   Querying the Registry

Since the entry representation is RDFS it is straightforward to use SPARQL, as a query language of the registry. For example one may want to obtain rulesets related to *Finance and Insurance* (code 52-NAICS), written in Jena and for which at least one positive example is available. Then the following SPARQL query can be designed:

```
SELECT ?reID
WHERE {
  ?re rdf:type wrr:RuleSetEntry;
      wrr:id ?reID;
      dc:related "52-NAICS";
      wrr:ruleSetReference ?rr;
      wrr:positiveUsage ?kpu.
  ?rr dc:type <http://jena.sourceforge.net/>.
FILTER (count(?kpu) > 0))
}
```

For a number of cases, *when the vocabulary of the ruleset is expressed by using RDFS*, more advanced searches are available.

For the above example, we can check the vocabulary to find if the ruleset contains *Driver* objects. In such a registry, no rules, no vocabularies nor other direct information regarding rule sets are stored, but instead, it offers access to all these resources by using URI references. This way, access to all information needed is provided, and more advanced searches are then supported. The limit depends on the implementation of the search engines through that information.

4 Conclusion and Future Work

In this paper we defined the structure of a web-based rules registry. The registry entries provide both required and optional information allowing rulesets discovery. The information model proposes RDF(S) as a knowledge representation language of the registry entries and SPARQL as a registry query language. Future work will investigate an extension of the registry under the assumption that properties *positiveUsage* and *negativeUsage* are open and complementary. Unfortunately this can't be expressed by using RDF(S), therefore the actual reasoning process over it can't take this into consideration. A possible solution is to investigate the usage of ERDF [3] for representation of the registry entries. In ERDF, those properties collapse into a single one, namely *usage* represented as a *PartialProperty*, and its complement can be expressed by using *strong negation*. However such a solution is subject of discussion since it involves a potential SPARQL extension to query ERDF knowledge bases.

References

1. RDF Semantics. W3C Recommendation (February 10, 2004),
   http://www.w3.org/TR/rdf-mt/
2. RIF Basic Logic Dialect (October 2007), http://www.w3.org/2005/rules/wiki/BLD
3. Analyti, A., Antoniou, G., Damasio, C.V., Wagner, G.: Extended RDF as a Semantic Foundation of Rule Markup Languages. Journal of Artificial Intelligence Research 32, 37–94 (2008)
4. Beckett, D., Berners-Lee, T.: Turtle - Terse RDF Triple Language (January 2008),
   http://www.w3.org/TeamSubmission/turtle/
5. Brickley, D., Guha, R.V.: RDF Vocabulary Description Language 1.0: RDF Schema. W3C Recommendation (February 2004), http://www.w3.org/TR/rdf-schema/
6. Dan Brickley and Libby Miller. FOAF Vocabulary Specification 0.91 (November 2007),
   http://xmlns.com/foaf/spec/
7. Forgy, C.: Rete – A Fast Algorithm for the Many Pattern / Many Object Pattern Match Problem. Artificial Intelligence 19, 17–37 (1982)
8. Dublin Core Metadata Initiative. DCMI Metadata Terms (January 2008),
   http://dublincore.org/documents/dcmi-terms/
9. OASIS. ebXML Business Process Specification Schema Version 1.01 (May 2001),
   http://www.ebxml.org/specs/ebBPSS.pdf
10. OASIS. ebXML Technical Architecture Specification v1.0.4 (February 2001),
    http://www.ebxml.org/specs/ebTA.pdf
11. OASIS. UDDI Version 3.0.2, UDDI Spec Technical Committee Draft, Dated 20041019 (October 2004), http://uddi.org/pubs/uddi-v3.0.2-20041019.pdf
12. OMG: Production Rule Representation (PRR), Beta 1. Technical report (November 2007)
13. Prud'hommeaux, E., Seaborne, A.: SPARQL Query Language for RDF (November 2007),
    http://www.w3.org/TR/rdf-sparql-query/
14. Ross, R.G.: The Business Rule Book: Classifying, Defining and Modeling Rules, 2nd edn. Database Research Group, Inc., Boston (1997)
15. Ross, R.G.: Principles of the Business Rule Approach, 1st edn. Addison-Wesley, Reading (2003)

Large-Scale Data Dictionaries Based on Hash Tables

Sándor Juhász

Department of Automation and Applied Informatics,
Budapest University of Technology and Economics,
1111 Budapest, Goldmann György tér 3. IV. em., Hungary
juhasz.sandor@aut.bme.hu

Summary. Data dictionaries allow efficient transformation of repeating input values. The attention is focused on the analysis of voluminous lookup tables that store up to a few tens of millions of key-value pairs. Because of their compactness and search efficiency, hash tables turn out to provide the best solutions in such cases. This paper deals with performance issues of such structures and its main contribution is to take into consideration the effect of the multi-level memory hierarchies present in all the current computers. The paper enumerates and compares various choices and methods in order to give an indication how to choose the structure and the parameters of hash tables in case of large-scale, in-memory data dictionaries.

Keywords: hash table, cache, large data dictionaries, performance optimization.

1 Introduction

Hash tables store the matching input-output (key-value) pairs instead of performing different calculation steps on the input data to provide the corresponding result. This paper studies performance optimization possibilities of large in-memory hash tables. When analyzing the storage layout choices, hash functions and optimization methods, we focus our attention to minimizing the number of L2 cache misses. This study addresses problems originating from real life projects related to web log processing [4] and model transformation issues [5]. While the primary goal to achieve was the highest performance possible, seeking for memory-economic solutions is also beneficial as it spares resources for other cooperating tasks. The rest of the paper is organized as follows. Section 2 deals with performance analysis and enumerates the possible optimization choices and methods suggested by other authors. Section 3 applies various optimization methods one by one to the hash tables, and analyzes the effect they have on the real performance. Section 4 provides an evaluation and comparison of different methods presented in the paper. We conclude by giving a suggestion how to choose the structure and the parameters of large hash tables to achieve the best performance.

2 Related Works

Lookup tables are memory intensive with low computation power requirements, thus their performance is equally effected by the memory access pattern and by the number

C. Badica et al. (Eds.): Intel. Distributed Comput., Systems & Appl., SCI 162, pp. 257–262, 2008.
springerlink.com
© Springer-Verlag Berlin Heidelberg 2008

or instructions (comparisons) completed to find the element corresponding to the key. Thanks to optimization work of different researchers trying to find better hash functions or improve collision or overflow handling, good hash tables provide a constant access time to the elements, where the average can be pushed down nearly as low as a single step [6]. Different parameters such as initial size and the type of the hashing play crucial roles in this process. Two different approaches exist: open hashing and bucket hashing. Open hashing uses a single directly addressable flat structure, while bucket hashing distributes the values into groups (buckets) based on the hash function, and stores these groups in separate lists.

If more keys are mapped to the same value a collision occurs. Lum, Yuen and Dodd were among the firsts to analyze the accuracy of hash functions and collision solving methods on real data [1]. They came to the conclusion that bucket hashing outperforms open hashing if the buckets are small enough. Their work was continued by Ramakrisnha [3], who verified the predicted analytical performance of hashing techniques in practice, taking in account both the successful and unsuccessful search lengths and the expected worst case performance. Owolabi in [8] used open hashing to compare five common hash functions (division, multiplication, midsquare, radix conversion and random) along with two collision handling techniques (linear probing and chaining) and found that random and division methods perform the best on the tested real-life data sets. It is also suggested [1] to use prime numbers as the number of buckets, or at least to avoid values having small numbers (less than 20) between their prime factors, but our tests presented in Section 3 did not confirm the importance of applying such restrictions.

Unfortunately the above works concentrate on reducing the number of comparisons only, Mitzenmacher [7] was the first to mention the importance of memory access patterns and caching when adjusting the parameters of hash tables. In modern computers as a multilevel cache acts as a temporary storage place for frequently or recently accessed data [9][10], thus in practice the performance complexity of data intensive application is always measured in the number of the slow memory accesses (cache misses).

3 Comparing the Different Approaches

This paper focuses on finding a performance optimal structure and memory usage pattern for hash tables. Our main contribution lies in analyzing the effect the different storage structures have on the memory access and through this, on their overall performance. This effect is captured by measuring the number of L2 cache misses. Caches in current CPUs administer lines of usually 64 bytes. When designing the storage structure accordingly, this fact can be turned into an advantage. This section illustrates the effects of the different design and run-time parameters on the efficiency of the lookups. The theoretic proposals are supported by various experiments. A lookup table used in real-life web log compression will serve to support our suggestion and observations, where a field of 20 bytes is transformed to 4 bytes). The code table in the measurements includes 10 million pieces of these 24 byte long key-value pairs. All the hash table implementations are written in C++, compiled with *Microsoft Visual Studio 2005*, and executed on *Intel Pentium 4* CPU @ 3.2 GHz, 2 MB L2 cache, with 4 GB system

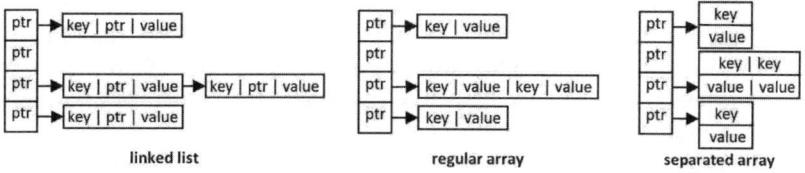

Fig. 1. Outline of the listed hash table structures

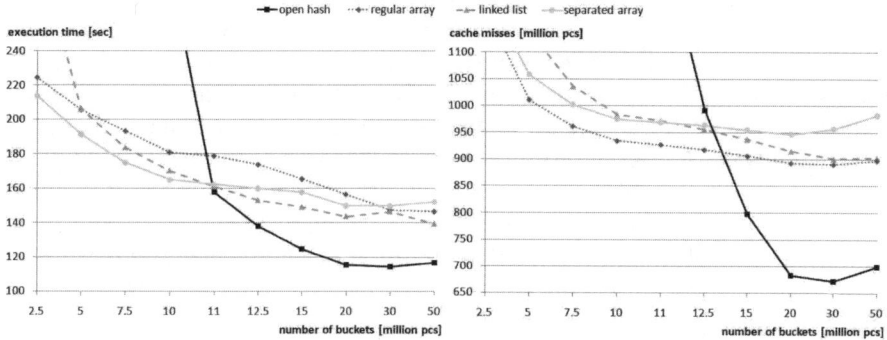

Fig. 2. Execution time and the number of cache misses of the hash tables

memory. The L2 cache usage was measured with *Intel VTune Performance Analyzer 9.0*. Each measurement starts after inserting the 10 million items and includes the execution time of 200 million lookups.

As the primary goal is to enable fast data conversions for voluminous data dictionaries, the first experiment was done with bucket hashing as suggested in [1]. The storage of the buckets themselves can be implemented with linked lists or with arrays. Lists can dynamically increase their size, while arrays require more administration (reallocation), but as the items are stored side by side arrays provide a cache friendlier behavior. Linked lists on the other hand do not have this advantage, as their items are spread all along the memory. If the average number of items per (non-empty) buckets falls bellow two, the possibility of taking advantage of cache line prefetches vanishes. Fig. 1 summarizes the different bucket hash structures handled in the paper. Next to the above mentioned array and linked list based implementations, we can see a special separated key-value approach.

Fig. 2 compares the performance of the above three storage structures to the traditional approach of open hashing by measuring the execution time and the number of cache misses. It is clearly visible that there is a strong correlation between the number of cache misses and the execution time.

As expected, arrays outperform linked at low bucket numbers (see Fig. 2). When increasing the number of buckets the average bucket size decreases and after a point linked lists become faster. This happens because arrays have higher initial cost (the size of the array has to be queried as well), but fetching items after getting the first one is cheaper. Fig. 2 also shows the curves of cache misses corresponding exactly to the execution time of the algorithms and we can see that the change in favor of linked lists is also verified. Our experience shows that arrays are advised when there are more than 2 items in the buckets, and linked lists are recommended below that bucket size.

Good data locality allows taking advantage of the prefetching. Reducing the size of the items increases the cache performance by allowing more items to fit in a single cache line. Although the size of key or the value cannot be modified, but their separation (putting them into separate arrays) is feasible, like it is done in case of the *separated array* version. Since searching uses the keys only for comparison, the values will not be needed until the key matches. This variation runs faster until the bucket number of 10 millions is reached. At this bucket number, the average number of items per bucket drops to nearly 1 thus the first item in the bucket is highly likely to be a match, and the value of that item will be requested immediately in both cases. With the modification, the value request will result in a cache miss, since it is stored at a different part of the memory; while in the original version, the key and the value is loaded in one step, and the value is directly accessible. The larger the buckets are, and the bigger part the values in the key-value pairs represent, the more significant improvement such modifications provide. As a consequence, the separation of keys and values is strongly advised when using arrays, especially when the number of buckets is smaller than the number of items.

The most significant factor determining the performance of hash tables is the number of buckets (or the expected size of the buckets). As a rule of thumb, the more buckets are used, the better the performance is. With the growing number of buckets, the average bucket size decreases, resulting in fewer items to be checked. However, the performance gain is limited. While the CPU time may decrease as the items in the buckets gets fewer and fewer, the number of memory reads stay constant after a point, therefore the time will not be reduced further, memory consumption on the other hand will continue to increase rapidly. The limit is reached around 20-30 million buckets for 10 million items, as visible in Fig. 2.

Having tuned the bucket hashing in the previous sections, now it is time to verify whether our initial choice of bucket hashing instead of open hashing was really a good one under the circumstances we face. The simplified description of both approaches was presented in Section 2. To do a fair comparison during the measurements, linear probing was used in the implementation of open hashing, as it takes the best advantage from the cache line prefetch, and is claimed to be best for medium saturation levels (40-70%) [11]. With the increasing amount of memory reserved for open hashing, the execution time decreases, while the allocated memory increases. Below 50% saturation (20 million buckets), the execution time does not change considerably showing same behavior that bucket hashing did. Open hashing is considerably faster in the high bucket regions, although, bucket hashing is still a viable choice for bucket numbers less than the number of items. Open hashing in these regions simply does not work, as by nature it requires more slots than items. Smaller bucket numbers also come with smaller memory need, thus for memory sensitive operations, bucket hashing is advised, otherwise open hashing is the suggested choice.

4   Comparison of Methods

An overview of the examined aspects is given in Table 1, showing the effect of the different options on the execution time and on the amount of the reserved memory. The best choice is often not obvious, but depends on the number of buckets (many buckets in the table means at least as many buckets as items, less buckets means between 10% and 75% of the number of items). Given that using bucket hashing makes only sense with small bucket sizes (otherwise open hashing is a clear winner), we did not consider cases where the number of buckets fall below 10% of the items.

The most controversial modification is the choice of storage structure in bucket hash tables. While arrays perform significantly better (up to 100%) for small bucket numbers by taking advantage of cache usage, linked lists will profit from higher bucket numbers. The decision is also influenced by the nature of the items. When the number of items is unknown, open hashing is out of the question, while linked lists will deteriorate rapidly with the growing number of items. In this case, bucket hashing is a safe bet as its performance deteriorates much slower.

The most dominant parameter is clearly the number of buckets. Estimating the best number of buckets is out of the scope of the current paper, but as a rule of thumb, increasing the number of buckets results in the better performance, but it no use going above 2-3 times of the number of items to store.

Table 1. Comparison of methods

		in favor of	performance gain	gain in memory
open vs. bucket hashing	many buckets	open	20-30%	-30%
	less bucket	bucket	open hash not applicable	
linked lists vs. arrays	many buckets	linked lists	20%	-
	less buckets	array	100%	5-10%
separation of keys	many buckets	key-value together	7%	-
with arrays	less buckets	separated key	10%	-
bucket number		high numbers	20-30%	-30%

5   Conclusion

This paper presented methods and aspects of hash table design when storing tens of millions of key-value pairs. The main focus was put on the performance of searching that was sped up by choosing the most suitable storage structures and methods taking advantage of L2 cache. During our work we found that bucket hashing does not out-perform open hashing even if the bucket sizes are kept small. As the difference in high performance region is small (below 10%), bucket hashing can prove to be useful when we ignore the number of items to be stored in advance.

Considering the memory usage, lookup performance and parameter sensitivity bucket hash tables turned out to have two optimal working points. The first version is recom-mended for known-number of items, where the lookup speed is a primary concern. In this case linked lists are advised; and the bucket number should be greater than the number of items. As the buckets store list pointers only, overestimating (or increasing) the number of buckets has a slight additional memory cost only. The second version is a tradeoff between memory usage and lookup speed. Being less sensitive to the number of items this version is advised when the number of items is not well-known in the beginning. In this case arrays with separated keys and values are used, and the number of buckets is chosen between one tenth and one fourth of the number of items. This version is very robust, the execution time declines only slowly when inserting more items than what the hash table was designed for.

Acknowledgment

This work was completed in the frame of Mobile Innovation Centre's integrated project Nr. 3.2. supported by the National Office for Research and Technology (Mobile 01/2004 contract).

References

1. Lum, V.Y., Yuen, P.S.T., Dodd, M.: Key-to-address transform techniques: A fundamental performance study on large existing formatted files. Communications of the ACM 14(4), 228–239 (1971)
2. Lum, V.Y.: General performance analysis of key-to-address transformation methods using an abstract file concept. Com. of the ACM 16(10), 603–612 (1973)
3. Ramakrishna, M.V.: Hashing in Practice, Analysis of Hashing and Universal Hashing. In: Proc. of ACM SIGMOD Int. Conf. on Management of Data, pp. 191–199 (1988)
4. Juhász, S., Iváncsy, R.: Tracking Activity of Real Individuals in Web Logs. International Journal of Computer Science 2(3), 172–177 (2007)
5. Lengyel, L., Levendovszky, T., Charaf, H.: Normalizing OCL Constraints in UML Class Diagram-Based Metamodels - AND/OR Clauses. In: Proceedings of the IEEE EUROCON 2005, Belgrade, November 21-24, pp. 579–582 (2005)
6. Litwin, W.: Linear hashing: A new tool for file and table addressing. In: Proceedings of the Sixth International Conference on Very Large Data Bases, New York, pp. 212–223 (1980)
7. Mitzenmacher, M.: Good Hash Tables & Multiple Hash Functions. Dr. Dobbs Journal 336, 28–32 (2002), http://www.ddj.com/dept/architect/184405046
8. Owolabi, O.: Empirical studies of some hashing functions. Information & Software Technology 45(2), 109–112 (2003)
9. van der Pas, R.: Memory Hierarchy in Cache-Based Systems, Technical report, High Performance Computing, Sun Microsystems, Inc. (2005), http://www.sun.com/blueprints/1102/817-0742.pdf
10. Wulf, W.A., McKee, S.A.: Hitting the Memory Wall: Implications of the Obvious. Computer Architecture News 23, 20–24 (1995)
11. Pagh, A., Pagh, R., Ruzic, M.: Linear probing with constant independence. In: Proceedings of the 39th ACM Symp. on Theory of Computing, San Diego, pp. 318–327 (2007)

Undo in Context-Aware Collaborative Ubiquitous-Computing Environments

Marco P. Locatelli and Marco Loregian

Dipartimento di Informatica, Sistemistica e Comunicazione,
Università degli Studi di Milano Bicocca,
viale Sarca 336,
20126 Milano (Italy)
{locatelli,loregian}@disco.unimib.it

Summary. A comprehensive approach to the design of Ubiquitous Computing systems must deal with the issues related to the restoration of an earlier or acceptable state of the system, if possible, when users intentionally want to undo some previous actions. Systems supporting collaborative Ubiquitous Computing environments should provide a default undo function, but also provide users and applications with awareness information to correctly decide which (compensative) actions should be undertaken. In this paper we describe how to achieve undo in distributed, dynamic, context-aware systems. We present a general approach to undo in collaborative Ubiquitous Computing environments in terms of the CASMAS model: part of the approach relies on the notion of active coordination artifacts, as defined also in CSCW literature.

1 Introduction

We have studied the problem of undo from various perspectives: from Business Process Management [2] to Human-Computer Interaction [6, 7]. Several approaches and theoretical frameworks for undo have been proposed in literature [3] and implemented in prototype systems as well as in commercial applications. However, even if undo can no longer be considered just an accessory function, the topic has not been treated systematically for ubiquitous-computing (ubicomp) systems.

In a previous work, we have analyzed experimentally the perception users have of undo in ubicomp environment [7]. Our research showed that the average users already have a complex idea about what undo should be like in ubicomp environments: context-aware, smart (semi-proactive), and with compensation mechanisms (e.g., to be able to deal with the consumption of resources).

In this paper, we present *how an undo function satisfying the expectations of users can be designed for cooperative ubiquitous-computing environments*.

We adopt the CASMAS model (Community-Aware Multi-Agents Situated Systems [4]) as a reference for the design context-aware ubicomp systems, and extend the notion of active coordination artifact [8, 4] to achieve *dynamic, distributed undo for context-aware systems*.

C. Badica et al. (Eds.): Intel. Distributed Comput., Systems & Appl., SCI 162, pp. 263–268, 2008.
springerlink.com

2 Undo in Context-Aware UbiComp Environments

A command can be defined as the high-level (*user*) action that causes the execution of a set of lower-level operations by the system, and the scope of a command can be defined considering the atomic operations it triggers, and their effects, as a whole. The term **undo** is appropriate when there has previously been the *intention* to do something [1]. As a consequence, undoing a command can be seen as the result of *the capability of a system to perform a set of actions to fulfill the will of the users to reach a state as if the previous command had never been given.*

Ubiquitous Computing systems are composed of many different technological elements (hardware and software) setup to behave coherently [4]. Regardless of the architecture of the system, each technological entity (device or service/application) is characterized by specific features and functionalities, thus by a distinctive set of low-level operations and/or a set of high-level commands. It is an established practice, acquired from the design of PC applications, to implement "general purpose" undo commands to trigger inverse operations [9] according to the specific user requests. Undo is designed as a meta-command that works only with the command history and with other commands [11, 10]. In this case, there must be a direct and explicit association between a command and its inverse or compensation one, and the undo command has only to execute such inverse command. All the data needed by the entity to perform the undo must be directly accessible: current data, history, and possibly also the previous versions of data. This generally means to have everything within the local scope, since independent applications are the reference case.

In the absence of inter-entity interactions requiring coordination for the execution of commands (even if there is some shared data), the undo process is generally straightforward and limited in scope to the single entity (with its mechanisms and data). Also, the involvement of the user is limited only to the initial undo request, and there is no way she can further affect or customize the execution. The only alternative to the predefined undo is to intentionally perform a different set of commands or operations [1].

2.1 Issues with Cooperating Entities

In systems where different entities are actively cooperating (influencing each other and possibly not only accessing the same data) it is necessary to achieve coordination: the execution of actions must be concerted to reach a common goal and to avoid conflicts, data must be passed along (sometimes more than once, in an iterated communication process), and so on. When the undo of a command that had involved various entities is requested through or by a specific entity, also the undo mechanism must be able to span across entities and to deal with all the effects of the original command. Among the problems that may arise, it is worth considering:

- how to start and to distribute the undo mechanism, i.e., how to enact communication between entities so that the enacted process is acknowledgeable to be an undo, e.g., to make also other entities aware of what is happening;
- how to make the different entities coordinate, e.g., to avoid conflicts and possibly to make all of them converge on the current goal;

- how to detect or to infer that the undo process has terminated, i.e., all the required commands or operations have been performed causing the system to be in a (stable) state meeting the expectations of users;
- how to deal with the inability of the undo process to terminate successfully, e.g., supporting interaction with human or technological entities to proceed by providing additional information.

These issues could be avoided by exploiting coordination information [4] to enforce the so-called must policies or prescriptive interaction: the set of operations to be performed must be defined uniquely and based only on the information kept by the entities involved in the operation. In other words, it is necessary to rely on a *complete* description of the undo process in terms of entities involved, information exchanged and the effects of the interaction with respect to the whole system. This technique is commonly adopted to deal with undo, rollback and even exceptions in Business Process Management (BPM) and Workflow Management systems [2] but it is not suitable for systems that do not rely on an explicit description of the supported processes.

2.2   History as a Coordination Artifact

In systems with cooperating entities — interacting in a dynamic, spontaneous, and not predictable way, — history can provide more than just an *a posteriori* trace (log) of the system: entities can adopt it as a coordination artifact [4] both to do and undo. The information in the history can be used by entities to coordinate the execution of their actions: all the actions performed are stored in the history to be available to all the entities. In particular, history can support specific undo mechanisms such as the non-linear ones by making temporal or logical dependencies between commands explicit. Entities rely on the sequence (order) of the already performed commands to define the sequence of action to be performed to achieve undo, e.g., waiting for the completion of other commands (in order to get correct data). In other words, even if there is no explicit description of a *do* process, the history can be used to define the undo process according to the selected undo strategy.

In collaborative ubicomp environments, usually there is no unified control over the overall environment and on the operations executed therein. This fragmentation of control — i.e., distributed coordination on a local basis — generally introduces an intrinsic inconsistency between the actual state of the whole environment and the state of the same environment as perceived by the participating entities. Hence, the consistent temporal sequence of commands can not be granted to be available because the access to the resources of the environment can not be forced to be done in a way that is purely functional to having a consistent history. As a consequence, the possibility of having inconsistencies in the history must be taken into account when choosing, designing and implementing an undo mechanism: other information regarding the environment must be exploited to achieve the desired goals (Section 2.3).

In CASMAS, history is modeled as a coordination artifact whose content is the list of commands, asserted and shared in the community fulcrum (which is the space where entities that are members of the community share information) to which the cooperating entities are linked. In this way, the coordination artifact is available to all the community members and they can add the *command record* to the history artifact each time they execute a command. The command record is the set of information related to a command and that contains at least the command name and the ID of the entity that executed it. When a generic undo request is sent by an entity — this event may be considered equivalent to the generic PC undo, or 'ctrl+Z' — the command to be undone is identified according to the adopted undo strategy, and a consequent assertion is made in the community fulcrum. As a reaction, if the input conditions match, the rules of some entities fire to perform what is needed to undo the specific command. The entity performing the undo might be the same that had performed the command to be undone, but this is not always the case. Consider for example a simple consumption case, where the entity consuming a resource within the system is generally not the same having produced or provided the same resource. Similarly, if a selective strategy is adopted, entities can also directly ask for the undo of some specific command. In this case, if a global undo is performed — meaning that the actions performed by all entities constitute the scope of the undo — it is sufficient to assert the undo request in the community fulcrum and the process continues as above. Also if local undo is performed, it might be necessary to make other entities aware of the process, especially if the entity is a constitutive element of the context, as we are describing next.

2.3   Context-Aware Undo

The adoption of an explicit construct like the history helps distributed systems to adapt to unexpected situations. For example, if some entity performs an undo locally, and the history of the system is changed accordingly then also other entities might need to react — this can be due to the fact that the designer of the individual entities of a ubicomp systems have only a partial knowledge of what the other elements of the system will be like.

Context-aware (ubicomp) systems need to be able to process information also at a further level in order to flexibly adapt and effectively support the different user (work) practices. Entities should be able to enact undo mechanisms according to context information, e.g., to implicitly or explicitly select between different complex strategies according to different context (trigger) conditions, and not only to restore a previous system or entity state or set of data: the possibility to sense various kinds of context-related information can give great flexibility, and can open many other possibilities.

By modeling (at design time) and processing (at runtime) awareness information, also may policies can be enacted, i.e., all those mechanisms that depend on the many facets of the current context, and that entities can discretionally enact according to their subjective perception of the environment.

The awareness module of CASMAS, with its awareness graphs [5] to model the various aspects of the context (location, roles, ...), can provide entities with the information they need to adapt to the current context and to perform the most appropriate set of commands or operations to fulfill the undo request. If the input conditions of no undo specification are completely met, the undo request might not take place as expected (by the requesting entity).

2.4   Active Coordination Artifacts as Carriers of Undo Policies

Even if active coordination artifacts are not necessary in principle to achieve entity coordination, they can provide an effective way to concert the coordinated execution of "do" mechanisms [4]. Similarly, active coordination artifacts can be effective when they transport undo policies, i.e., the specification of how and when to execute coordinated undo, and the information that is necessary to achieve that goal. To have some active coordination artifacts to enact the coordination can be especially useful in cases like the last in the list in Section 2.1 (i.e., undo impossible or unsatisfactory termination).

In CASMAS, active coordination artifacts (can) carry into the community the undo mechanisms that are possibly missing (e.g., when they had not been foreseen or designed) or introduce further elements also to solve critical cases (e.g., additional data that can help avoiding system deadlocks). The active coordination artifact shares the undo mechanisms by using the "share behavior" feature of CASMAS to assert the rules that implement the undo mechanisms in the community fulcrum; in this way undo rules are available to the entities that takes part to the community and they can acquire them. This means that these entities are now able to properly achieve coordination also during the execution of an undo command.

3   Conclusion

In this paper, we have discussed the characteristics of undo for context-aware ubiquitous-computing environments. While the undo function is commonly available for PC applications, and while there exists approaches to deal with undo in distributed processes such as workflows [2], the problem of undoing actions in ubicomp environments has not been investigated systematically. Starting from an experimental analysis of the problem [7], we outlined the implication of undo on the design of ubicomp systems: (*a*) it is necessary to provide interacting entities with alternative undo mechanisms, exploiting context information (awareness), instead of just a general purpose solution relying on fixed inverse specifications or compensation; (*b*) history can be adopted as a mean to enrich cooperation between entities when no process specification can be given; (*c*) if a support for coordination, such as active coordination artifacts, is introduced in the reference model, the same can be used to dynamically transport undo policies within and between ubicomp environments.

References

[1] Abowd, G.D., Dix, A.J.: Giving undo attention. Interact. Comput. 4(3), 317–342 (1992)
[2] Agostini, A., De Michelis, G., Loregian, M.: Undo in Workflow Management Systems. In: van der Aalst, W.M.P., ter Hofstede, A.H.M., Weske, M. (eds.) BPM 2003. LNCS, vol. 2678, pp. 321–335. Springer, Heidelberg (2003)
[3] George, B., Leeman, J.: A formal approach to undo operations in programming languages. ACM Trans. Program. Lang. Syst. 8(1), 50–87 (1986)
[4] Locatelli, M.P., Loregian, M.: Active coordination artifacts in collaborative ubiquitous-computing environments. In: Schiele, B., Dey, A.K., Gellersen, H., de Ruyter, B.E.R., Tscheligi, M., Wichert, R., Aarts, E.H.L., Buchmann, A.P. (eds.) AmI 2007. LNCS, vol. 4794, pp. 177–194. Springer, Heidelberg (2007)

[5] Locatelli, M.P., Vizzari, G.: Awareness in collaborative ubiquitous environments: The multilayered multi-agent situated system approach. ACM Transactions on Autonomous and Adaptive Systems 2(4), 13 (2007)

[6] Loregian, M.: Undo for Mobile Phones: Does your Mobile Phone Need an Undo Key? Do You? In: Proceedings of the 5th Nordic Conference on Human-Computer Interaction 2008, Lund, Sweden, October 20-22. ACM, New York (2008)

[7] Loregian, M., Locatelli, M.P.: An Experimental Analysis of Undo in Ubiquitous Computing Environments. In: Sandnes, F.E., Zhang, Y., Rong, C., Yang, L.T., Ma, J. (eds.) UIC 2008. LNCS, vol. 5061, pp. 505–519. Springer, Heidelberg (2008)

[8] Schmidt, K., Simone, C.: Coordination mechanisms: Towards a conceptual foundation of cscw systems design. Computer Supported Cooperative Work (CSCW) 5(2), 155–200 (1996)

[9] Sun, C.: Undo as concurrent inverse in group editors. ACM Trans. Comput.-Hum. Interact. 9(4), 309–361 (2002)

[10] Yang, Y.: Undo support models. Int. J. Man-Mach. Stud. 28(5), 457–481 (1988)

[11] Zhou, C., Imamiya, A.: Object-based nonlinear undo model. In: COMPSAC 1997: Proceedings of the 21st International Computer Software and Applications Conference, pp. 50–55. IEEE Computer Society, Washington (1997)

Understanding Distributed Program Behavior Using a Multicast Communication Scheme

Mihai Mocanu[1] and Emilian Guţuleac[2]

[1] University of Craiova, Software Engineering Dept., Bvd.Decebal 107, Craiova,
   RO-200440, Romania
   mocanu@software.ucv.ro
[2] Technical University of Moldova, Computer Science Dept., 168 Bd. Ştefan cel Mare,
   Chişinău, MD-2004, Republic of Moldova
   egutuleac@mail.utm.md

Summary. Events in a distributed global computation framework, unlike those in a sequential local computation, form a partially ordered set with respect to the causality relation revealed by timestamps. This paper describes a new logical timestamping mechanism based on multicasting, called *Collective Logical Time*, and compares it with other known schemes that have been developed in the domain mainly to help in detecting undesired (global) properties of distributed computations (such as deadlock). Unfortunately, due to excessive complexity and some unrealistic restrictions (such as a fixed number of processes), these schemes have produced limited results. Some of the benefits in using our scheme are revealed, together with the possibilities for direct applications in the development of low-level communication protocols.

1 Introduction

Tracing event execution in a distributed computing environment (DCE) might seem a simple idea if we want to analyze a programs' behavior, its efficiency, or with respect to unusual event occurrences, but the formation of correct global time measurements is a difficult task. It is hard to understand an execution using a set of traces, due to the non-deterministic duration (Non-DD) of processes, as a consequence of their distributed nature and multiple interactions. Time is not absolute and the events in DCEs, unlike those in a sequential computation (SC), form a partially ordered set with respect to the causality relation, revealed by timestamps. Real time clocks are not relevant here, more useful to track causal dependencies between global events is the "logical" time, based either on scalar [1] or vector clocks [2] [3]. Difficulties here are in mapping partial order of distributed events into total ordering, or in the need to setup a constant, known-in-advance number of processes in the DCE [4][5]. Moreover, a distributed program may be easily perturbed by a metric code inserted, or too sensible to the application architecture and deployment decisions.

We present here a new logical timestamping mechanism - named *Collective Logical Time* (CLT), and a way to build it on top of a multicast scheme. We show also some of its advantages over other known schemes and how can it be directly applied in the construction of concurrency domains of execution, defined and denoted here as *Collective Work Domains* (CWD). These can in turn be used to understand the behavior and the implications of alternative implementations for the overall performance.

C. Badica et al. (Eds.): Intel. Distributed Comput., Systems & Appl., SCI 162, pp. 269–274, 2008.
springerlink.com   © Springer-Verlag Berlin Heidelberg 2008

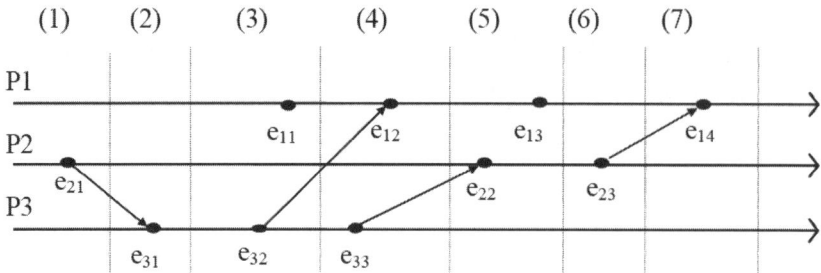

Fig. 1. Time diagram corresponding to an execution, as reported by an arbitrary observer

2   A Distributed Model of Execution

In our model, a set of sequential, "spatially" separated asynchronous processes cooperate to achieve a common task and communicate only by passing messages over a communication network. Process *Non-DD, unpredictable communication delays* are characteristic. The types and relative order of *events* (instantaneous and atomic actions occurring in each process) describe the model. Message transmission delays cannot be neglected. Communication between processes is point to point and FIFO order of message delivery is assumed. A global clock, or equivalently, perfectly synchronized local clocks are not available. If we denote by $E_1, E_2, \ldots, E_N$ the set of events occurring in processes $P_1, P_2, \ldots, P_N$, then $E = E_1 \cup E_2 \cup \ldots \cup E_N$ is the total set of all events occurring in the DCE. It is convenient to index the events in a process P_i in their order of occurrence, so let $E_i = e_{i1}, e_{i2}, \ldots, e_{ik}$, be a local sequence of events, usually denoted as (process P_i) *trace*. In terms of type, we need to distinguish between: **send** events, reflecting the fact a message was sent, **receive** events, denoting the fact a sent message has been received and state has been changed in consequence, and **internal** events, which affect only the local process state.

Unlike things happen in SC, in DCEs we cannot preserve a linear, unique sequence of events. We can see this if we examine closely the structure of a distributed computation, using a time diagram similar to the one in figure 1, and keep in mind that time scale may not be the same (for each process, a directed line representing local time whose flow is left-to-right, shows the progress). Events are symbolized here as dots on the process lines, according to their relative order of occurrence. Messages are drawn as arrows connecting send events with their corresponding receive events.

In the absence of a global system clock, computation in a DCE differs essentially from a sequential computation with respect to the definition of the temporal relationship between global events. Fortunately, almost all problems in understanding the execution of either one or many tasks in a DCE may be embedded in the fundamental problem of obtaining a consistent (non-contradictory) *global view of distributed computations*, and downsized to the detection of a consistent set of traces from event executions, with respect to cause-effect relationship. Therefore, most solutions proposed are based on causality mechanisms and use "timestamping" of events [8].

For the so-called *observability and consistency problems*, very well summarized in [3], the origin is in process non-DD duration, the lack of availability of absolute "real-time" and the unpredictable notification delays that any observation of the global computation may suffer. What is really important for different observations of a system is not to keep the real-time ordering of the events (difficult to obtain, may easily change) but to preserve the consistency of the observations, with respect to the causality relations. Put simply, observation consistency is directly derived from computation consistency, and equivalent to the requirement that, in the global linear event ordering, a cause must precede its effect (the *consistency criterion*).

Logical timestamping mechanisms are essentially based on event counting. In a system which uses logical timestamping, every process maintains a logical clock as a counter which is always advanced (never going back) using specific rules. Two of the most used systems of logical clocks are the scalar logical time (Lamport) [1] and vector time (Mattern-Fidge) [3][6]. A knowledge-based interpretation of vector time is that the component $V_j[j]$ of P_j's current vector time reflects the accurate logical time of P_j, as the "number of past events" at P_j, while $V_j[i]$ is the best estimate P_j was able to derive about P_i's current logical time value, from the messages received at P_j. Thus, if V(e) is the vector timestamp of an event occurring in P_j, then V(e)[i] is the number of events in P_i which e "knows about" i.e. which are in the causal history of e [5]. But, unlike it seems, vectors of dimension N are not mandatory in order to track N-distributed computations, if we need a simple mechanism consistent with causality. On the other hand, with respect to the size of vector clocks, a definite result would require some statement about the minimum amount of information that must be contained in timestamps in order to define an N-dimensional partial order on them. Stating this is still an open problem [4][6].

Finally, we must stress here that the causality relation introduced is potential, not real. For instance, events occurring in the same process are totally ordered, although some of them are not causally related (the absence of an event occurrence does not imply automatically the absence or modification of all the other events in its observed trace). This leads to the idea of relaxing the causality relation and to the aim of using other logical timing mechanisms to characterize the modified relation.

3   Collective Logical Time

A new timestamping mechanism, which relaxes the total (and thus artificial) order derived from Lamport logical time, seems to yield a partial order somewhat stricter, although in a more "natural" way than the order induced by vector time. We called it *Collective Logical Time* and based its construction on a multicast scheme. According to this, on each event occurrence in a process, which has potential to affect others, a notification is sent to all the other co-operative processes. It contains only time information and must use the fastest communication channels; the only condition needed here is that multicast communication of short notification messages is faster than longer messages exchanged between pairs of processes. This condition is realistic for today's distributed systems technology - where fast links or high priorities are dedicated to signals. As a consequence, a new category of events appear in a process: *collective* events, which are receive notification events which carry information from another process, about the occurrence there of a potentially-affecting event.

Definition 1. *(CLT). The* collective logical time *is a mapping $C : E \rightarrow N$ from events to integers, defined as follows:*

1. *If e is an* internal *or* send *event without (immediate) local predecessor, $C(e) = 1$.*
2. *If e is an* internal, send *or* receive *event and has as local predecessor the* internal *or* collective *event e', then $C(e) = C(e') + 1$.*
3. *If e is a* collective *event, it either has the timestamp of a local predecessor e', or carries the timestamp of corresponding send event e": $C(e) = max(C(e'), C(e"))$.*

Theorem 1. *The collective timestamping order $(C, <)$ is consistent with causality.*

Proof. Let e, e' be two events such as $e < e'$. If e and e' occur in the same process, the logical clock C is incremented once with each event between (exclusive) e and (inclusive) e', and never set back. This means $C(e) < C(e')$. If e and e' occur in different processes P_i and P_j, $e < e'$ implies there is a path from e to e', with two corresponding send-receive events s in P_i and r in P_j, $s < r$. But for local events we have $e < s$ and $r < e'$, and from the transitivity of causality relation it follows $e < e'$.

The diagrams in figure 2 illustrate similarities of our mechanism with Lamport timestamping, its simplicity, but also an increase in its potential to construct "time zones" denoted here as concurrency domains for the distributed processing. Comparing Lamport and Collective timestamping reveals that: a. our assignment for logical time, with concurrency domains marked, is quite similar to Lamport's and therefore, simple; b. the multicast communication scheme which assigns CLT works; c. the collective logical time assignment can be used in building concurrency domains, clearly marked in figure 2c.

An algorithm to build CWDs will be introduced in the next section, together with a discussion of a possible protocol.

4   Collective Work Domains

The collective timestamping mechanism transmits a sort of "global knowledge" in the form of chronological information in groups of cooperative processes. This information can be used by synchronization algorithms and protocols. Better grouping of collective logical time values, as opposed to the dispersion of Lamport logical time values, suggest the idea of defining concurrency domains. These should be able to put together distributed events that can be executed in the same time frame (horizon). What it important here is to base the decision of events immediate execution or postponement strictly on local information.

Definition 2. *(CWD) The* collective work domain *(concurrency domain) is a subset of distributed events having adjacent timestamps, which can be executed when they are ready as follows: sequentially, within the frame of the same process, and concurrently, in any order, if they belong to different processes.*

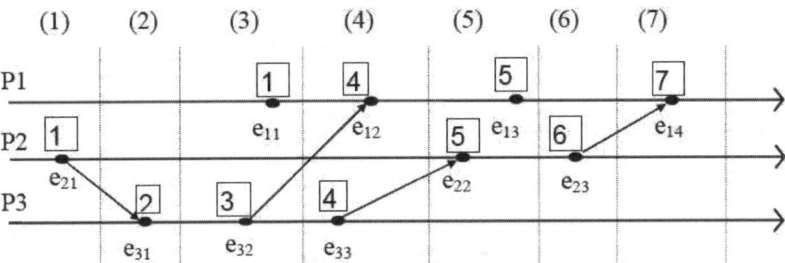

a. Lamport assignment for logical time, with concurrency domains marked.

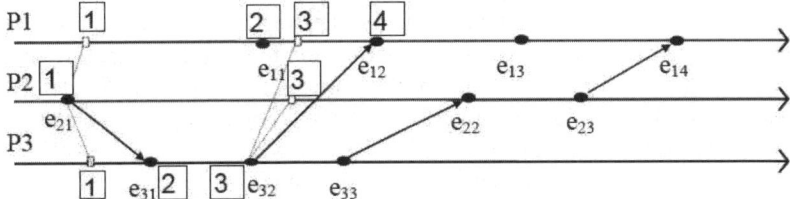

b. Multicast communication scheme which assigns collective logical time.

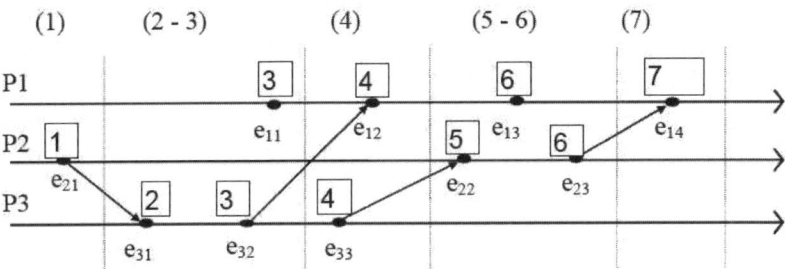

c. Collective logical time assignment, with collective work domains marked.

Fig. 2. A comparison between Lamport and Collective Timestamping

Algorithmic steps in constructing CWDs:

1. a) Initialize C=1 and b) first CWD to include all events with timestamp C.
2. a) If there is a causal communication between an event $e \in P_i$ and another event $e' \in P_j$, neither e' or another distributed event with timestamp C=C(e') will be included in CWD1, but will form the first timestamp value for the next CWD.
   b) Else, in the absence of a causal communication, C is incremented and the CWD is extended with distributed events timestamped with this new time value.
3. Repeat steps 2a and b until a causal communication is detected.
4. Repeat steps 1b, 2 and 3 until all distributed events have been processed.

The iterative construction of CWDs is based on the identification of causal dependencies among distributed events and their expression in collective logical time values. An application of this algorithm is illustrated by an improved construction of CWDs in figure 2c, as compared to the concurrency domains in figure 2a that can be obtained based on a Lamport scheme.

5 Concluding Remarks

Time is a multi-shape concept, not yet fully understood in DCE. Real (absolute) time is less important than logical time, whose flow is only determined by the events happened locally and by the knowledge acquired for global events.

We discussed in this paper different schemes to assign logical clocks, including a new, original scheme. CLT, and the separation of distributed events in CWD, exhibit some direct advantages over other known schemes. Reflecting in each local process knowledge on the global events more precisely than Lamport time, and propagating complete global knowledge with high probability in every moment, they serve in the detection of simultaneous events and in the exploitation of the maximum degree of concurrency in execution. This mechanism may also have others straightforward and important applications: it may be used to avoid race conditions and other synchronization errors, or in detecting the termination of a distributed computation; to accurately replay concurrent activities in distributed systems for the purpose of debugging and monitoring; for reasoning about the properties of asynchronous systems, since it reflects both causal and temporal structure of such a system; or in parallel and distributed simulations, either by conservative and optimistic protocols.

References

1. Lamport, L.: Time, Clocks, and the Ordering of Events in a Distributed System. Comm. ACM 21(7), 558–565 (1978)
2. Charron-Bost, B.: Concerning the Size of Logical Clocks in Distributed Systems. Information Processing Letters 39, 11–16 (1991)
3. Fidge, C.J.: Fundamentals of Distributed System Observation. IEEE Software 13(6), 77–83 (1996)
4. Landes, T.: Dynamic Vector Clocks for Consistent Ordering of Events in Dynamic Distributed Applications. In: Arabnia, H.R. (ed.) Proc. Intl. Conference on Parallel and Distributed Processing Techniques and Applications, Las Vegas, Nevada, vol. 1, pp. 31–37 (2006)
5. Raynal, M., Singhal, M.: Logical Time: Capturing Causality in Distributed Systems. IEEE Computer 29(2), 49–56 (1996)
6. Schwarz, R., Mattern, F.: Detecting Causal Relationships in Distributed Computations: In Search of the Holy Grail. Distr. Computing 7(3), 149–174 (1994)
7. Basten, T.a.o.: Vector Time and Causality among Abstract Events in Distributed Computations, TR-NSERC, University of Waterloo, Canada, pp.1-33 (1996)
8. Santoro, N.: Design and Analysis of Distributed Algorithms. John Wiley and Sons, Chichester (2007)

Multi-agent Conflict Resolution with Trust for Ontology Mapping

Miklos Nagy[1], Maria Vargas-Vera[2], and Enrico Motta[1]

[1] The Open University, Knowledge Media Institute (Kmi), Milton Keynes, MK7 6AA , UK
mn2336@student.open.ac.uk, e.motta@open.ac.uk
[2] The Open University, Computing Department Milton Keynes, MK7 6AA , UK
m.vargas-vera@open.ac.uk

Summary. Software agents that operate on the Semantic Web have to deal with scenarios where the discovery and combination of the relevant information from a variety of heterogeneous sources becomes contradicting. One such application area of the Semantic Web is ontology mapping where different similarities have to be combined into a more reliable and coherent view, which might easily become unreliable if trust is not managed effectively between the different sources. In this paper we propose a solution for managing trust between contradicting beliefs in similarities for ontology mapping based on the fuzzy voting model.

1 Introduction

Managing content related trust on the Semantic Web has not received too much attention from the research community so far. The reason for that is most probably the relatively small number of successfully deployed applications that can be tested by users of communities. However since Semantic Web applications become more and more adolescent, issues related to the very nature of this environment will likely emerge such as trustworthiness of the information available in this media or the uncertain nature of the deducted information as a direct consequence of the heterogeneous environments. If we assume that in the Semantic Web environment it is not possible to deduct an absolute truth from the available sources then we need to evaluate content dependent trust levels by each application that processes the information on the Semantic Web e.g. how a particular information coming from one source compares the same or similar information that is coming from other sources. The main contribution of this paper is a novel trust management approach for resolving conflict between agent's belief in similarities which is the core component in the DSSim[7, 8] ontology mapping system. DSSim addresses the uncertain nature of the ontology mapping by considering different similarity measures as subjective probabilities for the correctness of the mapping which is represented as an individual belief of a mapping software agent . It employs the Dempster Shafer theory of evidence in order to create and combine beliefs that has been produced by the different similarity algorithms.

The paper is organized as follows. Section 2 provides the description of the problem and its context. Section 3 describes the voting model and how it is applied for determining trust during the ontology mapping. Section 4 gives and overview of the related work. Finally, section 5 describes our future work.

C. Badica et al. (Eds.): Intel. Distributed Comput., Systems & Appl., SCI 162, pp. 275–280, 2008.
springerlink.com

2   Problem Description

The problem of trustworthiness in the context of ontology mapping can be represented in different ways. In general, trust issues on the Semantic Web are associated with the source of the information i.e. who said what and when and what credentials they had to say it. Consider an example from ontology mapping. When software agents assess similarity between two terms they can use different linguistic and semantic information in order to determine the similarity level e.g. background knowledge or concept hierarchy. The problem is that any similarity assessment can perform differently depending on domain because the context is determined by the available backgroud knowledge . In reality any similarity algorithm will produce good and bad mappings for the same domain depending of the actual context of the terms in the ontologies e.g. different background knowledge descriptions or class hierarchy. In order to overcome this shortcoming the combination of different similarity measures are required. In general any similarity combination can improve the overall result except if there is a conflict between the assessments. To handle this conflict in our ontology mapping method we propose using trust in the provided beliefs in similarities, which is assessed between the ontology entities and associated to the actual understanding of the mapping entities. These beliefs can differ from case to case e.g. a similarity measure can be trusted in one case but not trustful in an another case during the same process.

3   Trust Management for Belief Combination

In ontology mapping the conflicting results of the different beliefs in similarity can be resolved if the mapping algorithm can produce an agreed solution, even though the individual opinions about the available alternatives might vary. We propose a solution for reaching this agreement by evaluating trust between agents' established belief through voting which is a general method of reconciling differences. Voting is a mechanism where the opinions from a set of votes are evaluated in order to select the alternatives that best represent the collective preferences. Unfortunately deriving binary trust like trustful or not trustful from the difference of belief functions is not so straightforward since the different voters express their opinion as subjective probability over the similarities. For a particular mapping this always involves a certain degree of vagueness hence the threshold between the trust and no trust cannot be set definitely for all cases that can occur during the process. Our argument is that the trust membership value which is expressed by different voters can be modeled properly by using fuzzy representation as depicted on Fig. 1.

In fuzzy logic the membership function $\mu(x)$ is defined on the universe of discourse U and represents a particular input value as a member of the fuzzy set i.e. $\mu(x)$ is a curve that defines how each point in the U is mapped to a membership value (or degree of membership) between 0 and 1. For representing trust in beliefs over similarities we have defined three overlapping trapezoidal membership functions which represents high, medium and low trust in the beliefs over concept and property similarities in our ontology mapping system.

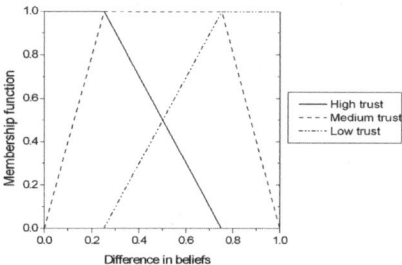

Fig. 1. Trust representation

3.1 Fuzzy Voting Model

The fuzzy voting model was developed by Baldwin [1] and has been used in Fuzzy logic applications. However, to our knowledge it has not been introduced in the context of trust management on the Semantic Web. In this section, we will briefly introduce the fuzzy voting model theory using a simple example of 10 voters (agents) voting against or in favor of a similarity measure for ontology mapping. According to Baldwin [1] a linguistic variable is a quintuple $(L, T(L), U, G, \mu)$ in which L is the name of the variable, $T(L)$ is the term set of labels or words (i.e. the linguistic values), U is a universe of discourse, G is a syntactic rule and μ is a semantic rule or membership function. We also assume for this work that G corresponds to a null syntactic rule so that $T(L)$ consists of a finite set of words. A formalization of the fuzzy voting model can be found in [6]. Consider the set of words { low-trust (L_t), medium-trust (M_t) and high-trust (H_t) } as labels of a linguistic variable trust with values in $U = [0, 1]$. Given a set "m" of voters where each voter is asked to provide the subset of words from the finite set $T(L)$ which are appropriate as labels for the value u. The membership value $\chi_{\mu_{(w)(u)}}$ is taking the proportion of voters who include v in their set of labels.

We need to introduce more opinions to the system i.e. we need to add the opinion of the other agents in order to vote for the best possible outcome. Therefore we assume for the purpose of our example that we have 10 voters (agents). Formally, let us define

$$V = A1, A2, A3, A4, A5, A6, A7, A8, A9, A10 \tag{1}$$
$$\Theta = L_t, M_t, H_t$$

Let us start illustrating the previous ideas with a small example - By definition consider our linguistic variable L as TRUST and TL the set of linguistic values as $TL = (low\_trust, medium\_trust, high\_trust)$. The universe of discourse is U which is defined as $U = [0, 1]$. Then, we define the fuzzy sets $\mu(Low - trust), \mu(Medium - trust)$ and $\mu(High - trust)$ for the voters where each voter has the overlapping trapezoidal membership functions as depicted on Figure 1. The random set L=TRUST is defined by the Table 1. Note that in the table we use a short notation L_t means Low_trust, M_t means Medium_trust and H_t means High_trust. Once the fuzzy sets (membership functions) have been defined the system is ready to assess the trust memberships for the input values. Based on the difference of beliefs in similarities the different voters will

Table 1. Possible values for the voting

A1	A2	A3	A4	A5	A6	A7	A8	A9	A10
L_t	L_t	L_t	L_t	L_t	L_t	L_t	L_t	L_t	L_t
M_t	M_t	M_t	M_t	M_t	M_t				
H_t	H_t	H_t							

Table 2. Voting

A1	A2	A3	A4	A5	A6	A7	A8	A9	A10
H_t	M_t	L_t	L_t	M_t	M_t	L_t	L_t	L_t	L_t

select the words they view as appropriate for the difference of belief. Assuming that the difference in beliefs(x) is 0.67 the voters will select the following labels:

Then we compute the membership value for each of the elements on set TL.

$$\chi_{\mu(Low\_trust)(u)} = 1 \tag{2}$$

$$\chi_{\mu(Medium\_trust)(u)} = 0.6 \tag{3}$$

$$\chi_{\mu(High\_trust)(u)} = 0.3 \tag{4}$$

and

$$L = \frac{Low\_trust}{1} + \frac{Medium\_trust}{0.6} + \frac{High\_trust}{0.3} \tag{5}$$

A value x is presented and voters pick exactly one word from a finite set to label x as depicted in Table 2.

Taken as a function of x these probabilities form probability functions. They should therefore satisfy:

$$\sum_{w \in TL} Pr(L = w|x) = 1 \tag{6}$$

which gives a probability distribution on words:

$$\sum Pr(L = Low\_trust|x) = 0.6 \tag{7}$$

$$\sum Pr(L = Medium\_trust|x) = 0.3 \tag{8}$$

$$\sum Pr(L = High\_trust|x) = 0.1 \tag{9}$$

As a result of voting we can conclude that given the difference in belief $x = 0.67$ the combination should not consider this belief in the similarity function since based on its difference compared to another beliefs it turns out to be an untrustful assessment. The before mentioned process is then repeated as many times as many different beliefs we have for the similarity i.e. as many as different similarity measures exist in the ontology mapping system.

4   Related Work

Dominantly the existing approaches for ontology mapping that address the problem of the trustworthiness of the available data on the Semantic Web are reputation and context based e.g. using digital signatures that would state who the publisher of the ontology is. For ontology mapping there are different methods to combine similarities. GLUE [2] for example uses Bayesian probabilistic model to combine the results of different learners, which exploit information in concept instances and taxonomic structure of ontologies. Other system COMA [3] applies the variations of max, min, weighted or threshold aggregation methods in order to assess the best possible similarity measures. In practice considering the overall results these combination methods will perform relatively well under different circumstances except when contradictory evidence occurs during the combination process. Trust is important in applications where the human-computer interaction is necessary in order to support the users' task with mimicking intelligent behavior. Ontology mapping is one of these areas and such there is a possibility to improve these systems if the algorithms can be enhanced with cognitive support [4]. To date this perspective of trust has not been investigated in the context of ontology mapping. Ongoing research has mainly been focusing on how trust can be modeled in the Semantic Web context [10] where the trust of user's belief in statements supplied by any other user can be represented and combined. Considering multi-agent systems on the Web existing trust management approaches have successfully used fuzzy logic to represent trust between the agents from both individual[5] and community[11] perspective. However the main objective of these solutions is to create a reputation of an agent which can be considered in future interactions.

5   Conclusion

In this paper we have shown how the fuzzy voting model can be used to evaluate trust, and determine which belief is contradictory with other beliefs before combining them into a more coherent state. We have proposed new levels of trust in the context of ontology mapping, which is a prerequisite for any systems that makes use of information available on the Semantic Web. Our system is flexible because the membership functions for the voters can be changed dynamically in order to influence the outputs according to the different similarity measures that can be used in the mapping system. There are many areas of ongoing work, with our primary focus being additional experimentation to investigate different kind of membership functions for the different voters and to consider the effect of the changing number of voters and the impact on precision and recall. In our future research we also intend to investigate different conflict detection methods. Our aim to measure how different levels of conflict can affect the overall performance of our solution.

References

1. Baldwin, J.F.: Mass assignment Fundamentals for computing with words. In: Selected and Invited Papers from the Workshop on Fuzzy Logic in Artificial Intelligence. Lecture Notes In Computer Science, pp. 22–44. Springer, Heidelberg (1999)

2. Doan, A., Madhavan, J., Domingos, P., Halevy, A.: Learning to map between ontologies on the semantic web. In: Proceedings of the 11th World Wide Web Conference, Honolulu, Hawaii (2002)
3. Do, H.H., Rahm, E.: COMA - A System for Flexible Combination of Schema Matching Approaches. In: Proceedings of 28th International Conference on Very Large Databases (VLDB), Hong Kong, China (2002)
4. Falconer, S., Storey, M.: A cognitive support framework for ontology mapping. In: Proceedings of the 6th International Semantic Web Conference, Busan, Korea (2007)
5. Griffiths, N.: A Fuzzy Approach to Reasoning with Trust, Distrust and Insufficient Trust. In: Proceedings of the 10th International Workshop on Cooperative Information Agents, Edinburgh, UK (2005)
6. Lawry, J.: A Voting Mechanism for Fuzzy Logic International Journal of Approximate Reasoning. 19, 315–333 (1998)
7. Nagy, M., Vargas-Vera, M., Motta, E.: DSSim - Managing Uncertainty on the Semantic Web. In: Proceedings of the 2nd International Workshop on Ontology Matching, Busan, Korea (2007)
8. Nagy, M., Vargas-Vera, M., Motta, E.: Multi-agent ontology mapping with uncertainty on the Semantic Web. In: Proceedings of the 3rd IEEE International Conference on Intelligent Computer Communication and Processing, Cluj, Romania (2007)
9. Shafer, G.: A Mathematical Theory of Evidence. Princeton University Press, Princeton (1976)
10. Richardson, M., Agrawal, R., Domingos, P.: Trust Management for the Semantic Web. In: Proceedings of the 2nd International Semantic Web Conference, Florida, USA (2003)
11. Rehak, M., Pechoucek, M., Benda, P., Foltyn, L.: Trust in Coalition Environment: Fuzzy Number Approach. In: Proceedings of The 4th International Joint Conference on Autonomous Agents and Multi Agent Systems - Workshop Trust in Agent Societies, The Netherlands (2005)

Algorithmic Trading on an Artificial Stock Market

Daniel Paraschiv[1], Srinivas Raghavendra[2], and Laurentiu Vasiliu[1]

[1] CIMRU/DERI National University of Ireland, Galway
   daniel.paraschiv@nuigalway.ie, laurentiu.vasiliu@nuigalway.ie
[2] Department of Economics National University of Ireland, Galway
   s.raghav@nuigalway.ie

Summary. This work introduces algorithmic trading on artificial stock markets and describes past and existing approaches. A proposed framework of the artificial stock market approach is presented, together with the used agent types. Then the simulation results' analyses are discussed. Conclusions and future work directions are presented, showing where the MACD algorithm and some rules can be used. The human behavior influence over the market is highlighted.[1]

Keywords: Artificial Stock Market, Double Auction, Back Testing, MACD.

1 Introduction

Algorithmic trading and artificial stock markets have been in the last decade of high interest for business, IT research and academia. The emergence of algorithmic trading has created a new environment where the classic way of trading requires new approaches. High trading speed and automated algorithms have accelerated the trading process beyond human capabilities, moving brokers in a new area that can be called 'micro-second economics'. In order to tackle this, new tools theories and approaches need to be created. Thus artificial stock markets have emerged as simulation environments where to test, understand and model the already complex human behaviours and also to analyse the impact in the system of algorithmic trading where humans and software agents may compete on the same market. Considering this, the purpose of this paper is to create a framework to test and analyse various trading strategies in a dedicated artificial environment.

2 Related Work

There exists a vast literature on the computer-simulated, artificial financial markets following the pioneering work done at the Santa Fe Institute [1]. Some studies proposed

---

[1] This research has been funded through Enterprise Ireland Research Grant CFTD/05/312, M2MN project.

C. Badica et al. (Eds.): Intel. Distributed Comput., Systems & Appl., SCI 162, pp. 281–286, 2008.

artificial markets populated with heterogeneous agents endowed with learning and optimising capabilities with an aim to mimic the performance of the real world markets. Other studies looked at various trading rules attributed to agent types and its implication for the market outcome. Given the vastness of this literature and for want of space, we will discuss a representative model from the literature and its relation to the present work.

In [2] is presented an agent-based artificial financial market with heterogeneous agents who trade one single asset through a trading mechanism, to study the process of price formation. The price-formation process of the market was built around a mechanism for matching demand and supply of market orders. In this market, agents are endowed with limited resources with the global-amount of cash in the economy in time-invariant. There are N agents and at each simulation step each agent issues a buy order with probability p_i or a sell order with probability 1-p_i. The orders are generated in the following way: Suppose the ith agent issues a sell order of quantity $a_i{}^s$ at time h+1. The quantity of stocks offered for sale at time step h+1 is a random fraction of the quantity of stocks owned at time step h according to the rule: $a_i^s = [r_i A_i(h)]$ where r_i is a random number drawn from a uniform distribution in the interval [0,1] and $A_i(h)$ is the amount of assets owned by the ith agent at time h. In addition, a limit sell price s_i is associated to each sell order. The limit prices are computed by $s_i = (\frac{p(h)}{N_i(\mu,\sigma_i)})$ where $N_i(\mu,\sigma_i)$ is a random draw from a Gaussian distribution with $\mu = 1.01$ and a standard deviation is proportional to the historical volatility computed through the equation $\sigma_i = k\sigma(T_i)$ with k being a constant and $\sigma(T_i)$ is the standard deviation of log-price returns. The buy orders are generated in a fairly symmetrical way with respect to sell orders, where the buy order, c_i, at time h+1 is a random function of available cash at time h, i.e., $c_i = r_i C_i(h)$ with r_i being a random number drawn from a uniform distribution in the interval [0, 1] and $C_i(h)$ is the amount of cash with i^{th} agent. The price formation process is set at the intersection of the demand and supply curves with the former is a decreasing step function of price and the latter is an increasing step function of price and the equilibrium price is computed by the system at which the two functions intersect.

Even though the results of this model and other subsequent models seem to capture some of the stylized facts of Herding, Bubbles, Crashes, Fat tails and Volatility Clustering that we observe in real financial markets, limiting the model to only random or uninformed agents is too simplistic and there is a scope for generalising this model with different types of agents to test the robustness of the results [3]. Nevertheless, in terms of comparing various artificial agent models that generate various stylized facts of the financial markets from, it is pertinent to develop an artificial agent environment that has the capability of back-testing whereby it lends itself as a standard for comparing the empirical verifiability of this class of models. It is in this sense, the model presented here is a contribution to the literature of artificial agent models.

3   Model Framework of Artificial Market with Algorithmic Trading

The framework of the artificial market proposed in this paper is presented in figure 1. The artificial stock market uses a double auction [4] system or if preferred uses real data from time series of real stocks. Using algorithmic trading the agents place buy or sell stock orders on the Artificial Stock Market that makes the connection between buy orders and sell orders of the same stock.

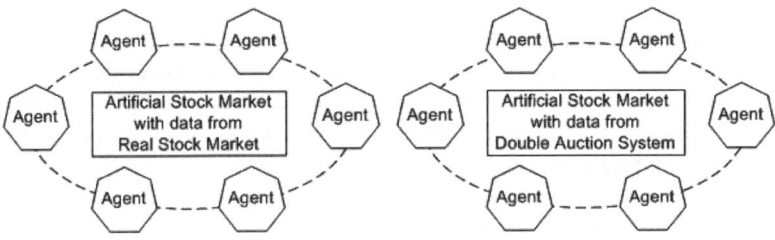

Fig. 1. The framework of the artificial environment

The algorithm of the Artificial Stock Market is:

1. while (market is open) do
2.   IF time to apply tick passed THEN
3.   Change all Stocks Price with a tick
4.   Process Orders for a cycle
5.   IF a day has passed THEN
6.   Change time period to all stocks

The aim of a double auction system is to maximize the number of shares processed [5]. To do so the price of each stock is changed every 100ms with a tick of value 0.005. The formula of the new price for each stock is·

$$P_{new} = \begin{cases} P_{old} + Tick \ IF \ V_B - V_S > 10 \\ P_{old} - Tick \ IF \ V_S - V_B > 10 \\ \quad P_{old} \qquad\qquad otherwise \end{cases}$$

Where V_B is the total volume of buy shares and V_S is the total volume of sell shares of the stock with price P. 1000 shares of each stock are matched at step 4. Shares from the same order can be processed with different prices and the final price of the shares is given by the volume weighted average price formula [6].

When a day has passed (step 6) the current data is added to a list of old data in case of double auction mode or real data are loaded from database and added to this list of old data in case of back testing mode. The real data were downloaded from Yahoo Finance [7], being the historical data from 1st of March 2005 to 24th of March 2008 for ten companies: Microsoft, Yahoo, IBM, Google, Apple, Sony, General Motors, Ford, Honda Motor and NISSAN.

3.1 Agent Types

The tested and analyzed classes of agents are random agents, human agents, market makers agent and three strategic agents.

1) Random Agents: They place, for each stock, buy orders with probability 0.5 and sell orders with probability 0.5. After a random number of placing orders they check the total volume of buy orders and sell orders and they place an order that makes the number of buy shares equal with the number of sell shares.

2) Human Agents & Market Makers agent: Anybody can place an order manually using a human agent. The orders can be limit orders or market orders. The human agent can place buy or sell orders for each company or for all companies at once. The Market Makers Agent automatically makes sure that buy volume and sell volume (demand and supply) are almost balanced.

3) Strategic Agents: These agents implement a rule of the form:

IF condition THEN buy/sell stock(s)

a) Rule 1 Agent type: IF the agent has stock X in portfolio AND Y is not in portfolio AND X is up UpThreshold AND Y is down DownThreshold in NoOfDays THEN sell stock X and buy Y with money from stock X.

b) Rule 2 Agent type: IF the agent has stock X in portfolio AND X is up UpThreshold OR X is down DownThreshold THEN sell stock X and buy Y that is not in portfolio and has the highest fall in NoOfDays, with money from stock X.

c) MACD Agent type: IF X is in portfolio AND the histogram changed from positive to negative THEN sell X; IF X is not in portfolio AND the histogram changed from negative to positive THEN buy X.

Moving Average Convergence Divergence - MACD [8] is a method to identify a trend for a stock. The formulas for MACD are:

MACD = EMA[D2] of price - EMA[D1] of price

signal = EMA[D3] of MACD

histogram = MACD - signal, where EMA is the exponential moving average [9] with the give number of days (Di), and D1 >D2 >D3.

4   Results and Discussion

There were tested 10 stocks and it was created an index as the arithmetic average of these stocks. The charts were implemented with JFreeChart [10]. For each stock and index are candlestick charts [11], a chart with logarithmic return and a chart with the distribution of return. Also for each stock can be seen the excess demand, current price, price change, current volume, average volume and other indicators, see figure 2. For each agent can be seen the initial wealth, current wealth, portfolio wealth, the percentage of profit or lose, no of bought/sold orders and no of bought/sold shares for each stock and per total. The agents are sorted by profit.

The random agents, also known as uninformed agents, are useful to guarantee a volume of shares on the artificial stock market. They place buy and sell orders without

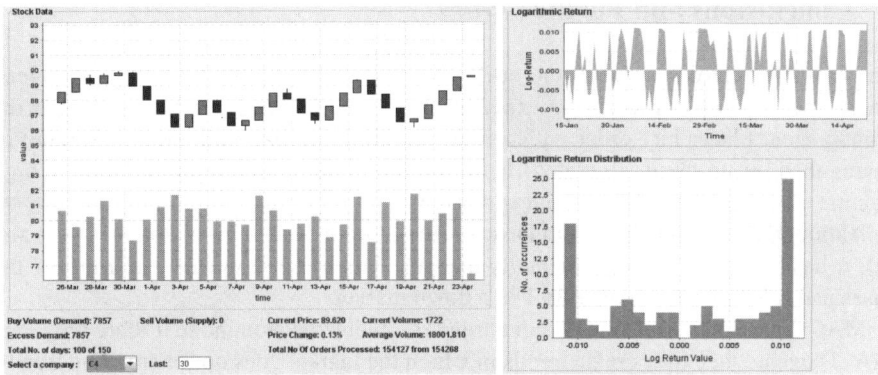

Fig. 2. The stock chart, the logarithmic return and logarithmic return distribution

Table 1. The profit and lose of agents from different types of agents

	Double Auction			Back Testing		
	\multicolumn{3}{c}{Double Auction}			\multicolumn{3}{c}{Back Testing}	
\multicolumn{7}{l}{Random Agents, Human agent and the agents with rule 1 and rule 2 (150 days)}						
Top:	**Rule 1**, Profit: 10.42%			**Rule 1**, Profit: 11.17%		
	Rule 2, Profit: 8.49%			Rule 2, Profit: 7.3%		
Loser:	Rule 2, Lose: 4.15%			Rule 2, Lose: 3.13%		
\multicolumn{7}{l}{Random Agents, Human A., rule 1 (R1) and rule 2 (R2) A. and MACD (M) A.}						
	250 days	500 days	750 days	250 days	500 days	750 days
Top:	**R1**: 11.8%	**R2**: 19.1%	**R2**:35.2%	**R1**: 15.2%	**R1**:23.8%	**R1**: 26%
	R2: 9.2%	R1: 13%	R1:16.8%	R2: 11.4%	R2: 21.4%	M: 22.6%
	M: 3.99%	M: 1.62%	M: -0.37%	M:11.01%	M: 15.19%	R2: 13.5%
Loser:	R2: -2.9%	M: -5.2%	M: -8.26%	R2: -4.4%	R2: -8.37%	R2: -8.6%

checking the price. Usually their profit is negative. If only these agents are active, in double auction mode, then the stocks follow a sinusoidal move.

There were four main simulations with different types of agents and different modes, as presented in table 1. The first two simulations were with 10 random agents, a human agent, 98 rule 1 agents and 98 rule 2 agents. The last two simulations were with 8 more MACD agents in double auction mode and 121 more MACD agents in back testing mode. In back testing mode the market maker agent was activated. Each agent had different values of parameters. In figure 2 is presented a stock with the first simulation. In table 1 are presented the best profit of agents from each type of agent and the type of agents with the highest lose. With human agents there were made available the shares that the strategic agents buy initially. When there are many MACD agents in double auction mode they create trends. In the simulation with MACD agents and double auction mode was a small down trend and all 8 MACD agents had lost wealth after 750 days. In back testing mode all 121 MACD agents had profit after 750 days.

5 Conclusions and Further Work

The human agent described in the artificial market can represent a group of traders. There had been observed during experiments in double auction mode that when the human agent places buy order with a high volume (more than 50% daily volume) of shares then an up trend is created. If the human agent places sell order with a high volume of shares then a down trend is created.

Although rule 1 looks very similar with rule 2 (as presented in section 3.1, 3) a) and b)), it has been proved experimentally that rule 1 is better than rule 2 in most of the cases and rule 2 is better than rule 1 in a down market.

MACD method is suited to detect a trend. In double auction mode if there are many MACD agents then they create trends in which the market goes up for small up trends, and for small down trends they create mass selling which produces down trends.

MACD has been proved to be useful in back testing mode and it was not so useful in double auction mode. This showed that MACD is not useful for traders that place big orders and influence the market, it is useful for small orders that do not influence the market.

As future work there is intended to create and test more types of agents. There will be created agents that chose their strategies based on index as some strategies are better suited than others when the index goes up or down. There will be created agents that forecast the prices of stocks. There will be automated more technical indicators, e.g. momentum indicators, William's and so on [12]

References

1. LeBaron, B.: Building the Santa Fe Artificial Stock Market. Brandeis University (2002)
2. Raberto, M., Cincotti, S., Focardi, S.M., Marchesi, M.: Agent-based simulation of a financial market. Physica A 299(1-2), 320–328 (2001)
3. Samanidou, Cf. Zschischang, E.E., Stauffer, D., Lux, T.: Agent-based models in financial markets, arXiv:physics/0701140v1 [physics.soc-ph] (January 11, 2007)
4. Double Auction,
   http://www.investorwords.com/1550/double_auction_market.html
5. Equity Trading Systems in Europe by Marianne Demarchi and Thierry Foucault
6. Electronic and Algorithmic Trading Technology by Kendall Kim
7. Yahoo Finance, http://finance.yahoo.com/
8. MACD, http://www.investopedia.com/terms/m/macd.asp
9. Exponential Moving Average,
   http://www.decisionpoint.com/TAcourse/MovingAve.html
10. JFreeChart, http://www.jfree.org/jfreechart/
11. Candlestick chart, http://www.investopedia.com/terms/c/candlestick.asp
12. Algorithmic trading indicators,
    http://www.programtrading.com/DAT/algorithm-trading-indicators.htm

Towards a Visual Definition of a Process in a Distributed Environment[*]

Dragoslav Pešović, Zoran Budimac, and Mirjana Ivanović

Department of Mathematics and Informatics, Faculty of Science, University of Novi Sad
Trg Dositeja Obradovića 4, 21000 Novi Sad, Serbia
{dragoslav,zjb,mira}@im.ns.ac.yu

Summary. Workers, Inc., a workflow management system implemented using the technology of mobile agents, is especially suited for highly distributed and heterogeneous environments. This paper discusses the issues related to visual definitions of new processes, and their translation into execution contexts of the system.

1 Introduction

The usage of mobile agents [3] in modeling and implementation of a workflow [5] simplifies the workflow management. Workers, Inc. consists of individual agents with autonomous behavior. Mobile agents carrying out workflow instances (the so-called workers) have the ability to move to different users, where they can interact with them locally, autonomously taking care of their current position, state, and further itinerary. In order to achieve the flow of work, workers split the work in logical parts, cooperate together, and synchronize themselves.

However, in the proposed system, new agent classes describing brand-new work-flows had to be hard-coded manually. In order to employ the workflow system in a real working environment, end-users of the workflow system should be enabled to create brand-new worker classes. This goal could be achieved by the means of a declarative language, which can be used for describing workflow definitions. Moreover, a software tool, which would enable the visual definition of a new worker, could be used for process modeling while extracting the necessary data.

The Workflow Management Coalition (WfMC) developed and proposed XML Process Definition Language (XPDL) [6] with the intention for it to become a common language in the workflow domain considering the definition and exchange of process definitions. The language allows for the import and export of process definitions between a variety of tools ranging from workflow management systems to modeling and simulation tools.

By adopting the proposed standardized language (XPDL), our system becomes capable not only to use a number of existing graphical workflow editors, but also to exchange process definitions with various other workflow products.

In order to comply with XPDL, the system first had to be modified to conform to the basic constructs of XPDL and the underlying meta-model. The structure of workers'

---

[*] This work was supported by project "Abstract Methods and Applications in Computer Science" (no. 144017A), of the Serbian Ministry of Science and Environmental Protection.

itineraries (previously supporting only sequential routing) was adapted to support more refined control-flow patterns, like parallel routing, alternative paths, or iterations. The system which developed in such a way has been named Workers, Inc. to emphasize the collaborative nature of agents.

Moreover, a system-specific import layer had to be provided to allow the translation of process definitions, generated using a visual modeling tool, into worker execution contexts, their internal system representations.

2   System Architecture

Workers, Inc. is envisioned as a community of cooperative agents, its main characteristics being full decentralization and distribution of workflow functions. It is built on top of the Mole mobile agent system [1]. The current architecture is essentially two-part, consisting of work-agents (workers) and host-agents (worker hosts).

2.1   Workers

A worker is the key system component encapsulating both the process definition and the execution state of a workflow. While performing a workflow, a worker itinerates among distributed resources carrying process-specific information and autonomously taking care of its execution state. In that way, workers manage not only to perform workflow activities locally with respect to assigned resources, but to avoid the need to consult a central server or the originating machine at every step.

A workers behavior is entirely defined by its execution context. A worker context is an executable process definition, a worker being just a medium through which its context is transmitted and accomplished. When a worker migrates, its entire execution context as an object net is being encompassed by object serialization, and then transported and reconstructed at the target location.

The most important part of a context is the worker itinerary, which represents a flow of a worker through a network. By representing itineraries with directed graphs we are able to support complex flow patterns that could be needed by workflow applications.

To allow concurrent activity execution, agent social abilities are employed. When a single thread of control needs to split into two or more threads, which can be executed in parallel, the worker context is cloned and multiple worker instances are allowed to be executed simultaneously. On the other hand, when multiple parallel threads of execution need to converge into a single thread, agent coordination mechanisms and synchronization techniques are employed.

To strengthen security of the system, Mole mobile agents and thus workers are forbidden to access any system resources directly. Critical resources can be accessed only by communicating with stationary system agents, i.e. worker-hosts.

2.2   Worker Hosts

Every node in the network contains a worker host, which is implemented as a stationary system agent, having special privileges for the access to host system resources. A worker host is a passive entity, which spends most of its lifetime receiving requests from workers or users and coordinating their actions. There are the three main subcomponents of a worker host: an application manager, a participant manager, and a user interface.

3   Execution Contexts

The design of an execution context is done so as to comply with the workflow meta-model specification [6]. According to the specification, a process definition is the main meta-model entity, while a package represents a container for the grouping of common data entities from a number of different process definitions whose scope may be wider than a single process definition. To conform to the meta-model specification, the two distinct kinds of contexts have been set up: a worker context representing a process definition, and a package context representing a package definition.

3.1   Participant Specification

The participant declaration represents an abstraction level between the actual performer and the activity which has to be performed. Declared participant identifiers figure in activity performer expressions. At run time, performer expressions are evaluated, and activities are assigned to concrete humans, programs, and/or machines.

In Workers, Inc., human or system users are distinguished by their network locations. The participant resolution process will therefore result in the location of the user being determined and assigned to the declared participant.

3.2   Application Declaration

In XPDL, applications are barely declared. In fact, they are just named. Additionally, a list of formal parameters may be supplied. The real definition of an application is not required at the process definition level, and may be handled by an application manager at run time. The reason for this approach is the support for heterogeneous environments, where a different program has to be invoked for each platform.

In Workers, Inc., an application manager of a worker host maintains the list of locally available applications (and their calling mechanisms) registered by the owner of the host. In case that its current activity refers to an application declaration, a worker will use locally available information to map the declaration to the concrete application and its calling mechanism.

Besides a generic application declaration, Workers, Inc. also allows for a richer application declaration, where the concrete application and its calling mechanism may be identified already at the level of the process definition.

3.3   Data Fields

Workflow relevant data represents the data created and used within each process instance during process execution. Workflow relevant data may be referenced from activities or transitions, and it may be passed to invoked applications. Furthermore, data fields may be used to pass persistent information or intermediate results between activities. Package data may be used for communication between multiple process instances originating from the same ancestor, while the process data may be used only for passing information between activities within the particular process instance.

Workers, Inc. adopts this strategy with certain limitations imposed by the fact that the system is highly distributed and lacks a centralized control. If the process involves parallel execution paths, there is a potential risk for concurrency problems to arise: a worker context is cloned as a whole, and concurrently executed activities do not point to the common data storage any more. Whenever multiple execution threads converge into a single thread, the data consolidation mechanism is used to provide as much as possible data consistency of the merged context. Corresponding data fields of the two merging contexts are compared by the time of their last modification, and the lastly modified value is taken. However, this may lead to inconsistent results depending on the activity execution sequence that is taken.

3.4   Itinerary

The itinerary has the structure of an arbitrary complex directed graph, where vertices of the graph represent process activities, and edges of the graph correspond to process transitions.

An activity is the smallest, atomic unit of work in a business process. The three main properties of an activity specification, which can be seen as answers to the accompanied questions, are:

- Performer assignment (Where?) — It specifies the performer of the activity. In the process of participant resolution, the actual location of a participant is determined. By evaluating a performer expression, a worker knows where its activity needs to be carried out, and will transfer itself over the network accordingly.
- Implementation specification (What?) — It specifies what the concrete realization of the activity is. It can be a call to a declared application, another workflow process, or an embedded activity set. Also, the activity may have no implementation at all, in which case it supports complex flow transitions or manually performed activities.
- Automation modes (How?) — Information on whether the activity is to be started/ finished manually by the user or automatically by the worker itself.

Transitions connect individual activities. A transition may contain a condition which must be fulfilled for the worker to start performing the target activity. If the performer assigned to the target activity is different than the one of the source activity, the worker will first transfer itself to the appropriate node in the network before it actually starts the activity.

3.5   Transition Restrictions

Transition restrictions of an activity specify how a worker should operate when encountered with multiple incoming or outgoing transitions of the activity. When faced with multiple outgoing transitions, a worker can take one, several, or all of the encountered paths, depending on the split type of the activity and the conditions associated with the transitions. On the other hand, when an activity has multiple incoming transitions, worker should know whether to start the activity immediately or to wait for the other active flows to complete, depending on the join type of the activity.

Any transition restriction may be of one of the three differed types:

1) XOR – representing alternative paths (only one path is taken)
2) AND – representing concurrent paths (all paths are taken)
3) OR – representing alternative paths with the potential of concurrency (several paths are taken, ranging from one to all)

Split points generating multiple concurrent threads are achieved by cloning a worker context into the appropriate number of copies, and by creating the same number of separate workers to carry and interpret those copies.

On the other hand, supporting join points that involve convergence of multiple threads has a necessary precondition: all merging workers must meet at the same node. This subject is directly related to the mechanism used for participant resolution. When participant locations are resolved in process initialization phase (as is the case in the current implementation), it is not an issue. If participant locations were evaluated dynamically during run time, it would become an issue raised by the same concurrency problems that apply to data fields.

4 XPDL Compiler

XPDL Compiler for Workers, Inc. takes an XPDL source code as its input, and produces appropriate context classes needed by workers to carry out defined processes. Once the system has been made compliant to the workflow meta-model, the translation itself becomes a routine: every XPDL element is translated into a block of Java code, in which the corresponding context member is constructed and associated with the appropriate context. The use of the compiler is required as the intermediate step between a modeling tool, which provides the XPDL process definition, and the actual process invocation in Workers, Inc. The compiler works through three phases:

1) XPDL document parsing. The XPDL source is validated against the XPDL schema for well-formedness, and a DOM tree is built.
2) Code generating for appropriate context classes. By traversing the DOM tree, the Java source code for the adequate number of context classes is generated. Process graphs are checked for correctness, and local methods are incorporated in the appropriate context classes.
3) Running Java compiler to produce Java bytecode.

The Graph Structure Checker component of the XPDL Compiler analyzes the process graph during compile time, so that potentially problematic points in the process graph can be detected prior to run time. A process definition containing problematic constructs may be rejected and asked for a revision, or the suspicious points may be marked and treated accordingly at run time.

5   Related Work

WADE [4] is another software platform that facilitates the development of distributed multi-agent applications where agent tasks can be defined according to the workflow metaphor. WADE is built on top of JADE [2], which is primarily a multi-agent system providing limited support for agent mobility. The current version of JADE supports only intra-platform mobility, i.e. an agent can move only within the same platform from one container to another.

WADE comes with a development environment called WOLF that provides support for the graphical definition of WADE workflows. WOLF does not use internally any standard workflow definition language. However, in order to facilitate import/export operations, WADE adopts the core elements of the workflow meta-model defined in the XPDL standard. Still, some meta-model elements, such as packages, type declarations, or block activities, are left unsupported. Also, there is no explicit support for process participants and activity performers in terms of the meta-model.

6   Concluding Remarks

A workflow modeler creates new processes by using a workflow process editor. He/she places the activities in lanes of process participants, connects them with transitions, and fills out all other process definition relevant data. The generated XPDL process definition is then taken as input to the XPDL Compiler which produces context classes necessary for workers to carry out the defined process.

The system is used experimentally within the department. First experiences proved all our expectations.

References

1. Baumann, J., Hohl, F., Rothermel, K., Strasser, M.: Mole – Concepts of a Mobile Agent System. Homepage of the University of Stuttgart (1997)
2. Bellifemine, F., Caire, G., Poggi, A., Rimassa, G.: JADE – A White Paper (2003), http://jade.tilab.com/
3. Green, S., Hurst, L., et al.: Software Agents: A review, IAG review (1997)
4. Wade User Guide, http://jade.tilab.com/wade/
5. Hollingsworth, D.: Workflow Management Coalition: The Workflow Reference Model. Homepage of Workflow Management Coalition (1995)
6. Workflow Management Coalition: Workflow Process Definition Interface – XML Process Definition Language, Version 1.0. Homepage of Workflow Management Coalition (2002)

A Multi-agent Recommender System for Supporting Device Adaptivity in E-Commerce

Domenico Rosaci and Giuseppe M.L. Sarné

University Mediterranea of Reggio Calabria
{domenico.rosaci,sarne}@unirc.it

Summary. E-Commerce recommender systems provide customers with useful suggestions about available products. However, in presence of a high number of interactions between customers and Web sites, the generation of recommendations could become a heavy task. Moreover, customers often navigate on the Web using different devices whose different characteristics may influence customer's preferences. In this paper we propose a new multi-agent system in which each device exploited by a customer is associated with a software agent which autonomously monitors the customer's behaviour. The use of device agents leads to generate recommendations taking into account the exploited device, while the fully decentralized architecture introduces a strong reduction of the time costs.

1 Introduction

In an E-Commerce environment, a customer generally spends a large amount of time to search interesting products while the sellers need to propose their products taking into account customers' preferences. A number of adaptive E-Commerce systems proposed in the last years [1, 6, 4, 2] face these problems by using a *profile* of the customer, which represents his interests and preferences, and exploit *software agents* to construct such a profile. Generally, when the customer accesses an E-Commerce site, his agent exploits the profile to interact with the site. In this interaction, the site can use both *content-based* and *collaborative filtering* techniques to provide recommendations to the customer's agent by adapting the site presentation. Moreover, nowadays customers can navigate on the Web by using different devices as traditional desktop PCs, notebooks, cellular phones, etc. The necessity that recommender systems consider the device in generating suggestions is becoming a key issue [5, 4]. Indeed, if a user accesses a site by a PC, the site can propose its recommendations by a presentation including graphics and other multimedia while, if the exploited device is a cellular phone, the site presentation has to be lighter. In order to tackle this issue, some approaches have been proposed in the past. As an example, a multi-agent recommender for E-Commerce, called EC-XAMAS, is presented in [4]. EC-XAMAS assists a customer in the search of products of interest, accordingly to his past interests and considering different possible devices. However, the computation for generating suggestions is entirely performed on the client-side, introducing high costs for those devices having limited resources. Another proposal is represented by the MASHA system [5].

C. Badica et al. (Eds.): Intel. Distributed Comput., Systems & Appl., SCI 162, pp. 293–298, 2008.

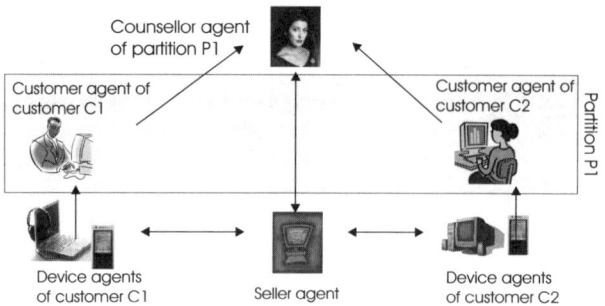

Fig. 1. The ARSEC Architecture

MASHA provides each device with a *client agent* that builds a user's profile associated to just that device. This profile is continuously updated by a *server agent* which collects the information provided by the different users' devices. A third component, called *adapter agent*, generates a personalized site's representation containing some recommendations derived by both an analysis of the user's profile and the suggestions coming from other users that in the past exploited the same device. However, although MASHA effectively takes into account the different devices and reduces the client-side time costs, it presents a significant cost for computing recommendations. In order to reduce both these costs, we propose in this paper a new multi-agent architecture, called *Adaptive Recommender System for E-Commerce* (ARSEC). ARSEC (see Figure 1) maintains the three MASHA agents, namely: (*i*) a *device agent*, (*ii*) a *customer agent*, associated with each customer, and (*iii*) a *seller agent*, associated with each E-Commerce site. However, differently from MASHA, the recommendations provided by ARSEC are not autonomously generated by the seller agent, but they are the result of a collaboration between the seller agent and a new agent type, called *counsellor agent*. The basic idea underlying ARSEC is grouping the customers in partitions of customers having similar profiles, where each partition is managed by a counsellor agent. Consequently, when a customer accesses a site, the seller agent does not compute recommendations, but it exploits the help of the counsellor agents associated with the clusters which the customer belongs to. We have experimentally evaluated ARSEC by comparing it with MASHA, and we have observed a significative improvements of the recommendation performances and a low time cost for generating recommendations. In Section 2 we provide the technical details of the ARSEC architecture. Finally, in Section 3, some experiments are presented and some conclusions are drawn.

2   The ARSEC Architecture

2.1   The Device Agent

A device agent is associated with each customer's device. It stores a *Device Profile DP* and a *Customer Profile CP*. In its turn, *DP* contains: (*i*) The set of the counsellor agents associated to the partitions which the customer belongs to; (ii) The maximum size (in Kbyte) of text (resp. audio, video) content of a product that the customer desires to handle when using the device; (iii) three parameters $\rho_1, \rho_2, \rho_3 \in [0,1]$, associated to the actions performable by the customer (i.e., visit, buy or add to favorite); (iv) an integer value T used to evaluate the customer's interest in a product; (v) three parameters k, z and r, that are exploited by the device agent in its interaction with the seller agent of each visited E-Commerce site and that respectively represent the maximum number of: (a) interesting categories belonging to the site that the customer desires to be considered; (b) similar agents that the customer desires to be considered in collaborative filtering recommendations; (c) recommendations to be considered for each similar agent. The customer profile CP stores the profile of the customer, based on the whole E-Commerce sessions history. More in detail, CP is a set of pairs (τ, IW), each one associated with a category τ, where IW (*Interest Weight*) is a measure of the customer's interest in the category τ by using the device. We define IW by using the actual time t spent by the customer when visiting the page containing τ. Moreover, the customer can buy τ, add it to favorite or visit the page containing τ, and this is taken into account by weighting IW with a coefficient ρ_a for each action a (where $a = 1, 2, 3$). More formally, for each new update, IW is computed as follows:

$$IW = \begin{cases} (IW + \frac{t}{T} \times \rho_a)/2 \text{ , if } t \leq T \\ (IW + \rho_a)/2, \qquad \text{elsewhere} \end{cases}$$

2.2   The Customer Agent

A customer agent is associated with each agent and collects information about the categories visited during the customer's activities. These information will be send to the counsellor agents of the customer's partitions. The customer agent contains two elements, namely the *Connection Setting* (CS) and the *Global Customer Profile* (GCP). In its turn, CS stores the the number ND of device agents associated with the customer and a vector PM of ND elements, where each element PM_i is the cost for Kbyte of the Internet connection of the i-th device. GCP stores a global representation of the customer's interests relative to the visited categories. In particular, it is a list of pairs $\langle \tau, GIW \rangle$, where τ identifies a category accessed by the customer and GIW is the *Global Interest Weight* shown by the customer, computed as the weighted mean of all the interest weights, relative to the different devices. That is $GIW = \frac{\sum_{i=1}^{ND} PM_i \times IW_i}{\sum_{i=1}^{ND} IW_i}$, where IW_i is the interest weight computed for the given category τ by the i-th device, $i = 1, ..ND$.

2.3   The Counsellor Agent and the Seller Agent

A *counsellor agent* is associated with a set of customers that are interested in the same domain. A *seller agent* is associated with a site to manage the products contained in the site. The data structure of the counsellor agent is composed of three elements called *Seller Catalogue* (*SC*), *Global Profile Set* (*GPS*) and *Profile Collector* (*PC*). The seller catalogue contains, for each site E that interacted with the counsellor agent in the past, all the products of E. The global profile set contains the global profiles of all the customers associated with the counsellor. The Profile Collector contains several data sections, each one relative to a site E and denoted by DS_E. Each DS_E contains in its turn the list of the profiles associated to the past visitors of E. We denote by $DS_E[c,d]$ each of these profiles, associated to a given customer c and his device d. The elements of $DS_E[c,d]$ are pairs (τ, IW) where τ is a category that c considers interesting in the site E, and IW is the interest weight of τ. The information relative to each visitor profile $DS_E[c,d]$ is provided to the counsellor agent by the site agent of E when c ends its session. Suppose that c visits the site E exploiting a given device d; then, the c's device agent sends to the seller agent the device profile DP. The customer c belongs to some customer partitions, each of which is associated to a counsellor agent. In this case, the seller agent contacts each counsellor agent and sends to it the device profile DP. In order to generate content-based recommendations, the counsellor agent has built a list CB that contains those products of E whose categories belong to the global profile of c (this global profile is contained in the Global Profile Set). Then, the counsellor agent orders CB in a decreasing fashion based on the coefficient IW of each category and maintains only the first k products deleting the remaining ones (remember that k is contained in DP). Moreover, to generate collaborative filtering recommendations, the counsellor agent compares the profile $DS_E[c,d]$ contained in the data section DS_E and relative to the customer c, with each profile $DS_E[q,d]$ of each other customer q, that has visited E in the past exploiting the same device of c. As a result, a list CF of the products belonging to those categories accessed by the z visitors less different to c is obtained (remember that also z is contained in DP). The difference between each pair of customers c and q that use the same device d is computed as follows. Let τ be a category that belongs both to the data sections $DS_E[c,d]$ of c and $DS_E[q,d]$ of q, and let $IW_c(\tau)$ (resp. $IW_q(\tau)$) be the interest rate assigned to τ in the profile of c (resp. q). The value $d(\tau) = |IW_c(\tau) - IW_q(\tau)|$ is assumed to be a reasonable measure of the difference between c and q in the evaluation of the category τ. We measure the global difference between the two customers c and q, denoted by $D(c,q,d)$ by summing all the contributions $d(\tau)$ relative to all the categories shared by the profiles of c and q. More formally, $D(c,q,d) = \sum_{\tau \in DS_E[c,d] \cap DS_E[q,d]} |IW_c(\tau) - IW_q(\tau)|$. Each counsellor agent of the customer c that is visiting an E-Commerce site returns to its seller agent the lists CB and CF, which contains products suitable to be recommended to c. Besides these lists, the counsellor returns to the seller agent also the *similar customer list* which contains the z customers more similar to c. These lists are used by the seller agent to generate an adapted presentation for the visiting customer. In particular, the seller agent generates a Web page that contains only elements compatible with the specification of the customer's device, contained in the device profile DP. Moreover, the Web page presents two section of recommendations, namely *The seller recommends* and *The other customers recommend*, containing the products contained in the lists CB and CF, respectively. A third section, called *Contact other customers*, gives the possibility to send a message to the customers that have been considered when generating the CF list.

3 Experiments and Conclusions

In this section, we present some experiments aiming to evaluate the advantages introduced by ARSEC, in terms of recommendation quality and time cost, with respect to both MASHA and EC-XAMAS. We used in our experiment 25 E-Commerce sites, each one provided with about 50 products. We monitored 53 customers during 42 days and in particular we exploited 10 of the 25 sites for building the customers' profiles. The remaining 15 sites have been used to test the systems. We recorded, for each customer, the customer's choices into a log file, containing a list of 200 elements $\langle a, b, t \rangle$, relative to 200 different products accessed by a customer, where a (resp. b) is the identifier of the *source* (resp. *destination*) product, and t is the timestamp associated to the choice to cross from a to b. We have implemented both the above systems in JADE [3]. In particular for ARSEC we have considered three device agents associated with three different devices, namely a desktop PC, a palmtop and a cellular phone. Finally, each ARSEC customer agents adopt the same parameters values: (*i*) $n = 3$, having only three types of device agents for each user; (*ii*) the prices per Mbyte (in euro cents) that we have considered are: $PM_1 = 0.9, PM_2 = 1.4, PM_3 = 1.8$. We considered four sets S_1, S_2, S_3, and S_4 containing 20, 40, 60, and 80 customers, respectively. The following experiment was repeated for each set of customers S_1, S_2, S_3, and S_4. For each customer, in correspondence of each triplet $\langle a, b, t \rangle$ belonging to the test-set, we generated a list of recommended product $R(a)$, for both of the evaluated systems. We checked if b belongs to $R(a)$ in order to measure the effectiveness of the different approaches and we stored the result in a value δ_a, where $\delta_a = 1$ if $b \in R(a)$, $\delta_a = 0$ otherwise. The Average Precision $(\overline{P})$ of each E-Commerce system is defined as the average of the δ_a values on all the triplets $\langle a, b, t \rangle$. The first 3 rows of Table 1 report the results obtained by the two approaches considering, in terms of Average Precision, the global performance, the content-based and the collaborative filtering component, respectively. We can note that, in all the three cases and for different size of the

Table 1. Experimental Results for ARSEC(A), MASHA(M) and EC-XAMAS(E)

	S_1			S_2			S_3			S_4		
	A	M	E	A	M	E	A	M	E	A	M	E
Global $\overline{P}$	0.80	0.72	0.69	0.85	0.75	0.71	0.89	0.79	0.74	0.91	0.82	0.78
$CB \, \overline{P}$	0.58	0.52	0.50	0.56	0.55	0.53	0.61	0.54	0.53	0.61	0.55	0.54
$CF \, \overline{P}$	0.47	0.40	0.38	0.54	0.45	0.39	0.67	0.49	0.48	0.71	0.55	0.52
$\overline{T}$	0.99	2.40	4.38	1.07	2.95	4.51	1.11	3.69	4.49	1.31	4.27	4.92

customer agent community (S_1, S_2, S_3, and S_4), ARSEC performs better than the other two systems. Finally, we have compared the impact of the different recommendation algorithms on the performances of the E-Commerce sites. The last row of Table 1 reports the average waiting time of the customers when accessing an E-Commerce site considered in the experiment above, computed on all the E-Commerce sites and on all the client accesses. The experiment shows that ARSEC introduces a waiting time significantly smaller than both MASHA and EC-XAMAS, and the advantage increases with the dimension of the customer set. We argue that this good performance is due to the presence of the counsellor agents that pre-compute recommendations, thus avoiding onerous computation for both the customer client and the seller server.

References

1. Badica, C., Ganzha, M., Paprzycki, M.: Mobile agents in a multi-agent e-commerce system. In: Int. Symp. on Symbolic and Numeric Algorithms for Scientific Computing, p. 8. IEEE, Los Alamitos (2005)
2. Alim, A., Siddique, S.A., Chisty, K.J.A., Rahman, G., Hossain, A.: Developing an agent-mediated e-commerce environment for the mobile shopper. In: Proc. of the 7th Int. Conf. on Computer and Information Technology
3. http://www.jade.tilab.org (2005)
4. De Meo, P., Rosaci, D., Sarnè, G.M.L., Terracina, G., Ursino, D.: An xml-based adaptive multi-agent system for handling e-commerce activities. In: Int. Conf. ICWS-Europe 2003, Erfurt, Germany, pp. 152–166. Springer, Heidelberg (2003)
5. Rosaci, D., Sarnè, G.M.L.: MASHA: A Multi Agent System Handling User and Device Adaptivity of Web Sites. UMUAI 16(5), 52–64 (2006)
6. Di Stefano, A., Pappalardo, G., Santoro, C., Tramontana, E.: A multi-agent reflective architecture for user assistance and its application to e-commerce. In: Int. Work. CIA, pp. 90–103. Springer, London (2002)

Dynamically Computing Reputation of Recommender Agents with Learning Capabilities

Domenico Rosaci and Giuseppe M.L. Sarné

University Mediterranea of Reggio Calabria
{domenico.rosaci,sarne}@unirc.it

Summary. The importance of mutual monitoring in recommender systems based on learning agents derives from the consideration that a learning agent needs to interact with other agents in its environment in order to improve its individual performances. In this paper we present a novel framework, called EVA, that introduces a strategy to improve the performances of recommender agents based on a dynamic computation of the agent's reputation. Some preliminary experiments show that our approach, implemented on the top of some well-known recommender systems, introduces significant improvements in terms of effectiveness.

1 Introduction

The issue of realizing recommender agents with learning capabilities has received a great deal of attention in the recent past. A key problem in such a context [1] is to realize an effective mutual monitoring among the agents, providing each agent with useful suggestions about other agents which could be contacted to obtain a fruitful cooperation. For instance, in [1] and in [5], it is proposed that the owner can integrate the individual knowledge of its agent with that of other agents that have similar interests in the community. A limitation of these approaches is given by the relatively simple cooperation mechanism, only based on a similarity measure. Instead, most of the recent cooperative approaches in multi-agent systems use reputation models to select the best candidates for collaboration [7, 2]. However, these reputation models are conceived for general multi-agent systems, and do not take into account the particular characteristics of a system composed by learning agents, that continuously update their knowledge bases and than can improve/worsen in time their capabilities. In this paper, we propose to face the problems above by an *evolutionary* framework, called *EVolutionary Agents* (EVA). In our approach, a user that is not satisfied by his/her agent can require the system to provide him/her with a new agent. In order to satisfy the request of the user, the system selects in the community that agent appearing as the best candidate to the substitution, based on an opportune *score*. Such a score is computed taking into account the reputation of each agent in the community. The selected agent is *cloned* and the created copy (clone) is sent to the requester user. The core of our method consists in

C. Badica et al. (Eds.): Intel. Distributed Comput., Systems & Appl., SCI 162, pp. 299–304, 2008.
springerlink.com   © Springer-Verlag Berlin Heidelberg 2008

the mechanism to compute the reputation of an agent, that takes into account the cloning mechanism, and considers a *genetic* component of the reputation. The paper is organized as follows. Section 2 describes the EVA framework, and Section 3 introduces our reputation model. Section 4 presents an evaluation of the framework and draws some final conclusions.

2   The EVA Framework

In our framework, each user u is assisted by a set A_u of information software agents able to provide him with recommendations. When u accesses a Web page, each agent $a_i \in A_u$ provides him with some recommendations, where each recommendation is a Web link. We denote by R_i the set of recommendations that a_i during its life suggested to u. To evaluate the quality of the set R_i, *precision* and *recall* are the best known measures used in information retrieval [3]. Precision is the fraction of the recommendations that are considered as relevant by u. Recall is the fraction of the links actually selected by u that have been recommended by the agent a_i. A well-known approach to take into account both recall and precision is represented by the F_β-*measure* defined in [4], i.e the harmonic mean of precision and recall, where β is a non-negative real that weights the recall with respect to the precision. In this paper we use the F_β-measure of R_i to compute the satisfaction of the user u for the recommendations provided by his agent a_i. Analogously, we can define the F-measure $F_\beta(A_u)$ for the whole agent-set A_u, considering that the recommendations provided by the agent-set is the union of the sets R_i relative to each agent $a_i \in A_u$. EVA framework introduces a dynamic strategy to improve the satisfaction of the agents composing the MAS. In particular, in the EVA architecture two types of agent are appositely conceived to manage such a strategy. First, each user u is provided with an agent called *Local Evolution Manager* (LEM_u). Moreover, the whole MAS is provided with a *Global Evolution Manager* (GEM). Each user u can fix a *satisfaction threshold* ρ_u for $F_\beta(A_u)$, under which the quality of the recommendations provided by A_u is unsatisfactory. In the case $F_\beta(A_u) < \rho_u$, LEM_u determines which agent a_i has a measure $F_\beta(a_i) < \rho_u$. We denote by UA_u the set of those unsatisfactory agents. If $UA_u \neq \emptyset$, then LEM_u de-activates the agents belonging to UA_u and sends a *help request* to GEM. This help request informs GEM that the agent u deactivated the k agents belonging to the set UA_u (where k is the cardinality of UA_u) and then it needs to substitute them with other k more satisfactory agents. As we will see below, GEM will determine a set of substitutes agents based on both their similarity with the unsatisfactory agents contained in UA_u and the *reputation* that the community gives to the agents. Therefore, in the help request the agent LEM_u provides to GEM, besides of the set UA_u and the threshold ρ_u, also a parameter ψ_u (a real value ranging in $[0, 1]$) that represents how much u weights the similarity w.r.t the reputation. The agent GEM maintains a similarity matrix $\Sigma = \{\Sigma_{ab}\}$, $a, b \in MAS$ where each element represents the similarity between two agents a and b belonging to the MAS. This similarity is computed as described in [5], and it is a real value belonging to $[0, 1]$.

Moreover, GEM also stores, for each agent a belonging to the MAS, a *reputation coefficient* r_a. This coefficient is a real value belonging to $[0,1]$, computed as described in Section 3, representing a measure of how much the whole community considers the performances of a as satisfactory. When GEM receives a help request $h_u = (UA_u, \rho_u, \psi_u)$ by u, it determines a substitute for each agent $\mu \in UA_u$. To this purpose, GEM examines as candidate to the substitution each agent a of the MAS to which is associated a F-measure greater than ρ_u. We denote by C_μ the set of these candidate agents. Then, GEM computes for each agent $a \in C_\mu$, the score: $s(a, \mu) = \psi_u \cdot \Sigma_{a,\mu} + (1 - \psi_u \cdot r_a)$ Finally, GEM chooses as substitute of μ the agent sub_μ to which is associated the maximum score $max(s(a, \mu))$, $a \in C_\mu$. For each agent μ belonging to UA_u, the agent GEM clones the substitute agent sub_μ. The clone agent sub_μ^* is then transmitted to the local evolution manager of u, that adds it to the agent-set of u in substitution of the unsatisfactory agent μ. From now on, the *evolution* of sub_μ^* will be independent on the *parent agent* sub_μ, since the model encoded in it will be applied to the environment of the user u. It is important to point out that the core of the strategy is that an unsatisfactory agent μ is *killed* and substituted by another agent sub_μ^*, provided via the activity of the global evolution manager GEM. The reason that leads to believe that this substitution will lead some advantages to the user u of μ is that sub_μ^* is the clone of an agent that (i) has a F-measure greater than the satisfaction threshold required by u and (ii) presents a good score, computed by taking into account both the similarity with the substituted agent and the global reputation in the community. Feature (i) alone is not sufficient to conjecture that sub_μ^* will produce an F-measure greater than the u's satisfaction threshold but only assures that the parent agent of sub_μ^* is goodly evaluated by its own user, that could obviously have a different perception of the satisfaction with respect to u. However, feature (ii) assures that the parent agent of sub_μ^* has a personal ontology similar enough to that of substituted agent μ and, in addition, it has a good reputation in the community. Considered together, the two features give a reasonable motivation to believe in a possible improvement of the u's satisfaction w.r.t. the original situation in which the unsatisfactory agent μ was present.

3   Agent Reputation in an Evolutionary Environment

It is widely recognized that the reputation on an agent, rather than being a single information, is a multi-dimensional concept, since it should take into account different aspects. Moreover, each user usually has a different way of combining the single aspects of the reputation associated with his agents, weighting each aspect by his/her personal point of view (this is called the *ontological* dimension of the reputation [7]). Finally, when an agent belongs to a group, besides the personal evaluation of the reputation (the *individual* dimension of the reputation), we need to consider the opinion of the whole community (the *social* dimension of the reputation). Our reputation problem presents a unique individual dimension, i.e. the reputation of providing good recommendations to the

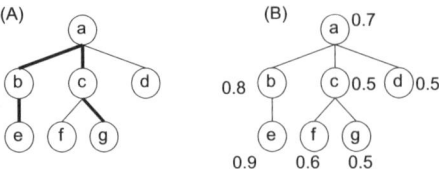

Fig. 1. An example of Descent Tree

agent's owner. Therefore, it is reasonable to consider, as possible ontological dimensions of the individual reputation, both the precision and the recall of the recommendations. Consequently, in our approach we assume the F-measure as a global measure of the individual reputation of the agent a, considering both the two ontological dimensions and where we denote as u_a the owner of a and as β_{u_a} the personal consideration of u_a for the precision with respect to the recall. Since the evolutionary strategy we have described above implies that an agent can be cloned and its clone can be moved in a new environment, it is necessary to define the agent reputation under this novel, particular perspective.

To study this problem, it is useful to represent the relationships the cloning process introduces in the set of the agents, by using some definitions directly derived from the terminology used to represent relationships between entities that share a genealogical origin. Figure 1-(A) helps us to introduce these definitions. It represents a set of agents, mutually involved in cloning activity, by using a sort of "genealogical" tree, where each node is associated to an agent, and each edge represents a cloning process between two agents, where we call *parent* the agent that is cloned and *child* the agent that results from the cloning. For instance, in Figure 1-(A), the agent a is the parent of the agent b. Obviously, while an agent has only one parent, it can have zero or more children. We call *siblings* two agents, like b and c in Figure 1-(A), that have the same parent. The parental relationship can be recursively applied, leading to introduce the notion of *ancestor* of an agent. For instance, a is an ancestor of e since it is the parent of b which is, in its turn, the parent of e. Already using the analogy with the human kinship system, we define *relatives* two agents that share a common ancestor, as all the agents in Figure 1-(A), and we can also define a kinship degree between two relatives, as the distance between them in the tree passing through the common ancestor. For instance, the agents e and g have kinship degree equal to 4, since the path linking them is composed by four edges. Obviously, if an agent a is cloned, then each of its clones $b \in children_a$, being identical to a, has to inherit at the cloning time the same reputation of a. Then, the inherited reputation of b can be considered as a sort of initial reputation. However, considering that b will moved to a user different from that of a, the reputation of b should evolve in time, taking into account, besides the initial inherited reputation, the individual satisfaction expressed by its current owner. It is also necessary to determine how these two components (i.e., inherited reputation and individual satisfaction) are weighted for determining a unique, global, measure of reputation. Moreover, due to the cloning process, each agent a belongs to a *family* of relatives, represented in the descent tree DT_a. Since a shares with all these relatives some similarities deriving from the cloning process, it seems reasonable to consider the performances of all these relatives to determine the reputation of a. This consideration introduces a social component (deriving from the relatives' performances) in the computation of the reputation. To take into account the observations above, in our approach we introduce a *reputation coefficient* r_a associated with each agent a, belonging to the interval $[0, 1]$, where $r_a = 1$ means complete reliability of a. Such a coefficient is computed as a weighted mean of n contributions, where each contribution is associated with one agent of the descent tree DT_a and where n is the number of relatives composing the descent number DT_a. The first contribution is associated with the agent a itself, and it is equal to the F-measure $F_{\beta_a}(a)$. Each of the other $n-1$ contributions is associated with one of the relatives of a, and it is equal to the F-measure $F_{\beta_a}(a)$. Each of this component, associated with an agent $x \in \mathcal{F}_a$, is weighted by a coefficient, equal to $k_{a,x} + 1$. This way, the contribution of $F_{\beta_a}(a)$ to the overall reputation is equal to $F_{\beta_a}(a)$, being $k_{a,a} = 0$, while the contribution of the satisfaction obtained by each other relative b is as smaller as higher is the kinship degree between a and b. As an example, consider the situation in Figure 1-(B), where to each agent of the descent tree is associated the F-measure w.r.t. its own owner. The F-measure of the agent e is equal to 0.9. The reputation of e is equal to 0.696.

4   Experimental Evaluation and Conclusions

To evaluate our approach, we have built some experiments that exploit EVA framework to improve some agent-based recommender system, i.e. MASHA [6], CILIOS [5] and SPY [1]. We have considered a set U composed by $n = 37$ real users, each provided with a set of 3-5 recommender MASHA (resp., CILIOS, SPY) agents, and we have exploited the same set of Web sites on Travel Agencies described in [6]. We have compared the performances, in terms of F-measure (using $\beta = 0.5$ and $\psi = 0.5$), in these two different cases: (i) the system does not use the EVA framework to substitute the unsatisfactory agents; (ii) the system uses the EVA framework. We have measured the percentage of improvement in the average satisfaction of the agents obtained in the case (ii) with respect to the case (i), where the average satisfaction has been computed as follows: $\overline{F} = \frac{\sum_{u \in U} F_\beta(A_u)}{n}$ and the percentage of improvement is defined as $P_i = \frac{|\overline{F}_E - \overline{F}_i|}{\overline{F}_i}$, where $\overline{F}_E$ is the average satisfaction in presence of EVA framework, where $\overline{F}_i$ is the average satisfaction of the system i, $(i = CILIOS, MASHA, SPY)$, without applying EVA strategies. Figure 2 shows how the improvement P significantly increases with the number of total sessions globally performed by the users, for all the considered systems, achieving for CILIOS a value of about 50 percent after 20000 sessions. Although promising, these experiments represents only a very preliminary evaluation of the framework. In our ongoing research, we are studying other evaluation methodologies, both theoretical and experimental, aiming at

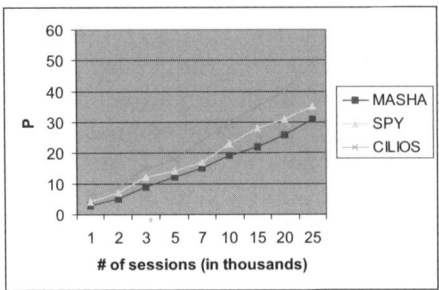

Fig. 2. Percent of improvement using EVA with different recommender systems

precisely characterizing the advantage of the evolutionary strategy in improving the global system performances.

References

1. Buccafurri, F., Palopoli, L., Rosaci, D., Sarné, G.M.L.: Modeling cooperation in multi-agent communities. Cognitive Systems Research 5(3), 171–190 (2004)
2. Carbo, J., Molina, J.M., Davila, J.: Comparing predictions of sporas vs. a fuzzy reputation agent system. In: Proc. of 3rd International Conference on Fuzzy Sets and Fuzzy Systems, Interlaken, pp. 147–153 (2002)
3. Raghavan, V., Bollmann, P., Jung, G.S.: A critical investigation of recall and precision as measures of retrieval system performance. ACM Trans. Inf. Syst. 7(3), 205–229 (1989)
4. Van Rijsbergen, C.J.: Information Retrieval. Butterworths, London (1979)
5. Rosaci, D.: Cilios: Connectionist inductive learning and inter-ontology similarities for recommending information agents. Inf. Syst. 32(6) (2007)
6. Rosaci, D., Sarné, G.M.: Masha: A multi-agent system handling user and device adaptivity of web sites. UMUAI 16(5), 435–462 (2006)
7. Sabater, J., Sierra, C.: Review on computational trust and reputation models. Artificial Intelligence Review 24(1), 33–60 (2004)

Topic Map for Medical E-Learning

Liana Stănescu, Dan Burdescu, Gabriel Mihai, Anca Ion, and Cosmin Stoica

University of Craiova, Software Engineering Department
Bvd.Decebal 107, Craiova, RO-200440, Romania
{stanescu_liana,burdescu_dumitru,mihai_gabriel,soimu_anca,
stoica_cosmin}@software.ucv.ro

Summary. The paper presents original ways of using a modern concept - topic map - in medical e-learning. The topic map is mainly used for visualizing a thesaurus containing medical terms. The topic map is built and populated in an original manner, mapping an xml file that can be downloaded free, to an xtm file that contains the structure of the topic map. Only a part of the MeSH thesaurus was used, namely the part that includes the medical diagnosis's names. The student can navigate through topic map depending on its interest subject, having in this way big advantages. The paper presents also how to use the topic map for semantic querying of a multimedia database with medical information and images. For retrieving the interest information this access path can be combined with another modern solution: the content-based visual query on the multimedia medical database. Combining these possibilities to access a database with medical data and images, allows students to see images and associated information in a simple and direct manner. The students are stimulated to learn, by comparing similar cases or by comparing cases that are visually similar, but with different diagnoses.

1 Introduction

In the last decade, the electronic learning became a very useful tool in the students' education from different activity domains. Among these domains is the medical domain, even if in this case it is adopted a hybrid process that combines the traditional learning methods with the electronic ones. The accomplished studies, including the ones focusing on medical domain, indicated that the students substantially appreciate the e-learning method, due to the facilities offered, but they don't consider it as a replacement of the traditional learning which has other advantages [5].

Taking into account all these, an e-learning platform, called TESYS, was designed and implemented for the University of Medicine in Craiova. It provides both traditional functionalities and original facilities that are included only in few advanced medical e-learning platforms [7, 8, 9]:

- A database with medical images acquired by doctors in the patients' diagnosis process. Along with these images are provided other relevant data: diagnosis, treatment, evolution. This database completes the small number of images presented in the specialty courses or books.
- Content-based visual retrieval at the image or region level, based on color or texture characteristics automatically extracted from images.This modern possibility of database query is used both in the learning and testing process.

C. Badica et al. (Eds.): Intel. Distributed Comput., Systems & Appl., SCI 162, pp. 305–310, 2008.
springerlink.com

This paper presents another original functionality included in the TESYS platform, namely a topic map based on a thesaurus with medical terms. The topic map represents a browsing tool for student, but also a means to enable semantic search. To build the topic map it was used a part of MeSH, the National Library of Medicine's controlled vocabulary thesaurus, especially the part that includes medical diagnosis.

The topic map concept is used not only to help student to understand the semantic context in which a collection and it's single items are embedded and to discover unexpected knowledge, but also to execute semantic queries on the multimedia database that was included in the TESYS platform. On the record set selected by the semantic query, content-based retrieval functions can be executed next. This type of query uses characteristics like color, texture or color regions that were automatically extracted from the medical images [6].

The student has the possibility to combine different access options, which is considered the most successful approach in image retrieval. Using content-based visual query with other access methods to medical imagistic database allows students to see images and associated information in a simple and direct manner. They only have to select a query image in order to find similar ones. The student is stimulated to learn by comparing similar cases or by comparing cases that are visually similar, but with different diagnoses [1].

2   Topic Maps

Topic maps represent a powerful tool for modeling semantic structures (associations) between any kinds of things (topics), which are linked to any kind of documents (occurrences) [3, 4, 10]. The available resources that can act as a source of input to auto-population of the topic map are identified: ontology, relational or object-oriented database, metadata about resources, index glossary, thesaurus, data dictionary, document structures and link structures or unstructured documents [3]. It can be observed among these resources, the presence of thesaurus and XML, which are used for the design and self- population of topic maps, solution adopted by us also.

When converting a thesaurus to a topic map, the thesaurus provides the topics and some basic associations: the superclass-subclass association, synonyms and related terms. In this software tool for medical learning, it is used a thesaurus that was designed based on MeSH that is the National Library of Medicine's controlled vocabulary thesaurus. It consists of sets of terms naming descriptors in a hierarchical structure that permits searching at various levels of specificity [2]. Taking into account that the most important information associated to medical images is the diagnosis, and the retrieval is based on this criterion, we used from the MeSH thesaurus only the "Diseases" part that defines the medical terms necessary in the diagnosis process. This thesaurus served as a fundament for the construction of the topic map. Starting from the information offered by MeSH, it is obtained an xtm file that contains the structure of topic map that presents only details for digestive diseases. Of course the extension of the topic map for other pathologies or categories of medical terms can be easily realized.

2.1   Building and Populating the Topic Map

MeSH 2008 has two important files:

- An xml file named desc2008 that gives information about diseases or different diseases categories.
- A txt file trees2008 that gives information about the organization of these diseases in a tree structure.

For generating the xtm file that contains topic map, the necessary operations are grouped in 2 steps:

1)Parsing the xml desc2008 file and generating the topics.

The desc2008 file has a series of XML tags of DescriptorRecord type. Each tag of this type has information about one disease or a diseases category. This file is parsed and analyzed and the content of each DescriptorRecord tag will be used to generate a topic. The xtm file will also contain two topic types: diseases_category and entryterm with the next structure:

```
<topicRef xlink:href="#diseases_category"/>
   </instanceOf>
   <baseName>
     <baseNameString>diseases category</baseNameString>
   </baseName>
</topic>

<topic id="EntryTerm">
   <baseName>
     <baseNameString>Entry Term</baseNameString>
   </baseName>
</topic>
```

The disease categories presented in the desc2008 file are organized into a hierarchy and the TreeNumber tag indicates the sub-tree where the disease or disease category is included, as in the next example:

```
<TreeNumberList>
    <TreeNumber>C04.588.274</TreeNumber>
    <TreeNumber>C06.301</TreeNumber>
</TreeNumberList>
```

The content of each TreeNumber type tag (that is unique) will be used when the association type tags for the topic map will be generated. The parsing and generation module uses a hashtable that stores (key, value) pairs as in the example:

h["C04.588.274"] = "D004067";
h["C06.301"] = "D004067";

2) Parsing the txt trees2008 file and associations generation.

The file trees2008 contains a set of records with the following form: (disease name or disease category; identification string of the tree where it is included), as in the example:

```
Carcinoma, Ductal, Breast; C04.588.180.390
Phyllodes Tumor; C04.588.180.762
Digestive System Neoplasms; C04.588.274

Pancreatic Fistula; C06.267.775
Digestive System Neoplasms; C06.301
Biliary Tract Neoplasms; C06.301.120
```

We can see that the category Digestive System Neoplasms appears in 2 sub-trees meaning that this category is part of two associations. To generate the association that includes the record: (Digestive System Neoplasms; C04.588.274), the identification string C04.588.274 must be analyzed. Three groups compose it: C04, 488 and 274. The parsing and generation module will search in the trees2008 file, records that have the identification string composed by 4 groups (C04, 488, 274, XYZ) with X, Y, Z as variables. To obtain the unique ID of the topics included in the association, the h hashtable that was obtained after parsing Mesh2008 file is used.

2.2   Topic Map View

By parsing these 2 files, as presented above, we identify the information needed to generate the file in xtm format that must be analyzed and displayed next in a graphical fashion (figure 1).

As a result, the students can use the topic map as a navigation tool. They can navigate through topic map depending on their interest subject, having in this way big

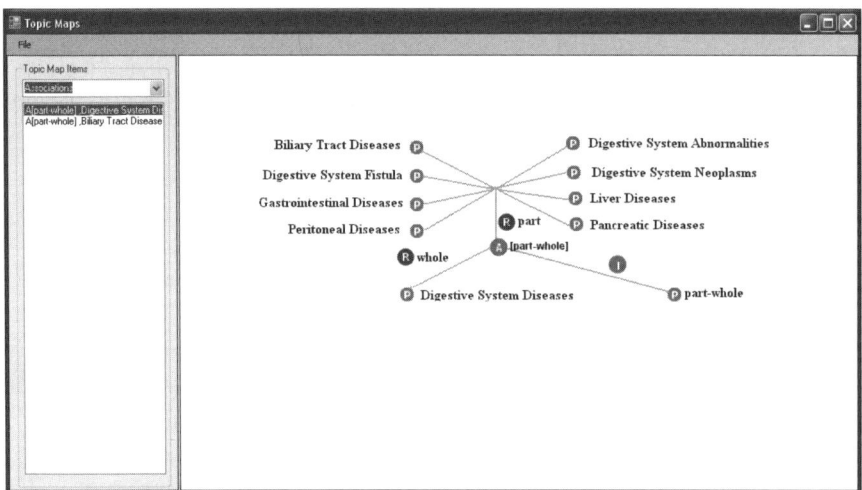

Fig. 1. Extract from the topic map

advantages. They don't have to be familiar with the logic of the database, they will learn about the semantic context, in which a collection and its single items are embedded and they may find useful items that they would not have expected to find them in the beginning.

The hierarchical structure of the descriptors from MeSH thesaurus, that has also multiple relationships between the medical terms, and each term may have a series of synonyms, can be properly visualized only by means of a topic map that offers to the student the opportunity to understand exactly these things.

3   Using the Topic Map for Querying the Medical Imagistic Database

Using the built topic map the student can launch a semantic query on the medical imagistic database. In our topic map, since the occurrences are stored in the database, every topic will be defined as a database query. This query may be simple: the topic "peptic ulcer", for example, will initiate a query for "peptic ulcer" in the diagnosis field of the table Images in the database. Consequently, every relevant image of "peptic ulcer" will be retrieved. The database search can be done in two ways: Using a single topic. In this case it is executed a Select command with the general form:

SELECT * FROM image where diagnosis="topic"
Example: Select * FROM image where diagnosis="peptic ulcer"

Using the topic and all the synonyms, if there are, and the Select command has the following form:

SELECT diagnosis FROM image where diagnosis="topic1" or diagnosis="topic2" or...
Example: Select * FROM image where diagnosis="peptic ulcer" or diagnosis= "gastroduodenal ulcer" or diagnosis= "marginal ulcer"

The second query modality is very useful in the learning process, because the images are introduced in the database by different specialists, and for diagnosis they can use synonyms, very known in the medical language, but less known by students.

This access path to the medical imagistic database can be combined with other modern modalities, the results being useful to the student in the learning process. For example, the student browses the topic map, learning about the hierarchical structure of diagnosis, and he decides to launch a query that uses the synonyms. As result, all the images corresponding to synonym diagnosis will be displayed. On the returned images set the content-based visual query can be executed.

4   Conclusions

The paper presents two original applications of the topic map concept in the medical e-learning: for browsing a thesaurus with medical terms and for semantic querying a multimedia database containing medical images and their diagnoses. To build and populate the topic map, taking into account that diagnosis is the main information attached to a medical image, we use a part of the MeSH dictionary. It has as component the medical diagnosis hierarchy, and also relationships between these and their synonyms. The topic map was build and populated identifying the useful information in an xml file and it's mapping to the xtm format corresponding to the topic map.

The topic map represents the most appropriate modality of visualizing this hierarchy in an educational way, capable to offer also contextual information. The topic map is also used to launch semantic queries on the multimedia medical database. This access path can be easily combined with other modern modalities like content-based image query or content-based region query. This combination is considered the most successful approach in image retrieval.

This new functionality added to TESYS medical e-learning platform was accepted by the teachers from Gastroenterology department of the Medicine and Pharmacy University from Craiova and appreciated as useful and original. During this year, the new functions will be tested in students training, in order to study the improvements of the learning process.

References

1. Muller, H., Michoux, N., Bandon, D., Geissbuhler, A.: A Review of Content-based Image Retrieval Systems in Medical Application - Clinical Benefits and Future Directions. Int. J. Med. Inform. 73(1), 1–23 (2004)
2. National Library of Medicine, Medical Subject Headings, MeSH Tree Structures, http://www.nlm.nih.gov/mesh/008/MeSHtree.C.html
3. Park, J., Hunting, S.: XML Topic Maps: Creating and Using Topic Maps for the Web. Addison Wesley, Reading (2002)
4. Rath, H.: The Topic Maps Handbook. In: Empolis GmbH, Gutersloh, Germany (2003)
5. Ruiz, J., Mintzer, M.J., Leipzig, R.M.: The Impact of E-Learning in Medical Education. Academic Medicine 81(3) (2006)
6. Smith, J.R.: Integrated Spatial and Feature Image Systems: Retrieval, Compression and Analysis. Ph.D. thesis, Graduate School of Arts and Sciences, Columbia University (1997)
7. Stanescu, L., Mihaescu, C., Burdescu, D., Georgescu, E., Florea, L.: An Improved Platform for Medical E-learning. In: Leung, H., Li, F., Lau, R., Li, Q. (eds.) ICWL 2007. LNCS, vol. 4823, pp. 392–403. Springer, Heidelberg (2008)
8. Stanescu, L., Burdescu, D., Ion, A., Panus, A.: An Original e-testing Method for Medical e-learning. In: 8th IEEE International Conference on Advanced Learning Technologies (ICALT), Santander, Spain (2008)
9. Stanescu, L., Burdescu, D., Ion, A., Panus, A.: Imagistic Database for Medical E-learning. In: 21st IEEE International Symposium on Computer-Based Medical Systems (CBMS), Jyvaskyla, Finland (2008)
10. TopicMaps. Org, http://www.topicmaps.org/

Author Index

Printing: Krips bv, Meppel, The Netherlands
Binding: Stürtz, Würzburg, Germany